Advances in Geographical and Environmental Sciences

Advances in Geographical and Environmental Sciences synthesizes series diagnostigation and prognostication of earth environment, incorporating challenging interactive areas within ecological envelope of geosphere, biosphere, hydrosphere, atmosphere and cryosphere. It deals with land use land cover change (LUCC), urbanization, energy flux, land-ocean fluxes, climate, food security, ecohydrology, biodiversity, natural hazards and disasters, human health and their mutual interaction and feedback mechanism in order to contribute towards sustainable future. The geosciences methods range from traditional field techniques and conventional data collection, use of remote sensing and geographical information system, computer aided technique to advance geostatistical and dynamic modeling.

The series integrate past, present and future of geospheric attributes incorporating biophysical and human dimensions in spatio-temporal perspectives. The geosciences, encompassing land-ocean-atmosphere interaction is considered as a vital component in the context of environmental issues, especially in observation and prediction of air and water pollution, global warming and urban heat islands. It is important to communicate the advances in geosciences to increase resilience of society through capacity building for mitigating the impact of natural hazards and disasters. Sustainability of human society depends strongly on the earth environment, and thus the development of geosciences is critical for a better understanding of our living environment, and its sustainable development.

Geoscience also has the responsibility to not confine itself to addressing current problems but it is also developing a framework to address future issues. In order to build a 'Future Earth Model' for understanding and predicting the functioning of the whole climatic system, collaboration of experts in the traditional earth disciplines as well as in ecology, information technology, instrumentation and complex system is essential, through initiatives from human geoscientists. Thus human geoscience is emerging as key policy science for contributing towards sustainability/survivality science together with future earth initiative.

Advances in Geographical and Environmental Sciences series publishes books that contain novel approaches in tackling issues of human geoscience in its broadest sense—books in the series should focus on true progress in a particular area or region. The series includes monographs and edited volumes without any limitations in the page numbers.

Praveen Kumar Rai · Ashutosh Singh ·
Bobby Beingachhi · Suraj Kumar Singh
Editors

Natural Resources Management and Sustainable Livelihoods in the Mountainous Region

Evidence, Gap and Future Strategies

Springer

Editors
Praveen Kumar Rai
Department of Geography
Khwaja Moinuddin Chishti Language University
Lucknow, Uttar Pradesh, India

Bobby Beingachhi
Department of Geography
Pachhunga University College
Aizawl, Mizoram, India

Ashutosh Singh
Department of Geography
Pachhunga University College
Aizawl, Mizoram, India

Department of Geography
HNB Garhwal University (A Central University), SRT Campus
Tehri, India

Suraj Kumar Singh
Centre for Sustainable Development
Suresh Gyan Vihar University
Jaipur, Uttar Pradesh, India

ISSN 2198-3542 ISSN 2198-3550 (electronic)
Advances in Geographical and Environmental Sciences
ISBN 978-981-97-2099-6 ISBN 978-981-97-2100-9 (eBook)
https://doi.org/10.1007/978-981-97-2100-9

This Springer imprint is published by the registered company Springer Nature Singapore Pte Ltd.
The registered company address is: 152 Beach Road, #21-01/04 Gateway East, Singapore 189721, Singapore

Contents

Chapter 1
Concept of Sustainable Livelihood: A Case Study of Laika Dadhiya Forest Village, Tinsukia, Assam

Ghanashyam Deka

Abstract Humans have been rapidly altering their interactions with the environment, a transformation that continues unabated. While the reciprocal influence between nature and humanity exists, it is ultimately nature that dictates the extent and manner of human activity. This underscores the prevalence of natural determinism. The world's population is burgeoning, and the scientific and technological revolution has accelerated developmental activities to the extent that they now pose a significant threat to the natural environment in some regions. This research aims to delve into consequential interactions between humans and their environment. One prominent example is the renowned Dibru-Saikhowa National Park, a habitat for migrating birds, wild horses, buffalo, and white-winged ducks. Situated within the low-lying active floodplain zone between the Brahmaputra and Dibru Rivers, the entire park faces unique challenges. Our study investigates the relationship between humans and the forest community within this context, as well as the impact of human actions on the forest and the corresponding adjustments made. Additionally, we explore how the habitat area expands due to the government's transparent strategy for delineating it and the park's conservation efforts, responding to the increasing population. Using both sample data and direct observations, we analyze how humans interact with and adapt to their environment. This study provides valuable insights into the complex interplay between humanity and the natural world within the forest community of Dibru-Saikhowa National Park.

Keywords Dibru-Saikhowa · Forest environment · Forest village · Man–environment relationship

---

G. Deka (✉)
Department of Geography, Pachhunga University College, Aizawal, India
e-mail: assameseguy@gmail.com

P. K. Rai et al. (eds.), *Natural Resources Management and Sustainable Livelihoods in the Mountainous Region*, Advances in Geographical and Environmental Sciences,
https://doi.org/10.1007/978-981-97-2100-9_1

1.1 Introduction

The notion that humans play a pivotal role in the dynamic relationship between humanity and the environment is not a novel concept. Human beings and their natural surroundings are intricately interconnected and mutually dependent. The capacity of humans to adapt to their environment and rely on the natural world has been a fundamental aspect of human civilization throughout history. Consequently, it is a two-way street, with humans shaping nature and nature influencing the timing and nature of human activities (Mathur 1988).

As environmental factors remain uncertain, influenced by various climatic events with a range of potential outcomes, humans exhibit diverse responses. This dynamic interaction even leads to humans assuming the role of "climate creators." Another significant determinant is the cultural fabric of society, which dictates how people interact with one another and adapt to their environment (Mishra et al. 2012; Rai et al. 2021; Singh et al. 2021; Brototi et al. 2022; Rai et al. 2022).

The study of how humans adapt to forest environments and engage with them is both important and captivating. This is because humans must strike a delicate balance between preserving their environment and making necessary modifications. In the context of current environmental challenges, the central question revolves around the manner in which adaptation occurs.

The exploration of the concept of the relationship between humanity and the environment is by no means a recent endeavor. As noted by Beyer (1974), this idea and its theoretical framework have roots dating back to the early days of modern geography. However, it gained renewed significance with the onset of the environmental awareness movement in the early 1980s. In contemporary scholarship, the study of human–environment dynamics is centered on both natural and human-induced hazards.

Geographers continually strive to bridge the longstanding notion that human actions influence nature with the latest insights and data. Pioneers in this field, including Burton, Beyer, Kates, Swell, Kayastha, Taher, Bhagabati, Gogoi, and others, have demonstrated their expertise in examining the intricate interactions and adaptations between humans and their environments.

1.2 The Study Area

Located in the far eastern region of Assam, Dibru-Saikhowa stands as one of the most expansive biodiversity hotspots. It spans between latitudes 27°30'N and 27°45'N and longitudes 95°01'E and 95°04'E, at an average altitude of 118 m above sea level. Dibru-Saikhowa's ecosystem encompasses various forest types, including semi-evergreen, deciduous, littoral, swamp, and pockets of wet evergreen forests. The area experiences an average annual rainfall ranging from 230 to 280 cm, while temperatures fluctuate between 7 °C and 34 °C. Within the confines of Dibru-Saikhowa

National Park, the landscape comprises approximately 35.84% moist mixed forests, 9.5% degraded forests, and 21.25% grasslands. The park boasts an impressive array of wildlife, with 36 mammal species identified. Its herpetofauna is equally diverse, featuring two monitor lizard species, eight snake species, 62 fish species, and a stunning variety of over 350 bird species. This sanctuary is particularly renowned as a natural haven for migratory birds.

The history of safeguarding Dibru-Saikhowa's natural treasures traces back to 1890 with the establishment of Dibru Reserved Forest, followed by the creation of Saikhowa Reserved Forest in 1929. In a significant development, the area was officially designated as the Dibru-Saikhowa Wildlife Sanctuary in 1986, encompassing not only the core forests but also the adjoining areas. Finally, in 1997, it achieved the prestigious status of a National Park, covering a core area spanning 340 square kilometers, complemented by a surrounding buffer zone of 425 square kilometers (Fig. 1.1).

Located at the western edge of the sanctuary are two forest villages, Laika and Dadhiya. Presently, these villages predominantly consist of Mishing residents. However, historically, Dadhiya was known for its substantial Nepali population, particularly renowned for dairy farming before the devastating earthquake of the 1950s. The name "*Dadhiya*" itself is derived from the Nepali word for milk (Das 2004). Following the earthquake, persistent flooding forced the Nepali community to relocate their dairy operations, leading to their departure from the area. Subsequently, in 1951–52, the Mishing community settled in the region, and at that time,

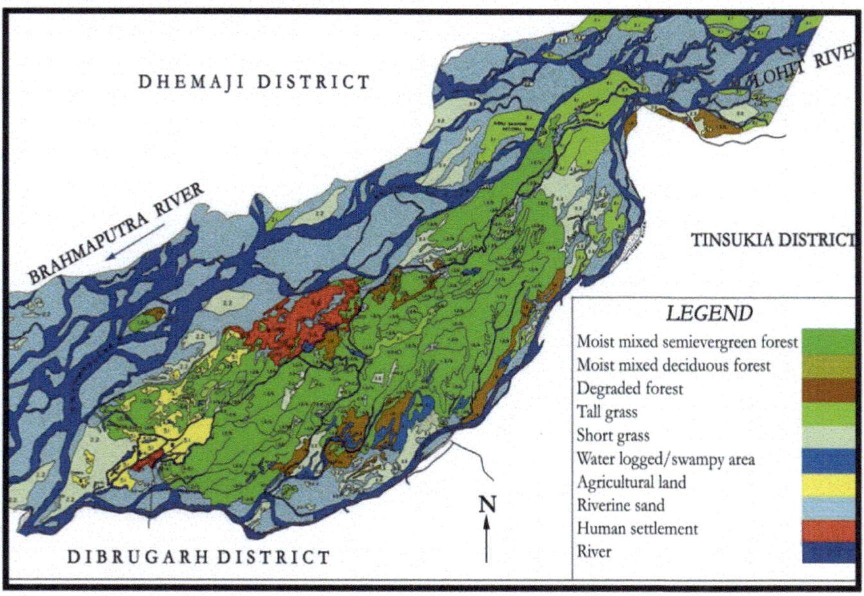

Fig. 1.1 Map of Dibru-Saikhowa national park

both villages were home to only 210 families. In this dynamic region, residents have long adapted to the challenges posed by floods and forests. In the 1950s, this area provided a suitable natural environment for the residents, facilitating the development of forest structures with relative ease. Presently, the population has grown to encompass around 1,500 families, typically consisting of 7 to 8 individuals per family.

At the socio-economic level, the surveyed area faces significant challenges, with a majority of the population lacking access to essential services such as health care, communication infrastructure, and transportation.

1.3 Objectives

The current study aims to address the following key objectives:

1. To explore the nature of human adaptation to the forest environment.
2. To examine the concept of sustainable livelihood concerning the forest environment.
3. To investigate the primary issue(s) affecting the national park in the surveyed region.

The research findings and analysis draw from a sample of 120 households located in two villages, specifically Laika and Dadhiya. In addition to this primary data collection, information from secondary sources has been gathered to facilitate comparative analysis. Additionally, maps and other essential data have been sourced from the DFO Office in Tinsukia. However, it is important to note that the primary discussion in this study primarily relies on the data collected from the sample households and personal observations.

1.4 Analysis

1.4.1 Family Profile

The average family size in the area exceeds seven persons, contributing to a total population that has grown significantly. Among the collected samples, it was observed that 53% of the total population consists of males, while 47% are females. In terms of age distribution, 61.64% of the total population is below 20 years old, indicating a high growth rate. Approximately 26% of the population falls within the 20–50 age range, with the remaining portion being above 50 years old. The area continues to experience a high population growth rate alongside a relatively low literacy rate.

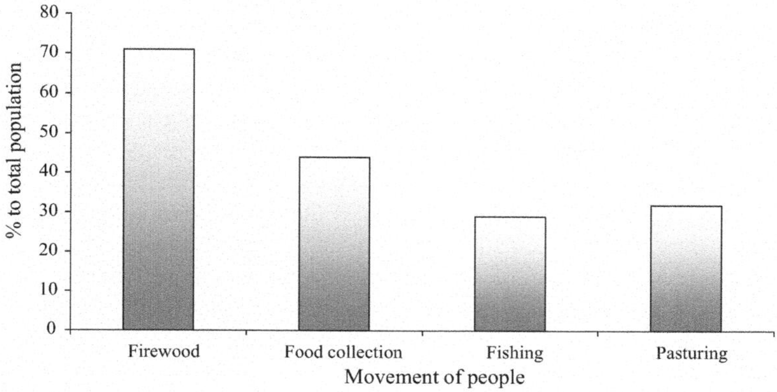

Fig. 1.2 Dependency on forest

1.4.2 Forest Dependency

In 1950–51, the two recently settled forest villages of Laika and Dadhiya were home to just 210 families. Presently, this figure has surged to 1,413 families, accommodating a population exceeding 10,000. This substantial increase underscores the significant population pressure in the neighboring forested areas. The initial survey reveals that the majority of the population relies heavily on the surrounding forests of Dibru-Saikhowa for their daily needs. They have developed a strong connection with the forest, utilizing it for various necessities such as food (including fish), firewood, animal fodder, and house building materials. Forest dependency is widespread in both towns (see Fig. 1.2), with 71% of the total population relying on the forest for firewood and building materials, 44% for food, and 29% for fishing. Additionally, more than 32% of the total population ventures into the forest daily for animal fodder. These exercises are not performed during the stormy season. In the stormy season the development is limited by the floods and submergence of available track. According to the forest authority report, inside the park area, there are in excess of 15,000 homegrown creatures. Cows, buffalos, and goats are by and large tamed by individuals living in the Dibru-Saikhowa. For their brushing, and any remaining exercises as referenced, clearly individuals move to the inside region of the timberland and gather the vital foods, etc.

1.4.3 Flood and Livelihood

Rivers have been the lifeblood of human civilization since ancient times. Dibru-Saikhowa serves as a compelling example of how people have adapted their way of life in the face of constant flood risks and waterlogging. The rainy season brings about floods that leave the local population vulnerable. Floods are a recurring event, happening multiple times each year within the park. Consequently, the entire National Park consists of low-lying floodplains, which yield highly fertile soil, ideal for agriculture.

In the aftermath of floods, people are compelled to venture out in search of essential supplies. During the rainy season, interactions among the residents are frequent due to the availability of water routes for boating and the scarcity of food for both humans and animals. In this manner, the people of Dibru-Saikhowa demonstrate their resilience as they adapt to their forest surroundings (Figs. 1.3 and 1.4).

When it comes to housing, the Mishing people, not just in this area but also elsewhere, typically construct Chang ghars (raised houses) and Chang bharals (raised food storage) as protective measures against floods of varying intensities. Additionally, every family owns at least one boat for self-rescue during floods and for fishing purposes. As soon as the floodwaters recede, people return to their agricultural fields with increased activity and urgency, taking advantage of the enhanced fertility of the soil.

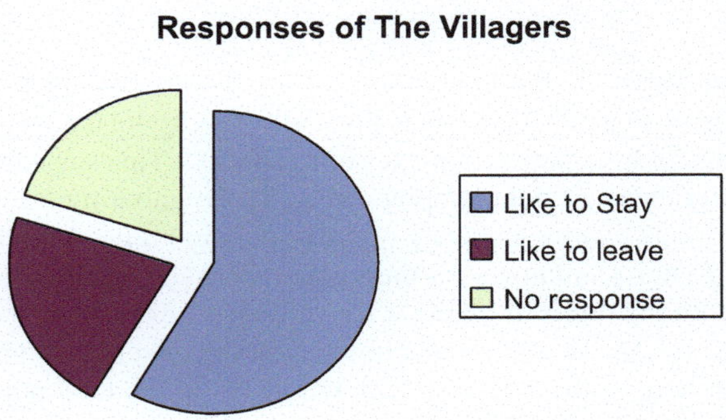

Fig. 1.3 Responses for forest village life

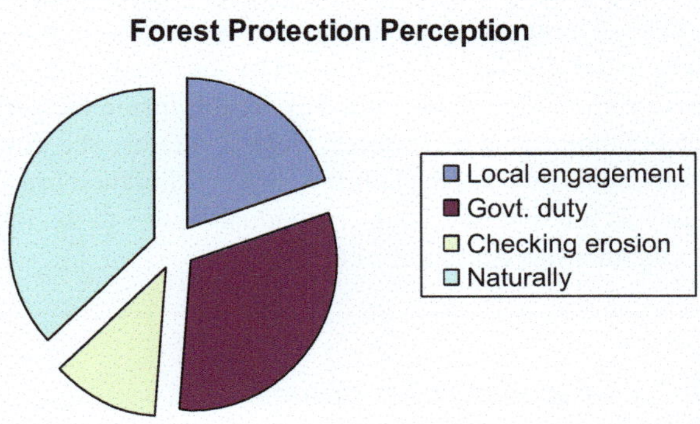

Fig. 1.4 Response regarding conservation

1.4.4 Responses for Living in Forests

The studied area faces significant challenges, including a low standard of living, limited development opportunities, and poor communication infrastructure. Despite these difficulties, the residents in the aforementioned villages exhibit a strong affinity for the forest environment, viewing it as the most suitable habitat for them. The majority of people in Laika and Dadhiya are determined to maintain their identity as forest dwellers, a sentiment clearly reflected in their perceptions.

The study of 120 households reveals that a significant portion of the villagers is resolute about continuing their lives within the forest area. More than 58% of the total population adamantly opposes leaving the forest village, even if offered improved survival facilities. Meanwhile, 22% are open to the idea of relocating for better opportunities, while the remaining 20% (primarily female respondents) did not provide a clear response.

When questioned about their reasons for not wanting to leave the forest, many residents cited their successful cultural and economic adaptation to the forest environment. However, the younger age group expresses reluctance to embrace this concept of adaptation. They are less inclined to remain in the forest village due to the absence of development opportunities and the recurring impact of annual flood hazards from all sides.

1.4.5 Perception Regarding Preservation

As the population continues to grow at a rapid rate, the reliance on forests has also increased, surpassing the sustainable exploitation levels nearly a decade ago. During the survey, a significant question arose concerning the protection, conservation, and preservation of the forest. Given that Dibru-Saikhowa is a national park, the issue of conservation is unavoidable. Therefore, the villagers were interviewed from a conservation perspective, as they reside permanently within the protected area. The researcher sought to understand their perception of the national park's protection.

Regarding the future of the park and forest protection, 20% of the total population expressed a willingness to actively contribute to forest protection through various means, such as becoming forest department employees, joining NGOs, or participating in semi-government organizations. Meanwhile, 31% argued that it is the government's duty to protect the forest (noting that several forest guard and security personnel positions remain vacant).

Among the residents, 12% believe that addressing bank erosion is a significant measure to protect the park, given that river meanders are expanding, pushing villages closer to the core areas. Furthermore, a substantial portion of the villagers (37%) have faith that nature itself will safeguard and sustain the forests from disasters. This belief reflects the deep-rooted tribal culture, as many tribal communities in Northeast India place their trust in Mother Nature, believing that she will restore balance in her own way.

In reality, the human interaction and adjustment toward sustainable development in the forest villages of Dibru-Saikhowa are still in the early stages of development. Activities such as tree felling, forest clearance for agriculture, overgrazing by domesticated animals, fishing, and hunting have collectively degraded and harmed the entire environment of the young National Park. Unfortunately, some individuals engage in illegal activities like poaching and hunting rare animal species under the guise of forest villagers' identities. However, it is important to note that not all residents of the two forest villages can be held responsible for these actions.

1.5 Summary of the Findings

The analysis and perception study based on the collected data yield the following insights:

1. The majority of the people in the study area express a strong desire to continue living in their forest village, despite facing numerous challenges in effectively adapting to the forest environment.
2. The government has yet to complete the demarcation of village boundaries within the forest through proper survey work. This could potentially assist the forest department in safeguarding the park from ongoing encroachments.
3. The lack of infrastructure development, effective implementation of Forest Act regulations, absence of rehabilitation programs, and recurring flood hazards have direct and adverse impacts on both the forests and the villagers residing within them.
4. Bank erosion and forest clearance activities pose severe and perilous threats to the overall environmental integrity of the national park.

1.6 Conclusion

To facilitate improved interactions, fair adjustments, and proper development of the forest villages in this area, innovative approaches are essential. This necessitates government involvement, rigorous adherence to Forest Act regulations, and the restructuring and reconstruction of the forest villages. Since both villages were already in existence when the Reserve Forest was declared, alternative rehabilitation methods become imperative. Alternative policies for rehabilitation or measures to prevent further encroachment, along with a reconsideration of the unrealistic pressure for land evacuations, could ameliorate the situation.

The present paper represents a partial investigative study focusing on the environmental challenges and human interactions within the forest village of Dibru-Saikhowa. This research aids in understanding how people adapt and interact within forest lands, along with the resulting issues and their potential solutions. Additionally, it is noteworthy that the Forest Department lacks a clear and comprehensive map of the National Park.

In summary, the mode of adjustment of the people residing in Laika and Dadhiya Forest villages remains a critical issue. While these residents are doing their best to minimize harm to the park's environment, the increasing population density in the future presents a looming challenge. The government must take responsibility for both the forest and the people living within it, as their mutual well-being is interlinked.

References

Altman I (1971) An ecological approach to the Functioning of Small Social Groups. In: Rasmussen, JE Individual and group behaviour in isolation, Aldine, Chicago

Beyer JL (1974) Global Summery of Human Response to Natural Hazard: Floods. In White FG (ed) Natural hazards: local, national, global, Oxford

Bhattacharya P (2004) Tourism in Assam: trends and potentialities, Bani Mandir, Guwahati, pp 277–278

Burton I (1968) The human ecology of extreme geographical events. Natural hazard research working paper 1:33

Brototi B, Singh A, Rai PK, Kumar J, Walker S (2022) GIS based study of reclamation of degraded semi-arid soil: A case study of Rajasthan from India. Indian J Environ Prot 42(3):302–315

Champ PE (1973) Approach to the visual environment. Arch Res Teach 2:153–160

Das R, Saikia P, Bhagabati AK (2004) Diversity of tropical grassland birds in Dibru-Saikhowa biosphere reserve. International conference on bird and environment, Nov 21–24

Goswami DC (1993) Assam's land resources: a satellite based ARSAC survey

Kaushik A, Kaushik CP (2004) The perspectives in Environmental Studies. New Age International Publishers, New Delhi, pp 211–236

Kumar HD (2001) Forest resources: conservation and management. Affiliated East West Press Pvt, Ltd

Mathur HS (1988) Essentials of biogeography. Anju Printers, Mahabeer Park Lane, Jaipur, pp 334–345

Mishra VN, Rai PK, Singh P (2021) Geo-information Technology in Earth Resources Monitoring and Management (edit. Book), Nova Science Publishers, USA, pp 1–395. ISBN 978-1-53619-669-6

Rai PK, Mishra VN, Singh P (2021) Recent technologies for disaster management & risk reduction-sustainable community resilience & responses (edit. Book). Springer Nature, Switzerland, pp 1–471. ISBN 978-3-030-76116-5. https://doi.org/10.1007/978-3-030-76116-5

Rai PK, Mishra VN, Singh P (2022) Geospatial Technology for Landscape and Environment Management: Sustainable Assessment & Planning (edit. Book), Springer Nature, Singapore, pp 1–431. ISBN 978-981-16-7373-3. https://doi.org/10.1007/978-981-16-7373-3.

Saharia VB (1982) Wildlife in India. Nataraj Publishers, Dehradun

Singh A, Rai PK, Deka G, Biswas B, Prasad D, Rai VK (2021) Management of natural resources through integrated watershed management in Nana Kosi micro watershed, District Almora, India. Ecol Environ Conserv 27 (February Suppl. Issue), pp S260–S268

Chapter 2
Integrated Hydrologic Assessment of Water Management Options for Regulated Wetlands Under Conditions of Rainfall Change: A Case Study of Bhandara District, India

Nanabhau S. Kudnar

Abstract The current study's examination of rainfall and groundwater exploration on biodiversity issues and mitigation was done in hard rock regions including granite, gneisses, and schist. Since 1991, the district has engaged in various phases of groundwater exploration. The method applies for IWD for 19 exploratory wells (EW) and 6 observation wells (OW) were drilled between 1991 and 1992. Additionally, between 1999 and 2011, three Piezometers (Pz) were built for the purpose of monitoring groundwater levels. Where there is a data gap, groundwater investigation has been undertaken down to a depth of 200 mbgl, and, as a result, seven exploratory wells have been built throughout the years 2021–2022. The district typically receives 1298.5 mm of precipitation per year, with a range of 48–93 wet days. Data on annual rainfall from 1998 through 2021 is analyzed. According to this, the wettest year was 2013 (1691.2 mm), and the driest was 2004 (765.3 mm). The investigation of rainfall trends reveals a decreasing tendency at 2.9 mm/year. Drinking or using polluted water for a long time can cause various diseases in humans and animals such as typhoid and cholera. When the toxic substance is mixed with water, it reduces the amount of oxygen in the water which creates problems for aquatic animals while respiring under water.

Keywords Climate change · Biodiversity issues and mitigation strategies · Bhandara district

N. S. Kudnar (✉)
C. J. Patel College Tirora, Gondia District, Maharashtra 441911, India
e-mail: nanabhaukudnar@gmail.com

P. K. Rai et al. (eds.), *Natural Resources Management and Sustainable Livelihoods in the Mountainous Region*, Advances in Geographical and Environmental Sciences,
https://doi.org/10.1007/978-981-97-2100-9_2

2.1 Introduction

Environmental problems, climate change, and global warming are no longer just topics for international seminars and conferences, but have affected every aspect of our lives. Be it a deadly heat wave or a dust storm from the Gulf. For the last 6 months, Maharashtra has been experiencing various changes in nature and seasons (Rai et al. 2018; Addisu et al. 2015; Aher et al. 2019; Bisen and Kudnar 2019; Mishra et al. 2019). Whether all these developments are related to global warming will be proven by research in the coming time, but this graph of changes in the last 6 months is certainly startling and introspective. These incidents should be taken seriously as nature's danger warning and efforts to conserve nature and environment should be increased (Fathian et al. 2014; Jain and Kumar 2012; Subash and Sikka 2014). Since we started using fossil fuels, gas, and coal to heat power plants, transport, and homes, the world has warmed by 1.2 °C.

It is very important to have two things for rain to have enough water vapor in the air and to have such a means by which the vaporized air can cool down and condense. Today we are going to give you detailed information about rain. Today, in this article, we will read how rain occurs, its types, how it is measured, how it is distributed in the world, etc. Convectional rain mainly occurs in the regions around the equator because due to excessive heat there, the air gets heated up and at the same time the water of the sea starts rising rapidly by turning into vapor. The air expands by going up and after cooling it becomes liquefied or condensed (Mishra and Rai 2016; Mishra et al. 2018; Kudnar et al. 2022; Ahmad et al. 2020; Ghosh et al. 2021; Gaikwad et al. 2023; Kudnar and Rajashekar 2023; Kudnar et al. 2023). Thus it starts raining. During the rains here, dark clouds are seen floating around. The rain is accompanied by thunder and torrential. Mountain rain (relief rain) or surface rain is the highest rainfall in the world, in which the air filled with vapor has to rise above the mountain and cool down. The air gets condensed after cooling and it starts raining. India's summer rainfall (by monsoons from the Arabian Sea and the Bay of Bengal) is of this type. When there is mountain rain, the leeward side of the mountains experiences more rain than the other, because the raised air starts to descend and pass the mountains after the rain, causing pressure to build and warming the ground. Due to coming in contact with water, it gets heated up and cannot rain. These parts of the mountain are called rain-shadow areas. For example, the Deccan Plateau of India falls in the rain-shadow of the Western Ghats in the rain-shadow of the Central Asian Himalayas. Cyclonic rains mainly occur in temperate countries (in the temperate zone where westerly winds blow). It rains due to cyclones. In cyclones, the air moves rapidly toward the center and is forced to rise. It is full of water vapor due to coming through the sea. When the cold polar air comes in contact with it, a kind of marginal region is formed in the middle and the vaporized hot air cools down and causes rain. This is cyclonic rain in which storms come more. This rain is not torrential, but it is in the form of light showers throughout the year. In tropical regions, cyclonic rains occur in summer because this season is favorable for the formation of cyclones. The lines that are drawn joining the places of equal rainfall on the map of the world are called isohyets. But the change of rainfall happens gradually from one place to another, so these lines are not being used to show the rainfall (Shortridge 2019; Rupa et al. 1992; Sivakumar et al. 2005; Anand et al. 2021; Bessah et al. 2021; Mishra et al. 2021; Rai et al. 2021, 2022; Rai 2023a and b). Nowadays underground storage tanks are used to transport petroleum and cooking gas through pipelines, sometimes these pipelines and storage tanks leak and contaminate the surroundings including water. When pollutants are released on the ground, these pollutants find their way to the ground water and contaminate it, it is called groundwater pollution.

The primary goals include a case study of Bhandara District, the influence of ground water and climate change on biodiversity issues, and mitigation techniques. Most of the rainfall here comes from the southwest monsoon. In India, the Tropic of Cancer passes from east to west. It has a direct impact on the climate of the country. Of the summer, winter, and rainy seasons, the rainy season is often called monsoon.

2.2 Study Area

Bhandara district is situated in Wainganga basin. It has a common border on North with Seoni and Balaghat districts of Madhya Pradesh and is surrounded by Chandrapur district on the south, Gondia district on the east and Nagpur district on the west. The total area of the district is 3716.65 Sq. Kms. which is about 1.21% of the total area of Maharashtra State. It comes under the area, there are 878 villages and there are also some gram panchayats whose information will be published soon. Bhandara 21°11′ north 80°00′ east Bhandara (Fig. 2.1) is situated at an altitude of 244 m above sea level, Bhandara is 855 km northeast of Mumbai, the capital of Maharashtra, on Highway 53 and 1188 km southeast of India's capital Delhi, on National Highway 44. In this district, the number of women for every thousand men is 982. The same ratio is 983 in rural areas and 981 in urban areas. Tumsar, Lakhandur Taluka has the highest ratio of 992 and Mohadi, Lakhani Taluka has the lowest ratio of 974. 83.8% people are literate in this district. Among them, the percentage of males and females is 90.4 and 77.1%, respectively. Literacy is 90.7% in urban areas and 82.1% in rural areas (Kudnar 2022). According to religion in the district, Hindus 84.1%, Buddhists 12.9%, Muslim 2.2%, Christian 0.2%, Jain 0.1%, Sikh 0.1%, other 0.3%, and unspecified population 0.2%. Wainganga is the major river in the district (Kudnar 2020a, b; Kudnar and Rajasekhar 2020). The flow of this river in the district does not dry up even in summer. Bavanthadi, Chulband, Kanhan, and Bagh are dams in this district. The district has three sub-divisions, namely, Bhandara, Tumsar, and Sakoli. Bhandara subdivision has 2 talukas, Tumsar subdivision has 2 talukas, Sakoli subdivision has 3 talukas, and has 878 villages (Bisen et al. 2022).

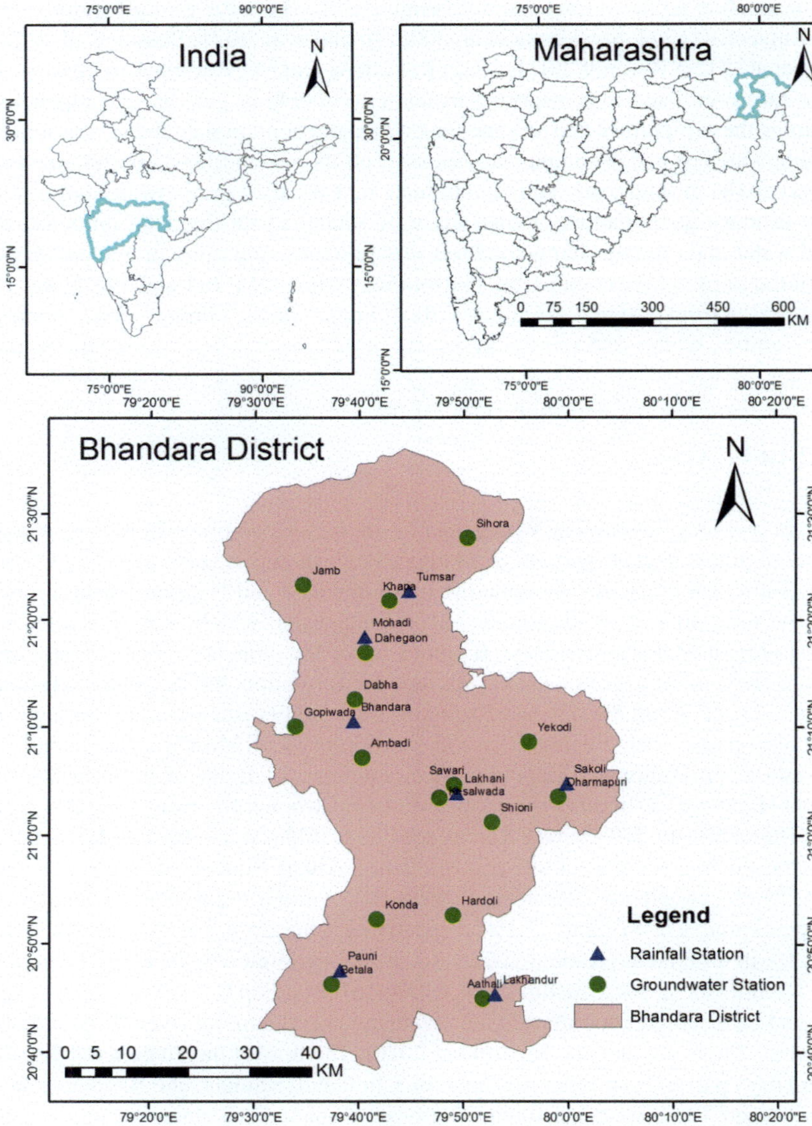

Fig. 2.1 Study area

2.3 Methodology

2.3.1 Standard Deviation

The standard deviation, which assesses how "spread out" the data are, is an often used indicator of dispersion. Although in practice it is less resilient than anticipated deviation or mean absolute deviation, standard deviation is more practical algebraically. Let X be a random variable with mean value μ:

$$E[X] = \mu$$

Here, the operator E stands for the mean or expected value of X. So, the standard deviation of X is

$$\sigma = E[(X - \mu)2].$$

In general, the standard deviation of a continuous real-valued random variable X with probability density function $p(x)$.

$$\sigma = \int (x - \mu)2\, p(x)\, dx$$

The root-mean-square (RMS) deviation of a discrete random variable's values from the mean is the standard deviation for that variable, which helps to explain why while the average is similar but the city near the coast has a narrower range, even though the average temperature of the two cities may be 15 °C. Contrary to expectations, inter-country variation is more likely to exist. The ability to calculate standard deviation enables us to understand that a city average with high standard deviation and large variance will not be as accurate at predicting temperature as a city average with lower standard deviation and less variation.

2.3.2 Coefficient of Variation

Data points scattered around the mean refer to the coefficient of variation. It represents the data dispersion across multiple series of measurement data. The coefficient of variation is a comparatively better and more reliable alternative than the traditional variance, which focuses only on the mean of the data. The former provides a plethora of effective tools that work wonders when comparing data points between multiple data series. Read more at: https://www.fincash.com/l/en/basics/coefficient-of-variation.

2.3.3 Ground Water and Hydrological Investigations

When hydrogeological research is mentioned, the following studies are mainly: determining underground geology using various sampling methods, determining and evaluating the flow and direction of ground water, determining the movement of water pollution in shallow underground sediments and aquifers, and studying its effects (Ahmad et al. 2020; Meshram and Ingle 2012; Rajasekhar et al. 2021). The contamination of groundwater, creation of techniques for treating water pollution, and implementation of these techniques. These are all hydrology-based investigations (Deshmukh and Aher 2016: Rajasekhar et al. 2020, 2021; Rajankar et al. 2011).

2.4 Results and Discussion

2.4.1 Rainfall Distribution

The average annual rainfall (Table 2.1) in Bhandara is about 180 cm, but there is great spatial variation. The climate of India is of "Tropical Monsoon" type. Monsoons are prominent within the tropics, especially on the eastern edge of the great land, but in Asia, it also occurs outside the tropics in China, Korea, and Japan. When the month of June begins, we hope for the arrival of rains. Being disturbed by the heat, we eagerly wait for the rain. We have been learning since childhood that our country gets maximum rainfall due to Southwest Monsoon winds. But how exactly do these winds form, how exactly do they travel across the country? "Monsoon set in Kerala," we see this every year before the rains start in newspapers or news channels. Monsoon makes its first appearance in the states bordering the Arabian Sea. As the monsoon winds flow from southwest to northeast, it first hits the southern state of Kerala on the Indian mainland. Monsoon generally enters Kerala on June 1. After March, as the sun moves north from the equator, the sun's rays fall perpendicularly to the northern hemisphere. Hence, a large low-pressure zone is formed there. But at the same time, the southern hemisphere has a cold climate due to the oblique rays of the sun. As a result, a zone of high pressure exists. High pressure constantly gives way to low pressure in winds. Currently, vapor winds are pouring from the high-pressure band to the south of India due to the spread of the Indian Ocean. So when they enter mainland India, these vaporous clouds collide with the north–south Western Ghats and give very heavy rainfall. That is why monsoon winds cause very heavy rainfall in the Konkan coast and some regions of the Ghats in Maharashtra. Monsoon season in India generally lasts from June to September. During this period, southwest monsoon rains in the country, while from September to December, northeast monsoon rains in some parts of the country. The period from June 1 to September 15 is considered as the southwest monsoon period in India. The period from September 15 to December 1 is the northeast monsoon. This is also called return monsoon. Northeasterly winds blow over land and are therefore drier, with less water vapor. Hence, India does

Table 2.1 Rainfall analysis 1998–2021

Sr. no	Year	Annual	Normal	% departure	No. of rainy days	Category
1	1998	1266	1439.8	−12	93	Normal
2	1999	1108.7	1439.8	−23	93	Normal
3	2000	1139.5	1439.8	−21	60	Normal
4	2001	1355.1	1439.8	−6	70	Normal
5	2002	1177.7	1439.8	−18	56	Normal
6	2003	1393.3	1439.8	−3	80	Normal
7	2004	765.3	1439.8	−47	56	Moderate
9	2005	1622.8	1439.8	13	68	Normal
10	2006	1031.9	1439.8	−28	62	Moderate
11	2007	1313	1439.8	−9	73	Normal
12	2008	974.4	1439.8	−32	70	Moderate
13	2009	983.6	1439.8	−32	65	Moderate
14	2010	1558.7	1439.8	8	79	Normal
15	2011	1113.1	1439.8	−23	79	Normal
16	2012	1363.2	1439.8	−5	72	Normal
17	2013	1691.2	1439.8	17	80	Normal
18	2014	799.9	1439.8	−44	48	Moderate
19	2015	1078.3	1439.8	−25	57	Normal
20	2016	927.9	1439.8	−36	64	Moderate
21	2017	850.5	1439.8	−41	62	Moderate
22	2018	1029.6	1439.8	−28	51	Moderate
23	2019	1286.5	1439.8	−11	79	Normal
24	2020	1368.3	1298.5	5	86	Normal
25	2021	1247.9	1298.5	−4	88	Normal

Source IMD, Pune

not receive large amounts of rain from the northeast monsoons. As only the winds blowing over the Bay of Bengal collect a small amount of water vapor, they bring rain over parts of Tamil Nadu and Andhra Pradesh. When the hot and humid winds are blocked by the mountainous parts, they start rising with the help of their slope, due to which the temperature of the air starts decreasing by 5.5 °F per 1000 ft due to the "dry adiabatic rate," due to which the wind gets colder and its relative humidity starts increasing. As one ascends higher, the air becomes saturated and begins to rise again. This causes condensation to begin, which mixes the latent heat of the condensation into the air. Eventually, the air rises once more and reaches the limit where the temperature of the air above it equals its own, as well as the air moist adiabatic temperature-fall rate (per 1000 feet). But after getting colder than 3.6 °F, it rains even on increasing height. The slope of the mountain on which the Guava collides is called the windward slope and it rains on this slope. It is also called Vrishti

Dhal. The slope of the mountain with the help of which the wind descends is called windward slope. The area situated on the leeward slope of a mountain range where the average rainfall is relatively less. The moist air rises on the windward side of the mountain range and its condensation causes heavy rainfall on this part. When these winds cross the mountain range and descend down the windward slope, they become warm and dry due to which there is very little rainfall in this part. The region located east of the Western Ghats in India is a prime example of rain-shadow region. The average annual rainfall is 300 cm at many places on the western slope of the Western Ghats. On its eastern slope (rain-shadow zone), it receives extremely little rainfall, and in certain locations, the average annual rainfall is reported to be less than 60 cm. According to meteorologists, this year, the return rains have started from October 3 to 4. The effect of return rains has also started to be seen in different parts of the state. In many places, crops like rice, urad, and onion are affected. Farmers are worried that if the sugarcane harvesting is delayed, the sugarcane will be damaged due to delay and the standing crop will be wasted. Farmers of different areas such as Bhandara, Gondia have suffered. According to meteorologists, anti-cyclones form on the border between Rajasthan and Pakistan. Anti-cyclone is the formation of high-pressure area. Once a high-pressure area is formed, the air is thrown toward the sea. But this process is happening gradually. So initially the return journey of monsoon starts from Rajasthan, then comes to Gujarat and then to Maharashtra. Currently, he is on the border of Maharashtra for the last 8–9 days. In such a case, the moisture that comes to us from the sea has no place to go, pressure builds up on it and it rains. The rain coming from Rajasthan cannot go further because of the moisture from the sea and it falls in Maharashtra, it is called return rain. The criteria currently used by the IMD for declaring return rainfall was adopted in 2006. But recently the rain has disappeared even in the rainy month. Therefore, the criteria formulated for return rains are considered in the last month, i.e., after September 1. No rain at all for 5 consecutive days in an area. An anti-cyclone condition appears in the last layer of the atmosphere (Fig. 2.2a–e).

The district typically receives 1298.5 mm of precipitation per year, with a range of 48 to 93 wet days. Data on annual rainfall from 1998 through 2021 is analyzed. According to this, the wettest year was 2013 (1691.2 mm) and the driest was 2004 (765.3 mm). The investigation of rainfall trends reveals a decreasing tendency at 2.9 mm/year. The investigation of the yearly rainfall reveals that the percentage difference between the annual rainfall and the typical rainfall ranged from −47 to +17%. The deviation % examined indicates the pattern of rainfall variation in relation to the period's average rainfall. As seen in Table 2.1, the region experienced moderate drought conditions and 16 times (67%) the usual rainfall. However, there were exceptional years 2004, 2008, 2014, 2016, and 2017 falling outside of this range in which the area received the less rainfall during the period (Nair and Mirajkar 2021; Reddy et al. 2000; Salunke et al. 2020; Sinha et al. 2019).

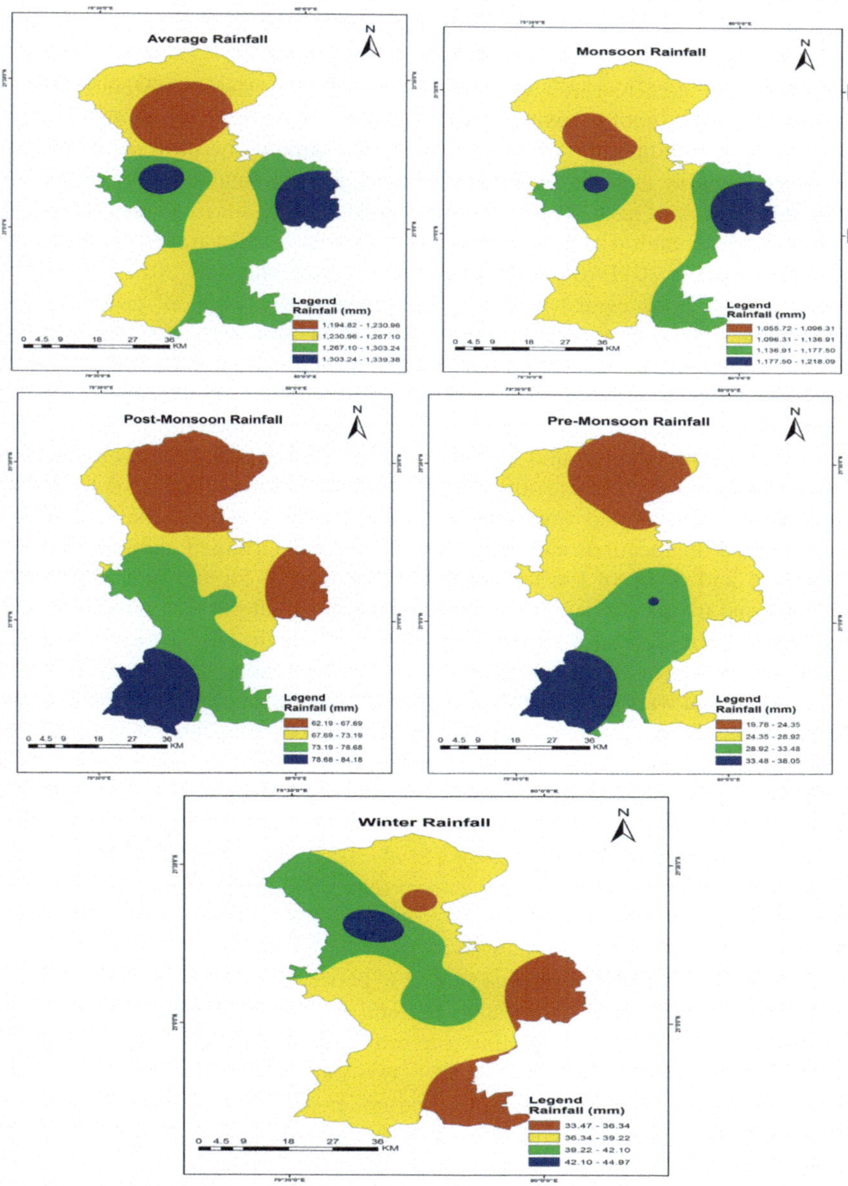

Fig. 2.2 Rainfall distribution. **a** Average. **b** Monsoon. **c** Post-monsoon. **d** Pre-monsoon. **e** Winter

2.4.2 Ground Water and Hydrological Investigations

When hydrogeological research is mentioned, the following studies are mainly: determining underground geology using various sampling methods, determining and evaluating the flow and direction of ground water, determining the movement of water pollution in shallow underground sediments and aquifers, and studying its effects. The contamination of groundwater, creation of techniques for treating water pollution, and implementation of these techniques. All these studies are hydrology-based studies (Singh et al. 2011; Srivastava et al. 2011; Kushwaha et al. 2022).

This method is necessary to check the infiltration capacity of ground water. Streams and water flows resulting from geological processes are controlled by different landforms. River-borne modern alluvial terraces, flood plains, moraine deposits, and glacial outcrops are some of the positive landforms for ground water. Geological units such as flood plains, soil types, and aquifers are the factors that control the occurrence of ground water (Table 2.2). Today, with the development of computer technology, various softwares are used in land modeling. In determining groundwater resources and water basins, a preliminary survey of the land is first carried out and the locations and depth of drilling are determined, thus determining the location and depth of the underground. In addition, groundwater flow rates, directions, and trends are determined. In the meantime, all the data combined with the geological structure and all other characteristics of the soil are entered into a program and thanks to the modeling of the ground, the trend of the role, the relationship of the ground water, the characteristics of the flow with different densities, and the aquifer parameters are determined, in short, solutions for groundwater management are prepared are done. The main rock formations in the district include Vindhyan/Penganga Shale and Limestone (1%), Gondwana Sandstone (3%), Alluvium (9% along the major rivers), and Archean Crystalline and Metamorphics (Kumar et al. 2022; Gosh and Gosh 2017; Kanga et al. 2017). Aquifer-wise characteristics have been identified based on the district's current groundwater exploration, and are displayed in Table 2.3.

Wetlands consist of large expanses of land that are regularly completely flat, reflecting the fact that their surface is subject to permanent and occasional flooding; The moment it is covered with water, the soil becomes saturated, reducing the aewze1mount of oxygen and creating a hybrid environment with the presence of aquatic and terrestrial species. River water is very important for plants, animals, and people, as it provides a variety of species to rural areas. For this type of ecosystem, when floods occur, they allow for the quality of water that sustains life around it, feeding plants and insects that form the basis for the food chain. Wetland ecosystems comprise a range of natural settings that characterize this biome, including peat bogs, an acidic wetland that gathers organic debris, and marshes, which are characterized as impure and relatively shallow layers of water that support a diversity of aquatic organisms. Among others, Pete. These surroundings were all formed organically without the help of humans. Anthropogenic wetlands can also be found, referring to environments that have been artificially modified or created by humans, the main

Table 2.2 Salient features of groundwater exploration

Block	Formation	Depth	SWL	Discharge	Zones	T	S
	Aquifer	(mbgl)	(mbgl)	(lps)	(mbgl)	m^2/day	
Bhandara	Gneiss	40.1–500.11	1.93–9.99	0.12–2.59	38–229	0.20–0.71	–
	Schist	73.86–201	2.2–12.75	Traces-3.19	24.8–49.8	62	8.7 × 10–4
Lakhandur	Gneiss	201–203.56	12.1–45.4	0.77	51–54	–	–
	Schist	146.7–147.3	3.43–5.66	5.7–6.42	43.9–156	23.86	1.49 × 10–4
Lakhni	Gneiss	172.2	3.76	0.33	108–112	–	–
Mohadi	Schist	179.3	3.67	0.16	61–62	1.72	–
	Gneiss	202	5.5	0.18	–	–	–
	Alluvium	9.1	2.8	Traces	–	–	–
Pauni	Schist	41–201.7	4.2–10.05	Traces-2.14	15–34.05	22	6.6 × 10–4
	Gneiss	201	14	0.17	29–180	–	–
Sakoli	Schist	104.6	8.7	Traces	23–35.90	–	–
	Granite	201.7	–	–	–	–	–
Tumsar	Gneiss	97.6–134	5.30–30.2	0.15–8.69	31–71	35.26	2.4 × 10–5
	Schist/ Quartzite	147–201	5.2–50.00	0.39–5.9	41–174	6.85	1.81 × 10–4
Total		9.1–500.09	1.90–50.00	Traces-8.78	0.79–180	0.22–62	1.49 × 10–4 to 2.4 × 10–5

Table 2.3 Chemical constituents (shallow aquifer)

Parameter	Drinking water standards (IS-10500–2012)		Shallow aquifer					
			Samples (<DL)		Samples (DL-MPL)		Samples (>MPL)	
	DL	MPL	No	%	No	%	No	%
pH	6.6–8.4	–	241	94.8	–	–	14	5.2
F (mg/L)	1.1	1.4	236	92.95	17	7.05	–	-
TDS (mg/L)	501	2001	218	85.00	39	14.8	–	–
Mg (mg/L)	31	101	18	82.63	5	17.38	–	0
TH (mg/L)	301	601	15	5.48	93	36.09	147	58.44
SO_4 (mg/L)	201	401	251	101	–	0	–	0
Ca (mg/L)	74	201	8	39.14	13	52.19	4	8.8
NO_3 (mg/L)	44	No relaxation	150	58.42	–	–	105	41.58
Cl (mg/L)	249	1001	192	75.67	61	23.52	3	0.79

purpose of which is to retain water and develop an environment rich in aquatic plants. This type of environment is developed in the open air, taking into account all the ideal environmental conditions for it, they can be widely viewed as natural reserves or natural benefits. Also in small expanses such as lakes or shallow ponds, deep pond, dam, etc. (Table 2.4).

Every year, during the pre-monsoon season (May), groundwater samples are taken from GWM wells. To understand the quality scenario of the shallow aquifer, the groundwater quality data from 255 monitoring wells operated by CGWB and GSDA

Table 2.4 Chemical constituents (deeper aquifer)

Parameter	Drinking water standards (IS-10500–2012)		Deeper aquifer					
			Samples (<DL)		Samples (DL-MPL)		Samples (>MPL)	
	DL	MPL	No	%	No	%	No	%
pH	6.4–8.4	–	122	97.59		0	3	2.43
F (mg/L)	1	1.6	81	84.67	14	15.32	–	0
TDS (mg/L)	501	2000	55	45.19	67	54.88	–	0
Mg (mg/L)	31	101	15	68.43	5	26.33	1	5.25
TH (mg/L)	302	601	60	49.12	53	43.56	9	7.24
SO_4 (mg/L)	202	400	100	100	–	0	–	0
Ca (mg/L)	73	201	19	94.75	1	5.27	–	0
NO_3 (mg/L)	44	No relaxation	54	56.13	–	–	44	43.89
Cl (mg/L)	252	1001	105	83.04	20	16.95	–	0

were used. The quality scenario of the deeper aquifer has been determined using data from 124 exploration wells, tubewells, and borewells that represent the CGWB and GSDA.

2.4.3 Water Pollution and Biodiversity Issues

The mighty Wainganga River flows through Bhandara District. There is a big dam at Gose on this perennial river. But lately the sewage from Nagpur gets mixed with Wainganga through Nag River. Due to this, the water of Wainganga has become polluted. Recently there has been a huge increase in this pollution and pollution has increased the turbidity. Also, ecornia, a water plant, and seaweed have increased in abundance. Sewage from the Nag River has contaminated the Wainganga River, which flows through the region, and the Gosekhurd Dam's holding water has become home to a vast number of aquatic plants and insects. The National Environmental Engineering Research Institute (NIRI) has reported that the turbidity of the river water has increased to a great extent and the water of Wainganga is unfit for drinking without purification. But even after that effective measures are not taken, it is likely to affect agriculture, humans, and animals. According to this, on the one hand, the amount of ecornia plant on the stagnant water has decreased the oxygen content of the water, while, on the other hand, Nag River is also increasing the pollution. The water is used for irrigation in the fields on the banks of Wainganga. But this water is feared to be dangerous to the crops and also to the person who comes in contact with this water. The river water has a normal pH and acidity level. Moreover, it has been disclosed in this report that the level of chemicals like calcium, magnesium, sodium, potassium etc. is normal and the amount of heavy metals except magnesium is also negligible. The turbidity of Wainganga water is 10 NTU at the top and 17 NTU at the bottom. A level of five NTU is considered normal. Nitin Tumane of Wainganga Rescue Abhiyaan Nagar Samiti has taken the initiative for the purification of Wainganga. Niri's suggestion is necessary to remove aquatic plants from the river bed to prevent organic pollution. Drinking water is only possible after proper purification. Safe irrigation practices are essential because of the risk of microbial contamination. It is necessary to stop the discharge of sewage of Nag River into Wainganga. The Wainganga River originates from the mountain ranges of Shivni District in Madhya Pradesh and enters the Bhandara after traveling through Balaghat, Gondia District. The river passes through the important towns of Tumsar, Bhandara, Pavani, and Lakhandur in the district and further into Gadchiroli District. The journey of Vainganga in Bhandara District is about 200 km. Wainganga is also joined by many tributaries on this journey. Indira Sagar Project was constructed on this river at Gosekhurd in Pavani Taluka. The drinking water problem started in the upper part of the project since the Waingange water was blocked in this project. Contaminated water of Nag River flowing from Nagpur City is responsible for this. Today, although the Wainganga vessel seems to be full for 12 months, this water is not potable. Today, although the Wainganga vessel seems to be full for 12 months, this water is not potable. Stagnant water causes annual growth of ecornia plants in the river bed. As a result, water supply schemes of villages and towns upstream of the dam have been affected. Bhandara City is the most affected. Due to blocking of water in Gosekhurd Dam, the "back water" of Wainganga has reached Bhandara Town. Every year Bhandara City gets flooded due to this water. However, no concrete measures have been taken yet to control the flooding caused by back water. Every year Wainganga is overgrown with ecornia plant. There is only discussion at the administrative level to eliminate this plant. Previously, grainy sand was seen in Wainganga. Now only silt is visible due to perennial water. The system has not yet paid attention to remove this sludge. The range of water quality is pH1.69, EC 2599.0, TDS 1701, TA 404, TH 547, Na^+ 490, Do 2.58, K^+ 399, COD 16.70, Ca^{2+} 219.960, Mg+2 101.00, HCO_3 412, Cl^- 3001.80, SO_2^{-4} 469.00, NO^-; 3 278.89, F^- 1.71.

Bhandara District was once known as the "District of Lakes," but gradually this identity is dying out. As the city expands, the extent of the lakes decreases and their pollution increases (Fig. 2.3). The water of the lakes has become polluted due to all the reasons like wastewater, sewage, animals, washing clothes, bathing in the city, and has covered them with algae. From that, they are moving toward degradation. Earlier, people used to go swimming in the lakes in the scenic surroundings at dawn. But at present, seeing the pitiful state of the lakes, people are staying away. Water is a natural resource. So it is our moral responsibility to keep the available water clean and pollution free. However, the reality is that except for a few or two social workers, most of them do not follow this moral responsibility. So it is time to start a movement to save water and make the available water potable. The relationship between nature and man is similar to that of other animals. Therefore, if we look at other animals and birds in nature, no element pollutes water. On the contrary, fish, crabs, herons, ducks, etc., all these organisms work to balance the ecological chain of water. However, the opposite is the case with humans. The water bodies that our forefathers consciously conserved near the village are now beset by the problem of pollution.

What should be done? Prohibition on bathing, washing clothes and animals during festivals, restrictions on disposal of impure water, and preventing sewage strict enforcement designing for continuous flow of water. Permanent measures to prevent water fouling, algae build-up, measures to maintain potable quality by testing the water once a month, pumping water in a limited manner as needed, so that there is room for constant mixing of freshwater. In addition to the increase in iron, zinc, and chromium metals that do not dissolve in water, faucal coliform, a toxic germ produced by human excrement, has been found in some lakes in an inspection. Due to the growth of blue green algae plants, the color of the water has changed, and if this poisonous water enters the mouth, many diseases will be invited.

Fig. 2.3 GPS location

2.5 Conclusion

A wide variety of living things on Earth have inspired human life for thousands of years. Additionally, between 1999 and 2011, three Piezometers (Pz) were built for the purpose of monitoring groundwater levels. Where there is a data gap, groundwater investigation has been undertaken down to a depth of 200 mbgl, and, as a result, seven exploratory wells have been built throughout the years 2021–2022. The district typically receives 1298.5 mm of precipitation per year, with a range of 48 to 93 wet days. Data on annual rainfall from 1998 through 2021 is analyzed. The district typically receives 1298.5 mm of precipitation per year, with a range of 48 to 93 wet days. Data on annual rainfall from 1998 through 2021 is analyzed. According to this, the wettest year was 2013 (1691.2 mm), and the driest was 2004 (765.3 mm). The investigation of rainfall trends reveals a decreasing tendency at 2.9 mm/year. The investigation of the yearly rainfall reveals that the percentage difference between the annual rainfall and the typical rainfall ranged from -47 to $+17\%$. Each animal can develop via its activities thanks to the support system created by the diversity of living things. It is also applied to development. Petroleum and cooking gas are now transported by pipelines via subterranean storage tanks; occasionally, these pipelines and storage tanks leak and contaminate the surrounding area, including the water. When pollutants are released on the ground, these pollutants find their way to the ground water and contaminate it, and it is called groundwater pollution. Agricultural activities such as the use of fertilizers and pesticides are the main causes of groundwater pollution. Many different chemicals are used in industrial and agricultural work. Chemicals used by these industries are released into the water bodies without treatment which contaminates the surface water as well as ground water. This is the most common type of water pollution.

References

Addisu S, Selassie YG, Fissha G, Gedif B (2015) Time series trend analysis of temperature and rainfall in lake Tana Sub-basin, Ethiopia. Environ Syst Res 4–25(1):1–12. https://doi.org/10.1186/s40068-015-0051-0

Aher S, Shinde S, Gawali P et al (2019) Spatio-temporal analysis and estimation of rainfall variability in and around upper Godavari River basin, India. Arab J Geosci 12:682. https://doi.org/10.1007/s12517-019-4869-z

Ahmad S, Israr M et al (2020) Spatio-temporal trends in snow extent and their linkage to hydro-climatological and topographical factors in the Chitral River Basin (Hindukush, Pakistan). Geocarto Int 35(7):711–734. https://doi.org/10.1080/10106049.2018.1524517

Anand A, Singh A et al (2021) Rainfall rate estimation over India using global precipitation measurement's microwave imager datasets and different variants of fuzzy information system. Geocarto Int. https://doi.org/10.1080/10106049.2021.1936208

Bessah E, Boakye EA, Agodzo SK et al (2021) Increased seasonal rainfall in the twenty-first century over Ghana and its potential implications for agriculture productivity. Environ Dev Sustain 23:12342–12365. https://doi.org/10.1007/s10668-020-01171-5

Bisen DK, Kudnar NS (2019) Climatology. Sai Jyoti Publication, Nagpur, pp 1–211

Bisen DK, Kudnar NS, Borude S et al (2022) Geo-spatial modeling in the assessment of environmental resources for sustainable water resource management in a semi- arid region: a case study of Bhandara district, India. Int J Sci Res Sci, Eng Technol 9(4), 286–299. https://doi.org/10.32628/IJSRSET229445

Deshmukh KK, Aher SP (2016) Assessment of the impact of municipal solid waste on groundwater quality near the Sangamner City using GIS approach. Water Resour Manage 30:2425–2443. https://doi.org/10.1007/s11269-016-1299-5

Fathian F, Dehghan Z, Bazrkar MH, Eslamian S (2014) Trends in hydrologic and climatic variables affected by four variations of Mann- Kendall approach in Urmia Lake basin, Iran. Hydrolsci J 61(5):892–904. https://doi.org/10.1080/02626667.2014.932911

Gaikwad V, Singh K, Salunke V, Kudnar N (2023) GIS-based comparative analysis of lineament extraction by using different azimuth angles: a case study of Mula river basin, Maharashtra, India. Arab J Geosci 16:538. https://doi.org/10.1007/s12517-023-11636-2

Ghosh S, Ghosh S (2017) Evaluating patterns of urban growth in Faridabad sub-region of NCR, India using remote sensing, GIS and entropy approach. Indian Cartogr 37:258–261

Ghosh S, Krishna Vidhata N, Kumar S, Midya K (2021) Seasonal contrast of land surface temperature in Faridabad. https://doi.org/10.4018/978-1-7998-2249-3.ch008

Jain SK, Kumar V (2012) Trend analysis of rainfall and temperature data for India. CurrSci 102(1):37–49

Kanga S, Tripathi G, Singh SK (2017) Forest fire hazards vulnerability and risk assessment in Bhajji Forest Range of Himachal Pradesh (India): a geospatial approach. J Remote Sens GIS 8(1):25–40

Kudnar NS (2020a) GIS-based assessment of morphological and hydrological parameters of Wainganga river basin, Central India. Model Earth Syst Environ 6:1933–1950. https://doi.org/10.1007/s40808-020-00804-y

Kudnar NS (2020b) GIS-based investigation of topography, watershed, and hydrological parameters of Wainganga River Basin, Central India, sustainable development practices using geoinformatics. Scrivener Publishing LLC, pp 301–318. https://doi.org/10.1002/9781119687160.ch19

Kudnar NS (2022) Geospatial modeling in the assessment of environmental resources for sustainable water resource management in a semiarid region: a GIS approach current directions in water scarcity research, vol 7, pp 133–151, https://doi.org/10.1016/B978-0-323-91910-4.00009-1

Kudnar NS, Rajasekhar M (2020) A study of the morphometric analysis and cycle of erosion in Waingangā Basin, India. Model Earth Syst Environ 6:311–327. https://doi.org/10.1007/s40808-019-00680-1

Kudnar N, Rajashekhar M (2023) Using geo-spatial technologies for land and water resource development planning: a case study of Tirora Tehsil, India. In: Balaji E, Veeraswamy G, Mannala P, Madhav S (eds) Emerging technologies for water supply, conservation and management. Springer water. Springer, Cham. https://doi.org/10.1007/978-3-031-35279-9_15

Kudnar NS, Mishra VN, Rajashekhar M (2023) Hydro-chemical characterization and geospatial analysis of groundwater for drinking and agriculture usage in Bagh River Basin, Central India. In: Rai PK (eds) River conservation and water resource management. Advances in geographical and environmental sciences. Springer, Singapore. https://doi.org/10.1007/978-981-99-2605-3_6

Kudnar NS, Diwate P, Mishra VN, Srivastava PK, Kumar A, Pandey M (2022) Spatio-temporal variability and trend analysis of rainfall in Wainganga river basin, Central India, and forecasting using state-space models. Theor Appl Climatol 150:469–488. https://doi.org/10.1007/s00704-022-04168-4

Kumar S, Ghosh S, Singh S (2022) Polycentric urban growth and identification of urban hot spots in Faridabad, the million-plus metropolitan city of Haryana, India: a zonal assessment using spatial metrics and GIS. Environ Dev Sustain 24:8246–8286. https://doi.org/10.1007/s10668-021-01782-6

Kushwaha N, Elbeltagi A, Mehan S et al (2022) Comparative study on morphometric analysis and RUSLE-based approaches for micro-watershed prioritization using remote sensing and GIS. Arab J Geosci 15:564. https://doi.org/10.1007/s12517-022-09837-2

Meshram RR, Ingle KB (2012) Mineralogy and origin of dumortierite from Girola area, Bhandara district, Eastern Maharashtra. J Geol Soc India 79:181–188. https://doi.org/10.1007/s12594-012-0020-4

Mishra VN, Rai PK (2016) A remote sensing aided multi-layer perceptron-Markov chain analysis for land use and land cover change prediction in Patna district (Bihar), India. Arab J Geosci 9:249. https://doi.org/10.1007/s12517-015-2138-3

Mishra VN, Rai PK, Prasad R, Punia M, Nistor MM (2018) Prediction of spatiotemporal land use/land cover dynamics in rapidly developing Varanasi district of Uttar Pradesh, India using Geospatial approach: a comparison of hybrid models. Appl Geomat 10(3):257–276

Mishra VN, Prasad R, Kumar P, Gupta DK, Agarwal S, Gangwal A (2019) Assessment of spatio-temporal changes in land use/land cover over a decade (2000–2014) using earth observation

datasets: a case study of Varanasi district, India. Iranian J Scie Technol Trans Civil Eng 43(1):383–401
Mishra VN, Rai PK, Singh P (2021) Geo-information technology in earth resources monitoring and management (edited book). Nova Science Publishers, USA, pp 1–395. ISBN: 978-1-53619-669-6
Nair SC, Mirajkar AB (2021) Spatio–temporal rainfall trend anomalies in Vidarbha region using historic and predicted data: a case study. Model Earth Syst Environ 7:503–510. https://doi.org/10.1007/s40808-020-00928-1
Rai PK (2023a) River conservation and water resource management. Springer Nature, Singapore, pp 1–352. ISBN No.: 978-981-99-2604-6. eBook ISBN: 978-981-99-2605-3
Rai PK (2023b) Advances in water resource planning & sustainability. Springer Nature, Singapore, pp 1–322. ISBN No. 978-981-99-3659-5
Rai PK, Mishra VN, Singh P (2018) Hydrological inferences through morphometric analysis of Lower Kosi River Basin of India for water resource management based on remote sensing data. Appl Water Sci (Springer) 8(15):1–16. https://doi.org/10.1007/s13201-018-0660-7
Rai PK, Mishra VN, Singh P (2021) Recent technologies for disaster management & risk reduction—sustainable community resilience & responses (edited book). Springer Nature, Switzerland, pp 1–471. ISBN: 978-3-030-76116-5. https://doi.org/10.1007/978-3-030-76116-5
Rai PK, Mishra VN, Singh P (2022) Geospatial technology for landscape and environment management: sustainable assessment & planning (edited book). Springer Nature, Singapore, pp 1–431. ISBN: 978-981-16-7373-3. https://doi.org/10.1007/978-981-16-7373-3
Rajasekhar M, Sudarsana Raju G et al (2021) Multi-criteria land suitability analysis for agriculture in semi-arid region of Kadapa District, Southern India: geospatial approaches. Remote Sens Land 5(2):59–72. https://doi.org/10.21523/gcj1.2021050201
Rajankar PN, Tambekar DH, Wate SR (2011) Groundwater quality and water quality index at Bhandara district. Environ Monit Assess 179:619–625. https://doi.org/10.1007/s10661-010-1767-y
Rajasekhar M, Gadhiraju SR, Kadam A et al (2020) Identification of groundwater recharge-based potential rainwater harvesting sites for sustainable development of a semiarid region of southern India using geospatial, AHP, and SCS-CN approach. Arab J Geosci 13–24. https://doi.org/10.1007/s12517-019-4996-6
Reddy GPO, Mouli K, Srivastav SK et al (2000) Evaluation of ground water potential zones using remote sensing data—a case study of Gaimukh Watershed, Bhandara district, Maharashtra. J Indian Soc Remote Sens 28:19. https://doi.org/10.1007/BF02991858
Rupa KK, Pant GB, Parthasarathy B, Sontakke NA (1992) Spatial and sub-seasonal patterns of the long term trends of Indian summer monsoon rainfall. Int J Climatol 12(3):257–268. https://doi.org/10.1002/joc.3370120303
Salunke VS, Bhagat RS, Kudnar NS (2020) Geography of Maharashtra. Prashant Publication, Jalgaon, pp 1–229
Shortridge J (2019) Observed trends in daily rainfall variability result in more severe climate change impacts to agriculture. Clim Change 157:429–444. https://doi.org/10.1007/s10584-019-02555-x
Singh PK, Kumar S, Singh UC (2011) Groundwater resource evaluation in the Gwalior area, India, using satellite data: an integrated geomorphological and geophysical approach. Hydrogeol J 19:1421–1429. https://doi.org/10.1007/s10040-011-0758-6
Sinha S, Chakraborty S, Shome D (2019) Mining footprint: a spatial indicator of environmental quality—a case study of a manganese mine in Bhandara district, Maharashtra. Arab J Geosci 12:96. https://doi.org/10.1007/s12517-019-4260-0

Sivakumar MVK, Das HP, Brunini O (2005) Impacts of present and future climate variability and change on agriculture and forestry in the arid and semi-arid tropics. Clim Change 70:31–72. https://doi.org/10.1007/s10584-005-5937-9
Srivastava PK, Mukherjee S, Gupta M et al (2011) Characterizing monsoonal variation on water quality index of River Mahi in India using geographical information system. Water Qual Expo Health 2:193–203. https://doi.org/10.1007/s12403-011-0038-7
Subash N, Sikka AK (2014) Trend analysis of rainfall and temperature and its relationship over India. Theor Appl Climatol 117(3–4):449–462. https://doi.org/10.1007/s00704-013-1015-9

Chapter 3
Measuring Sustainable Development of Cities Using Remote Sensing and Geospatial Technologies: A Review

Satyavati Shukla, Deeksha, Sasmita Chand, Praveen Kumar Rai, S. Sundeep Kumar, and Anoop Kumar Shukla

Abstract The United Nations (UN) established 17 Sustainable Development Goals (SDGs) to address global concerns such as poverty, inequality, and the consequences of climate change. Unsustainable socioeconomic growth connected to population expansion, urbanisation, and industrialization are putting tremendous ecological strain on Earth's ecosystems. Authorities throughout the globe are seeking long-term solutions to these problems. At various temporal and spatial dimensions, remote sensing enables the measurement, integration, and display of important information for successful governance. The research examines the resilience of Earth Observation data for SDG continuous planning, monitoring, and assessment. The scientific community has achieved significant advancements in delivering geospatial information at multiple spatial, spectral, radiometric, and temporal resolutions, allowing the information to be used for various purposes. This research also examines the use of big data from earth observation to accomplish the SDGs using a multidisciplinary approach. It examines publications from many scientific domains that use geospatial

S. Shukla
Key Laboratory of Geospatial Informatics, Guilin University of Technology, Guilin, PR China
e-mail: satyageomatics@gmail.com

Deeksha · S. Chand · S. S. Kumar · A. K. Shukla (✉)
Manipal School of Architecture and Planning, Manipal Academy of Higher Education, Manipal, Karnataka, India
e-mail: anoop.shukla@manipal.edu

Deeksha
e-mail: deeksha.msapma2@learner.manipal.edu

S. Chand
e-mail: chand.sasmita@yahoo.in

S. S. Kumar
e-mail: sundeep.k@manipal.edu

P. K. Rai
Department of Geography, Khwaja Moinuddin Chishti Language University, Lucknow, India
e-mail: rai.vns82@gmail.com

P. K. Rai et al. (eds.), *Natural Resources Management and Sustainable Livelihoods in the Mountainous Region*, Advances in Geographical and Environmental Sciences,
https://doi.org/10.1007/978-981-97-2100-9_3

data for mapping, monitoring, and analysing the earth's characteristics and processes and establishing the foundation for its use in achieving the SDGs.

Keywords Sustainable development goals · Remote sensing · Geospatial technology · GIS

3.1 Introduction

In a world that is becoming more linked and dynamic, sustainable development has evolved as a vital paradigm for tackling major global concerns while maintaining the well-being of current and future generations. The United Nations' Sustainable Development Goals (SDGs), launched in 2015, serve as a visionary roadmap that strives to combine the economic, social, and environmental components of human growth. The SDGs, which consist of 17 unique goals, constitute a worldwide call to action, uniting governments, organizations, and individuals in a collaborative effort to improve our planet.

The SDGs are committed to eradicating poverty, safeguarding the environment, and promoting prosperity for all. With an ambitious deadline of 2030, these goals serve as a comprehensive agenda spanning multiple disciplines, such as eradicating hunger and promoting gender equality, as well as developing sustainable cities and addressing climate change. The SDGs understand the delicate relationship between poverty, inequality, environmental degradation, and social injustice by including a comprehensive approach. To achieve these objectives, coordination across sectors and boundaries is required, reflecting the recognition that global issues demand united answers.

One of the SDGs' particular characteristics is its universality, which transcends geographical, economic, and cultural barriers. Every country, regardless of its level of development, is urged to participate in the objectives in a way that is appropriate for its circumstances and capabilities. Such an inclusive approach recognizes that sustainable development is a complex and adaptive framework that respects the diversity of human experiences and aspirations.

The SDGs, crucially, underline the notion of "leaving no one behind," recognizing the significance of addressing the most vulnerable and disadvantaged communities. This commitment emphasizes tackling systemic disparities and fostering social justice as essential components of long-term development. The SDGs aim to build a sense of ownership and agency by empowering individuals and communities and promoting revolutionary change from local to global scales.

The Sustainable Development Goals, in essence, provide a common vision of a more equal, resilient, and peaceful world. Their comprehensive scope and emphasis on collaboration highlight that no organization can address humanity's complicated concerns in isolation. As we move toward 2030 and beyond, adopting the spirit of the SDGs provides us with a road map for constructing a future in which prosperity, well-being, and environmental stewardship live in dynamic balance.

3.2 Overview of Sustainable Development Goals

The Sustainable Development Goals (SDGs), a series of 17 linked objectives developed by the United Nations in 2015, take a holistic approach to global concerns with the goal of creating a more sustainable and equitable society by 2030. Each aim focuses on a different development aspect, recognizing the complicated linkages between social, economic, and environmental elements.

No Poverty (Goal 1): The objective is to eliminate severe poverty and guarantee everyone access to necessities such as food, clean water, and sanitation. It also includes social security measures to assist needy individuals (World Bank 2018).

Zero Hunger (Goal 2): To eradicate hunger and malnutrition, this objective focuses on establishing food security, enhancing nutrition, and supporting sustainable agriculture.

Good Health and Well-Being (Goal 3): aspires to ensure healthy lifestyles and promote well-being for people of all ages. It contains goals for mother and child health, illness prevention, and access to high-quality treatment.

Quality Education (Goal 4): The goal here is to promote inclusive and equitable education for everyone while encouraging lifelong learning and skill development.

Gender Equality (Goal 5): aims to promote gender equality and empower all women and girls by eradicating discrimination, violence, and harmful behaviours.

Clean Water and Sanitation (Goal 6): This objective aims to prevent water shortages and pollution by guaranteeing the availability and sustainable management of water and sanitation for all (UNESCO 2021).

Inexpensive and Clean Energy (Goal 7): focuses on providing everyone access to affordable, dependable, sustainable, and contemporary energy while shifting to renewable sources.

Economic Growth and Decent Work (Goal 8): To foster long-term, inclusive, sustainable economic growth, full and productive employment, and decent work for all.

Industry, Innovation, and Infrastructure (Goal 9): This aim highlights the significance of resilient infrastructure, inclusive and sustainable industrialisation, and innovation.

Reduced Inequality (Goal 10): seeks to decrease inequality within and between nations by addressing income disparities, social disparities, and inequalities based on gender, age, disability, and other factors.

Sustainable Cities and Communities (Goal 11): The objective is to make cities and human settlements more inclusive, safe, resilient, and long-term, with an emphasis on urban planning, housing, and transportation.

Responsible Consumption and Production (Goal 12): encourage sustainable consumption and production patterns by addressing waste creation, resource efficiency, and environmental implications.

Climate Action (Goal 13): This objective emphasizes the importance of addressing climate change and its consequences, including mitigation, adaptation, and raising awareness (UNEP 2017).

Life Below Water (Goal 14): Goal 14 focuses on protecting and exploiting oceans, seas, and marine resources in a sustainable manner to reduce marine pollution, overfishing, and habitat destruction (SDNS 2019).

Life on Land (Goal 15): This objective strives to maintain, restore, and manage terrestrial ecosystems in a sustainable manner while addressing desertification, deforestation, and biodiversity loss (IPBES 2019).

Peace and Justice (Goal 16): aspires to foster peaceful and inclusive communities by ensuring access to justice, minimizing violence, and establishing effective, accountable, and inclusive institutions.

Partnerships for the Goals (Goal 17): The last objective underlines the significance of global partnerships for sustainable development, promoting collaboration among governments, industry, and civil society to accomplish the SDGs.

By tackling these 17 goals collaboratively, the world can move toward a more egalitarian, resilient, and sustainable future for all, ensuring that no one is left behind.

3.3 Sustainable Development Goals Focus Area

The Sustainable Development Goals (SDGs) contain a wide range of vital objectives, each addressing a different aspect of the world's difficulties. Goal 1 aims to eliminate extreme poverty and hunger by ensuring everyone can access basic essentials such as food, clean water, and sanitation. This objective recognizes that poverty is not just an economic issue but is also heavily influenced by social and environmental issues. It highlights the need to create inclusive economic opportunities and social safety nets to raise communities out of poverty and secure their long-term well-being.

Goal 2 is about establishing food security, encouraging sustainable agriculture, and putting a stop to malnutrition. With a rapidly rising global population, ensuring access to nutritious and diversified food supplies is critical while protecting agricultural techniques from environmental deterioration. To address the difficulties of hunger and malnutrition, this aim emphasizes the significance of empowering small-scale farmers, enhancing agricultural production, and developing resilient food systems.

Goal 5 prioritizes gender equality and empowerment. This aim recognizes the persisting gender inequities in terms of opportunities, rights, and representation. It works to end all types of discrimination and violence against women and girls and promote equitable access to education, healthcare, and economic opportunities. Societies may unlock their full potential and create more inclusive growth by empowering women and encouraging their active involvement in decision-making processes.

Goal 11 is concerned with the development of sustainable cities and communities. Congestion, inadequate housing, pollution, and inadequate infrastructure become increasingly severe as urbanization accelerates. This objective envisions inclusive, safe, resilient, and ecologically sustainable cities. To promote liveable and dynamic urban settings, it highlights the importance of smart urban planning, efficient transit systems, affordable housing, and effective waste management.

Climate action, as addressed by Goal 13, is one of the most pressing global imperatives. With the threat of climate change becoming more obvious, this aim necessitates rapid and concerted measures to cut greenhouse gas emissions, improve resilience, and promote sustainable behaviours. It is consistent with the Paris Agreement and emphasizes international collaboration's need to limit global warming and its destructive effects on ecosystems and communities.

Finally, Goal 16 promotes peace, justice, and healthy institutions. Recognizing that stable governance systems and the rule of law are inextricably related to sustainable development, this aim highlights the significance of eliminating violence, providing equal access to justice, and establishing responsible and transparent institutions. It aims to develop conditions devoid of fear, bigotry, and corruption, laying the groundwork for long-term success.

Finally, the Sustainable Development Goals include a comprehensive and integrated framework intended to steer mankind toward a future distinguished by diversity, equity, environmental stewardship, and prosperity. Nations, communities, and people may work together to build a society that preserves the principles of sustainability, social justice, and shared well-being by tackling particular issues within each aim while acknowledging their interconnection.

3.4 Role of Remote Sensing to Assess SDGs

In this book chapter, we shall discuss on use of remote sensing to assess SDG 6, 11, 13, and 15, as these goals are directly aligned with ecosystem services, such as water yield, carbon sequestration, and soil retention.

Water levels and flow in the Mekong River Basin, encompassing numerous Southeast Asian nations, are monitored using remote sensing satellites. This data assists countries in managing their water resources, forecasting floods, and ensuring sustainable water usage (Johnston and Smakhtin 2014; Lao et al. 2022; Markert et al. 2018; Rai 2023a and b). To improve transboundary collaboration and accomplish SDG 6 objectives, the Mekong River Commission uses remote sensing data.

Another case is Lake Victoria, which is polluted as a result of industrial and agricultural activity; by detecting contaminants and algal blooms, remote sensing aids in monitoring changes in water quality. Remote sensing is used by organizations such as the Lake Victoria Basin Commission to analyse pollution levels and execute suitable mitigation techniques (Mishra et al. 2021; Rai et al. 2021a, b; Rai et al. 2022; Njagi et al. 2022).

Assessing sanitary infrastructure is critical in quickly increasing metropolitan areas like Dhaka. To map sanitation facilities and assess their coverage and operation, remote sensing data, particularly high-resolution satellite images, is employed. This assists municipal planners in making educated sanitation decisions, contributing to SDG 6.

Remote sensing satellites monitor plant health and moisture levels in drought-prone areas like the Horn of Africa (Qu et al. 2019). Organizations such as the Food and Agriculture Organization (FAO) and local governments can take prompt efforts to alleviate the effects of drought on water supplies and agricultural productivity by examining this data (FAO 2016).

Land-based pollution, especially wastewater discharge, has an impact on the health of marine ecosystems such as the Great Barrier Reef. Remote sensing aids in monitoring contaminant dispersal from coastal regions into the ocean. This information is used by Australian authorities to control and manage wastewater discharge in order to safeguard the reef ecology.

Flooding is common throughout the Amazon Basin, harming populations and ecosystems. Remote sensing satellites monitor flood extent and river level fluctuations, assisting disaster response and preparedness operations (Parrens et al. 2017; Rosenqvist et al. 2020). These data are used by organizations such as NASA and the Brazilian government to monitor and control flood disasters.

3.5 Role of Remote Sensing to Assess SDGs in the Indian Scenario

3.5.1 Sdg 6

I. Water Bodies and Resources Mapping

India has been monitoring and mapping its water bodies, including rivers, lakes, reservoirs, and groundwater sources, via remote sensing satellites from the Indian Space Research Organization (ISRO). This aids in determining the availability of freshwater resources and understanding how water bodies vary over time (Rai et al. 2014; Rai et al. 2017; Rai et al. 2018; Anusha et al. 2022; Pande et al. 2021).

II. Water Quality and Pollution Monitoring

Water quality metrics such as turbidity, chlorophyll-a concentration, and sedimentation in bodies of water are monitored using remote sensing technologies. These indicators aid in assessing pollution levels and identifyingegions that require intervention to preserve water quality (Jha et al. 2020; Karnena et al. 2022; Sekharan et al. 2022).

III. Management of Floods and Disasters

Monsoon flooding is common in India. By detecting changes in river levels and analyzing flood extent using satellite images, remote sensing aids in flood monitoring and prediction. This data is essential for disaster response and management (Dash and Sar 2020; Sharma et al. 2017).

IV. Planning for Urban Water Infrastructure

Remote sensing data is often used to map and monitor sanitary infrastructure, wastewater treatment facilities, and water distribution networks in rapidly urbanizing areas. This helps city planners evaluate the coverage and functionality of sanitary services (Parry et al. 2018; Shukla et al. 2020).

V. Drought Monitoring and Management

Data from remote sensing satellites on vegetation health, land surface temperature, and soil moisture are crucial for drought monitoring and early warning systems. This contributes to reducing the effects of drought on water resources and agriculture (Mishra et al. 2020).

VI. Management of Coastal and Marine Ecosystems

Remote sensing technologies may be used to monitor India's vast coastline and marine habitats. This involves monitoring changes in marine ecosystems, tracking coastal erosion, and measuring water quality along shorelines (Nayak 2017).

3.5.2 Sdg 11

I. Infrastructure Development and Urban Planning

High-resolution pictures from remote sensing satellites may be utilized for urban planning and monitoring city infrastructure development. This involves examining changes in land usage, selecting suitable places for new developments, and monitoring urban expansion (Parry et al. 2018).

II. Slum Mapping and Improvement

Remote sensing can aid in the identification of informal settlements and slum areas in metropolitan settings. This data is critical for undertaking slum improvement projects and improving living circumstances for underprivileged communities (Thomson et al. 2014).

III. Disaster Management and Resilience

Remote sensing can help in disaster management and risk reduction. It may give information on sensitive areas, track changes following natural catastrophes, and assist in the development of resilient infrastructure to survive future calamities (Bera et al. 2021; Saxena et al. 2013).

IV. Transportation and Traffic Planning

Remote sensing can help cities monitor traffic congestion and analyze transportation trends. This data is useful for planning efficient transportation systems and minimizing traffic congestion (Bhagyaiah and Shrinagesh 2014; Verma and Dash 2020).

V. Environmental Monitoring and Green Space

Monitoring green space and plant cover is critical for constructing sustainable and habitable communities. Remote sensing may be used to monitor changes in urban vegetation, measure air quality, and support attempts to improve urban environmental quality (Deeksha and Shukla 2022; Sekharan et al. 2022).

VI. Infrastructure Maintenance and Monitoring

The state of infrastructure, such as roads, bridges, and buildings, may be monitored using remote sensing technologies. Regular monitoring assists in identifying maintenance needs and ensuring the safety of city dwellers.

VII. Housing and Land Policies

Remote sensing can help assess land tenure, property rights, and housing rules. It can aid in the identification of informal settlements and the implementation of inclusive housing policy.

3.5.3 Sdg 13

I. Climate Change Monitoring and Extreme Weather Conditions

Remote sensing satellites collect data on a variety of climate indicators, including sea surface temperature, land surface temperature, greenhouse gas concentrations, and ice cover variations. This data is crucial for tracking and comprehending climatic trends. It also aids in the monitoring and forecasting of extreme weather phenomena such as hurricanes, cyclones, and monsoons. This data aids disaster preparedness and response operations (Dash and Maity 2023; Halder and Bandyopadhyay 2021; Swain et al. 2022).

II. Monitoring Glaciers in the Himalayas

The Himalayan glaciers in India are sensitive markers of climate change. Remote sensing aids in monitoring glacier area, thickness, and movement. This data helps to analyse water availability and potential implications on downstream locations (Mishra and Rai 2020; Tewari and Misra 2019).

III. Sea Level Rise and Coastal Erosion

Satellite data on coastal erosion, coastline changes, and sea level rise is provided by remote sensing satellites. This data is critical for formulating plans to safeguard vulnerable coastal populations and infrastructure (Gopinath et al. 2023; Kumar Das et al. 2021).

IV. Monitoring of Forests and Vegetation

Changes in forest cover, deforestation rates, and vegetation health may all be tracked using remote sensing equipment. Monitoring these changes aids in the assessment of carbon sequestration and aids in the fight against deforestation (Ahmad et al. 2022; Das et al. 2023).

V. Agricultural Resilience and Adaptation

Data on crop health, soil moisture, and agricultural productivity may be obtained through remote sensing. This information contributes to the development of agricultural adaptation methods in response to changing climate conditions (Koppa and Amarnath 2021; Sun et al. 2022).

VI. Effects of Urban Heat Island

Remote sensing can assist in identifying urban heat islands, which are locations where cities have greater temperatures than adjacent rural areas. This data informs urban design in order to limit the effects of heat (Halder and Bandyopadhyay 2021).

3.5.4 Sdg 15

I. Biodiversity Mapping and Conservation

Remote sensing can help map and monitor biodiversity-rich regions. It aids in the identification of habitats and ecosystems in need of protection and conservation initiatives. It also contains information about forest cover, deforestation, and forest degradation. Remote sensing is used in India to monitor changes in forest regions and estimate the impact of land use changes on biodiversity and ecosystem health (Giri et al. 2015; Rai et al. 2021a, b).

II. Monitoring of Wetlands and Soil Erosion

There are various ecologically significant wetland habitats in India. Remote sensing can monitor changes in wetland areas, water quality, and plant cover, which can help with wetland conservation initiatives. It may also measure soil erosion rates and identify erosion-prone regions. This data is useful for establishing soil conservation strategies (Debanshi and Pal 2020; Sinclair et al. 2021; Singh et al. 2020).

III. Wildlife Habitat and Protected Area Monitoring

Remote sensing collects data on habitat features and changes, allowing wildlife management officials to monitor and conserve vital habitats for a variety of species.

It aids in the administration of protected areas by monitoring encroachment, unlawful activity, and changes in land use inside them (Ramachandra et al. 2018).

IV. Monitoring of Land Degradation and Desertification

Remote sensing technology aids in monitoring the processes of land deterioration and desertification. It allows for the identification of high-risk locations and supports steps to avoid and minimize these problems.

3.6 Conclusion

In India, the synergistic integration of remote sensing technology with SDGs 6, 11, 13, and 15 demonstrates the country's commitment to supporting sustainable development and environmental stewardship. The use of remote sensing to monitor water supplies, water quality, and sanitation infrastructure significantly improves SDG 6, which is focused on clean water and sanitation. The use of satellite imaging assists in the mapping of water bodies, the tracking of water pollution, and the prediction of water stress, allowing for evidence-based decision-making for equitable water management across different landscapes.

Moving on to SDG 11, which focuses on sustainable cities and communities, remote sensing is extremely useful in urban planning, disaster management, and infrastructure construction. By utilizing satellite data, India can monitor urban growth, decrease hazards connected with extreme weather events, and improve urban resilience. This technology identifies susceptible areas, informs strategies for long-term growth, and promotes inclusive and safe urban settings.

Remote sensing considerably helps India monitor climate patterns, identify climate-related hazards, and implement adaptation methods in the framework of SDG 13, which addresses climate action. Satellite observations make it possible to follow glacier movements, coastal erosion, and changes in plant cover. Such information enables the government to make educated decisions about combating climate change, strengthening resilience, and contributing to global climate mitigation efforts.

Finally, in keeping with SDG 15's emphasis on land-based life, remote sensing emerges as a critical tool for maintaining India's diverse biodiversity and ecosystems. Remote sensing allows for assessing biodiversity changes and supporting conservation activities by mapping forests, wetlands, and animal habitats. Furthermore, the technology contributes to the monitoring of land deterioration, the facilitation of sustainable land use practices, and the restoration of ecosystems.

Finally, India's use of remote sensing technologies to connect with SDGs 6, 11, 13, and 15 demonstrates the country's proactive attitude to sustainable development and environmental stewardship. The use of satellite data in policy formation, decision-making processes, and conservation initiatives demonstrates the country's dedication to accomplishing these worldwide objectives. The synergy between remote sensing and sustainable development will continue to be a cornerstone of India's journey toward a more resilient, fair, and ecologically aware future as technology progresses and capacities grow.

References

Ahmad F, Talukdar NR, Goparaju L, Biradar C, Dhyani SK, Rizvi J (2022) GIS-based assessment of land-agroforestry potentiality of Jharkhand State India. Reg Sustain 3(3):254–268. https://doi.org/10.1016/j.regsus.2022.10.004

Anusha BN, Babu KR, Kumar BP, Kumar PR, Rajasekhar M (2022) Geospatial approaches for monitoring and mapping of water resources in semi-arid regions of Southern India. Environ Chall 8:100569. https://doi.org/10.1016/j.envc.2022.100569

Bera A, Taloor AK, Meraj G, Kanga S, Singh SK, Đurin B, Anand S (2021) Climate vulnerability and economic determinants: linkages and risk reduction in Sagar Island, India; A geospatial approach. Q Sci Adv 4:100038. https://doi.org/10.1016/j.qsa.2021.100038

Bhagyaiah M, Shrinagesh B (2014) Traffic analysis and road accidents: a case study of Hyderabad using GIS. IOP Conf Ser: Earth Environ Sci 20(1):012026. https://doi.org/10.1088/1755-1315/20/1/012026

Das J, Mahato S, Joshi PK, Liou YA (2023) Forest fire susceptibility zonation in Eastern India using statistical and weighted modelling approaches. Remote Sens 15(5):1340. https://doi.org/10.3390/rs15051340

Dash SS, Maity R (2023) Effect of climate change on soil erosion indicates a dominance of rainfall over LULC changes. J Hydrol: Reg Stud 47:101373. https://doi.org/10.1016/j.ejrh.2023.101373

Dash P, Sar J (2020) Identification and validation of potential flood hazard area using GIS-based multi-criteria analysis and satellite data-derived water index. J Flood Risk Manag 13(3):e12620. https://doi.org/10.1111/jfr3.12620

Debanshi S, Pal S (2020) Modelling water richness and habitat suitability of the wetlands and measuring their spatial linkages in mature Ganges delta of India. J Environ Manag 271:110956. https://doi.org/10.1016/j.jenvman.2020.110956

Deeksha, Shukla AK (2022) Ecosystem services: a systematic literature review and future dimension in freshwater ecosystems. Appl Sci (Switzerland) 12(17):8518. https://doi.org/10.3390/app12178518

FAO (2016) The state of food and agriculture 2016. In: Climate change, agriculture and food security

Giri C, Long J, Abbas S, Murali RM, Qamer FM, Pengra B, Thau D (2015) Distribution and dynamics of mangrove forests of South Asia. J Environ Manage 148:101–111. https://doi.org/10.1016/j.jenvman.2014.01.020

Gopinath G, Thodi MFC, Surendran UP, Prem P, Parambil JN, Alataway A, Al-Othman AA, Dewidar AZ, Mattar MA (2023) Long-term shoreline and islands change detection with digital shoreline analysis using RS data and GIS. Water (switzerland) 15(2):244. https://doi.org/10.3390/w15020244

Halder B, Bandyopadhyay J (2021) Evaluating the impact of climate change on urban environment using geospatial technologies in the planning area of Bilaspur India. Environ Chall 5:100286. https://doi.org/10.1016/j.envc.2021.100286

IPBES (2019) The global assessment report of the intergovernmental science-policy platform on biodiversity and ecosystem services

Jha MK, Shekhar A, Jenifer MA (2020) Assessing groundwater quality for drinking water supply using hybrid fuzzy-GIS-based water quality index. Water Res 179:115867. https://doi.org/10.1016/j.watres.2020.115867

Johnston R, Smakhtin V (2014) Hydrological modeling of large river basins: how much is enough? Water Resourc Manag 28(10):2695–2730. Kluwer Academic Publishers. https://doi.org/10.1007/s11269-014-0637-8

Karnena MK, Konni M, Dwarapureddi BK, Satyanarayana Y, Saritha V (2022) GIS-based approach qualitative features of sub-surface water from coastal district in Andhra Pradesh. Appl Water Sci 12(3):45. https://doi.org/10.1007/s13201-021-01506-1

Koppa N, Amarnath G (2021) Geospatial assessment of flood-tolerant rice varieties to guide climate adaptation strategies in India. Climate 9(10):151. https://doi.org/10.3390/cli9100151

Kumar Das S, Sajan B, Ojha C, Soren S (2021) Shoreline change behavior study of Jambudwip island of Indian Sundarban using DSAS model. Egyptian J Remote Sens Space Sci 24(3):961–970. Elsevier B.V. https://doi.org/10.1016/j.ejrs.2021.09.004

Lao J, Wang C, Nie S, Xi X, Wang J (2022) Monitoring and analysis of water level changes in Mekong River from ICESat-2 spaceborne laser altimetry. Water (switzerland) 14(10):1613. https://doi.org/10.3390/w14101613

Markert KN, Schmidt CM, Griffin RE, Flores AI, Poortinga A, Saah DS, Muench RE, Clinton NE, Chishtie F, Kityuttachai K, Someth P, Anderson ER, Aekakkararungroj A, Ganz DJ (2018) Historical and operational monitoring of surface sediments in the lower mekong basin using landsat and google earth engine cloud computing. Remote Sens 10(6):909. https://doi.org/10.3390/rs10060909

Mishra AK, Rai SC (2020) Geo-hydrological inferences through morphometric aspects of the Himalayan glacial-fed river: a case study of the Madhyamaheshwar River basin. Arab J Geosci 13(13):533. https://doi.org/10.1007/s12517-020-05571-9

Mishra AK, Upadhyay A, Srivastava A, Rai SC (2020) Probabilistic groundwater recharge zonation in hard rock terrain using geospatial techniques in Veniar watershed South India. Ecohydrol Hydrobiol 20(3):456–471. https://doi.org/10.1016/j.ecohyd.2019.01.004

Mishra VN, Rai PK, Singh P (2021) Geo-information technology in earth resources monitoring and management (edit. Book). Nova Science Publishers, U.S.A., pp 1–395. ISBN: 978-1-53619-669-6

Nayak S (2017) Coastal zone management in India – present status and future needs. Geo-Spatial Inf Sci 20(2):174–183. https://doi.org/10.1080/10095020.2017.1333715

Njagi DM, Routh J, Odhiambo M, Luo C, Basapuram LG, Olago D, Klump V, Stager C (2022) A century of human-induced environmental changes and the combined roles of nutrients and land use in Lake Victoria catchment on eutrophication. Sci Total Environ 835:155425. https://doi.org/10.1016/j.scitotenv.2022.155425

Pande CB, Moharir KN, Khadri SFR (2021) Assessment of land-use and land-cover changes in Pangari watershed area (MS), India, based on the remote sensing and GIS techniques. Appl Water Sci 11(6):96. https://doi.org/10.1007/s13201-021-01425-1

Parrens M, Bitar AA, Frappart F, Papa F, Calmant S, Crétaux JF, Wigneron JP, Kerr Y (2017) Mapping dynamic water fraction under the tropical rain forests of the Amazonian basin from SMOS brightness temperatures. Water (switzerland) 9(5):350. https://doi.org/10.3390/w9050350

Parry JA, Ganaie SA, Sultan Bhat M (2018) GIS based land suitability analysis using AHP model for urban services planning in Srinagar and Jammu urban centers of J&K India. J Urban Manag 7(2):46–56. https://doi.org/10.1016/j.jum.2018.05.002

Qu C, Hao X, Qu JJ (2019) Monitoring extreme agricultural drought over the Horn of Africa (HOA) using remote sensing measurements. Remote Sens 11(8):902. https://doi.org/10.3390/rs11080981

Rai PK, Mishra S, Ahmad A, Mohan K (2014) A GIS-based approach in drainage morphometric analysis of Kanhar River Basin, India. Appl Water Sci (springer) 7:217–232. https://doi.org/10.1007/s13201-014-0238-y

Rai PK, Mishra VN, Mohan K (2017) A study of morphometric evaluation of the Son Basin, India using geospatial approach. Remote Sens Appl: Soc Environ 7:9–20. https://doi.org/10.1016/j.rsase.2017.05.001

Rai PK, Mishra VN, Singh P (2018) Hydrological Inferences through morphometric analysis of lower Kosi River Basin of India for water resource management based on remote sensing data. Appl Water Sci (springer) 8(15):1–16. https://doi.org/10.1007/s13201-018-0660-7

Rai P, Vineeta SG, Manohar KA, Bhat JA, Kumar A, Kumar M, Cabral-Pinto M, Chakravarty S (2021). Carbon storage of single tree and mixed tree dominant species stands in a reserve forest—case study of the eastern sub-himalayan region of india. Land 10(4):435. https://doi.org/10.3390/land10040435

Rai PK, Mishra VN, Singh P (2021) Recent technologies for disaster management and risk reduction-sustainable community resilience and responses (edit. Book). Springer Nature, Switzerland, pp. 1–471. ISBN: 978-3-030-76116-5. https://doi.org/10.1007/978-3-030-76116-5.

Rai PK, Mishra VN, Singh P (2022) Geospatial technology for landscape and environment management: sustainable assessment and planning (edit. Book). Springer Nature, Singapore, pp 1–431. ISBN: 978-981-16-7373-3. https://doi.org/10.1007/978-981-16-7373-3

Rai PK (2023a) River conservation and water resource management. Springer Nature, Singapore, pp 1–352. ISBN No.: 978-981-99-2604-6. eBook ISBN: 978-981-99-2605-3

Rai PKB (2023) Advances in water resource planning and sustainability. Springer Nature, Singapore, pp 1–322. ISBN No. 978-981-99-3659-5

Ramachandra TV, Bharath S, Gupta N (2018) Modelling landscape dynamics with LST in protected areas of Western Ghats, Karnataka. J Environ Manag 206:1253–1262. https://doi.org/10.1016/j.jenvman.2017.08.001

Rosenqvist J, Rosenqvist A, Jensen K, McDonald K (2020) Mapping of maximum and minimum inundation extents in the amazon basin 2014–2017 with ALOS-2 PALSAR-2 scan SAR time-series data. Remote Sens 12(8):1326. https://doi.org/10.3390/RS12081326

Saxena S, Geethalakshmi V, Lakshmanan A (2013) Development of habitation vulnerability assessment framework for coastal hazards: Cuddalore coast in Tamil Nadu, India-A case study. Weather and Climate Extremes 2:48–57. https://doi.org/10.1016/j.wace.2013.10.001

SDNS (2019) Sustainable development solutions network. www.pica-publishing.com

Sekharan S, Samal DR, Phuleria HC, Chandel MK, Gedam S, Kumar R, Sethi V, Supate AR, Karmakar S (2022) River pollution monitoring over an industrial catchment in urban ecosystem: Challenges and proposed geospatial framework. Environ Chall 7:100496. https://doi.org/10.1016/j.envc.2022.100496

Sharma VK, Mishra N, Shukla AK, Yadav A, Rao GS, Bhanumurthy V (2017) Satellite data planning for flood mapping activities based on high rainfall events generated using TRMM, GEFS and disaster news. Ann GIS 23(2):131–140. https://doi.org/10.1080/19475683.2017.1304449

Shukla AK, Ojha CSP, Garg RD, Shukla S, Pal L (2020) Influence of spatial urbanization on hydrological components of the Upper Ganga River Basin, India. J Hazard, Toxic, Radioact Waste 24(4):04020028. https://doi.org/10.1061/(asce)hz.2153-5515.0000508

Sinclair M, Vishnu Sagar MK, Knudsen C, Sabu J, Ghermandi A (2021) Economic appraisal of ecosystem services and restoration scenarios in a tropical coastal Ramsar wetland in India. Ecosyst Serv 47:101236. https://doi.org/10.1016/j.ecoser.2020.101236

Singh S, Bhardwaj A, Verma VK (2020) Remote sensing and GIS based analysis of temporal land use/land cover and water quality changes in Harike wetland ecosystem, Punjab India. J Environ Manag 262:110355. https://doi.org/10.1016/j.jenvman.2020.110355

Sun Q, Zhang Y, Che X, Chen S, Ying Q, Zheng X, Feng A (2022) Coupling process-based crop model and extreme climate indicators with machine learning can improve the predictions and reduce uncertainties of global soybean yields. Agriculture (switzerland) 12(11):1791. https://doi.org/10.3390/agriculture12111791

Swain S, Taloor AK, Dhal L, Sahoo S, Al-Ansari N (2022) Impact of climate change on groundwater hydrology: a comprehensive review and current status of the Indian hydrogeology. Appl Water Sci 12(6):120. https://doi.org/10.1007/s13201-022-01652-0

Tewari NK, Misra AK (2019) Landslide vulnerability assessment in Gangotri valley glacier Himalaya through GIS and remote sensing techniques. Appl Water Sci 9(4):115. https://doi.org/10.1007/s13201-019-0992-y

Thomson DR, Shitole S, Shitole T, Sawant K, Subbaraman R, Bloom DE, Patil-Deshmukh A (2014) A system for household enumeration and reidentification in densely populated slums to facilitate

community research, education, and advocacy. PLoS ONE 9(4):e93925. https://doi.org/10.1371/journal.pone.0093925
UNEP (2017) Environmental sustainability is essential for global equity.
UNESCO (2021) Valuing water. World Water Assessment Programme (United Nations)
Verma VR, Dash U (2020) Geographical accessibility and spatial coverage modelling of public health care network in rural and remote India. PLoS ONE 15(10):e0239326. https://doi.org/10.1371/journal.pone.0239326
World Bank (2018) Ending poverty investing in opportunity

Chapter 4
Eco-Entrepreneurship and Sustainable Development in Mizoram's Mountainous Landscape: Unleashing Potentials for Positive Change

Ashok Sharma

Abstract The pursuit of sustainable development in geographically challenged locations, like Mizoram's mountainous landscape, demands novel ways that balance economic growth and environmental preservation. This proposed study aimed to investigate the dynamic relationship between eco-entrepreneurship and sustainable development in Mizoram, with an emphasis on harnessing the hidden capacity for positive change. Mizoram, a northeastern Indian state, has a distinct geography and a diverse fauna, but it also faces development challenges due to its distant position and limited resources. The advent of eco-entrepreneurship, defined as environmentally responsible business endeavours, offers a way to solve these difficulties by capitalizing on the region's inherent assets and supporting long-term success. The purpose of this research is to determine the extent to which eco-entrepreneurship might serve as a catalyst for a revolutionary transformation of Mizoram's socioeconomic structure. The qualitative analysis for this conceptual paper is to identify essential theoretical frameworks, ideas, and best practices. By examining current policies, infrastructural support, and access to markets, the research aims to identify gaps that prevent the realization of eco-friendly entrepreneurial ventures in the region. There are several anticipated outcomes of this study. First, it aims to shed light on the underutilized potential of eco-entrepreneurship in the mountainous terrain of Mizoram, offering insights into how sustainable businesses can support community well-being and economic diversification. The second goal of the study is to offer valuable assistance to the government, development organizations, and entrepreneurs on how to foster and promote eco-entrepreneurial activities. Hence, the proposed research endeavours to bridge the gap between theory and practice by investigating the role of eco-entrepreneurship in driving sustainable development in Mizoram. By unleashing the potentials embedded within the mountainous landscape, this study aspires to foster positive change that is both economically viable and environmentally responsible.

A. Sharma (✉)
Department of Commerce, Pachhunga University College, Mizoram University, Aizawl 796001, India
e-mail: commerce.ashok@gmail.com

P. K. Rai et al. (eds.), *Natural Resources Management and Sustainable Livelihoods in the Mountainous Region*, Advances in Geographical and Environmental Sciences,
https://doi.org/10.1007/978-981-97-2100-9_4

Keywords Sustainable development · Eco-entrepreneurship · Mizoram · Mountainous landscape · Positive change

4.1 Introduction

Located away on India's eastern periphery, Mizoram's hilly terrain, with its verdant woods, pure rivers, and unrivalled biodiversity, is a monument to the magnificence of nature. The majority of Mizoram's geography is made up of mountains, with parallel ranges creating steep canyons that end in several streams and rivers (Rao et al., 2012). However, this breath-taking scenery hides a complicated web of difficulties and possibilities. Rich forests and a liveable climate are two examples of Mizoram's abundant natural wealth (Sati, 2022).

At a time when sustainable development is becoming increasingly important, states like Mizoram are in a position where decisions made now will have a significant impact on their future in terms of the environment, society, and economy. The combination of sustainable development and eco-entrepreneurship stands out as a fascinating focus point in this setting. Businesses have begun to prioritize environmental issues in an effort to ensure the sustainability of their operations and resources (Karagülle, 2012). According to Wei et al. (2023), environmental entrepreneurship has gained prominence as a crucial element that plays a major role in attaining sustainable development goals. The idea of eco-entrepreneurship captures the spirit of entrepreneurship, but it takes a special approach by emphasizing environmental sustainability. It aims to balance economic growth with ecological responsibility by bringing profit-making endeavours into line with environmental protection. The people of Mizoram are moving from agriculture to entrepreneurship for their primary source of income (Lalrokhawma & Singh, 2020). However, Mizoram may be considered as a prospective state for agribusiness (Kumar et al., 2023) for future sustainable entrepreneurship.

In the meantime, social justice, economic advancement, and environmental preservation must be balanced in order to achieve sustainable development, which is a global necessity (Mishra et al., 2021; Rai et al., 2021 and Rai et al., 2022). With its uneven landscape and diverse ecosystem, Mizoram serves as a microcosm of the greater global challenge, which depends on how we resolve the urgent problems of resource depletion, climate change, and socioeconomic inequality while promoting advancement. Entrepreneurship has the ability to significantly alter society (Stål & Bonnedahl, 2016). The importance of eco-entrepreneurship in helping Mizoram achieve sustainable development is highlighted in this conceptual research.

It is anticipated that the actions of sustainable entrepreneurs would align with the Sustainable Development Goals (SDGs), which were declared by the United Nations in 2015 (Akku & Çalıyurt, 2022). The inclusion of eco-entrepreneurship in development initiatives becomes a potent instrument for creating creative solutions to issues that appear insurmountable as countries throughout the world work to meet the UN-SDGs. According to Rajasekaran, (2013), entrepreneurship is gradually transitioning from a focus on meeting economic requirements in its early stages to including sustainable practises in later stages. By utilizing this strategy, we may unleash the possibility of positive change and redefine the parameters of development in the twenty-first century.

Using theories and current knowledge as a guide, this study explores the relationship between eco-entrepreneurship and sustainable development from a conceptual standpoint. It seeks to offer perspectives that are relevant to Mizoram as well as to other areas looking for a sustainable future. This paper hopes to shed light on the opportunities, difficulties, and revolutionary potential that lie ahead for those who dare to rethink development in a way that honours the environment, strengthens communities, and improves people's lives. This study begins with defining eco-entrepreneurship since it views ecopreneurship as a continuation of entrepreneurship. Subsequently, the study proceeds with the definition and concepts of sustainable development. This paper concludes with a discussion of eco-entrepreneurship's contribution to sustainable development.

4.2 Research Question

This conceptual study paper seeks to answer the following main query in relation to Mizoram:

How can eco-entrepreneurship be strategically utilised to drive sustainable development, promote environmental protection, and boost socioeconomic well-being in Mizoram's distinctive geographical and cultural setting?

The main question that directs this study into the relationship between eco-entrepreneurship and sustainable development is summed up in this research question. According to Sanal & Zare (2017), eco-entrepreneurs want to be change agents and issue solutions for the environment in addition to being more productive and environmentally conscious. This forces us to explore the complex dynamics of this relationship in an effort to identify opportunities, obstacles, and routes that can result in positive change in Mizoram and possibly act as a model for other areas confronting comparable developmental issues in the midst of delicate ecosystems.

4.3 Eco-Entrepreneurship

Eco-entrepreneurship refers to business ventures that prioritize environmental sustainability. It embodies the entrepreneurial spirit while also being deeply committed to preventing environmental deterioration and tackling important ecological challenges. Eco-entrepreneurs place a premium on environmental stewardship, striving to limit their businesses' negative environmental consequences through resource conservation, waste minimization, and the use of renewable and eco-friendly products. Eco-entrepreneurship thrives on creativity, looking for new ways to solve environmental problems. It frequently entails the creation of environmentally friendly products, services, or technologies that encourage sustainability. This method emphasizes the triple bottom line concept, which measures corporate success not only in terms of profit but also of social and environmental impacts. Entrepreneurs require fundamental entrepreneurial abilities to engage in sustainable operations, as observed by Wei et al. (2023), reaffirming the notion that eco-entrepreneurship is a thorough and complex endeavour.

As the global economy evolves, the importance of eco-entrepreneurship becomes clear. Nature is increasingly threatened by profit motives and rapid industrialization (Demir et al., 2015). As a result, eco-entrepreneurship is a proactive reaction to these difficulties. It is a commitment to a more responsible and sustainable business model, one that promotes economic progress while protecting the environment. This strategy has far-reaching implications for enterprises, society, and the environment, making it a critical component in the pursuit of a more balanced and harmonious future.

4.4 Sustainable Development

As a worldwide necessity, sustainable development is an all-encompassing strategy for advancing society. It seeks to strike a balance between social justice, environmental preservation, and economic progress (Singh et al., 2021). The goal of sustainable development is to satisfy current needs without endangering the capacity of future generations to satisfy their own. It acknowledges that progress must benefit all societal members, especially those who are marginalized, and it advocates for social inclusion, the provision of necessities, and the eradication of inequality. Operating in harmony with the environment is essential for sustainability, as the economy and society rely on the long-term balance of sometimes opposing components (Domanska et al., 2018). Hence, sustainable development is both a practical requirement and an ideal. It embodies the principles of social inclusion, sustainable development, and responsible resource management, all of which are necessary for a more just and sustainable world (Singh et al., 2021).

4.5 Eco-Entrepreneurship and Sustainable Development and Its Relevance to Mizoram

The biodiversity-rich hilly topography of Mizoram offers a special set of opportunities and difficulties that are especially pertinent to the ideas of eco-entrepreneurship and sustainable development. The mountains in the area are home to a diverse range of plants and animals, making them hotspots for biodiversity. Adopting sustainable development practises is crucial to protecting this priceless natural diversity. Due to its difficult geography and scarce resources, Mizoram needs to find new ways to boost its economy. A key factor in the shift to a more sustainable future is entrepreneurship, with an emphasis on balancing social, economic, and ecological goals (Raimundo & Cruz, 2022; Rai et al., 2023a and 2023b). In this situation, eco-entrepreneurship shows itself as a crucial instrument that may sustainably utilize regional resources and strengthen the economy of the area.

On the other hand, Mizoram's close-knit communities have strong cultural and social ties to their surrounding environment. Respecting these cultural values and actively involving local populations in decision-making processes are important aspects of sustainable development initiatives. In addition, the region is susceptible to the adverse consequences of climate change, including erratic weather patterns and landslides. Sustainable practises are not only desirable but also essential to mitigate and prepare for these environmental concerns. Mizoram has a promising future thanks to the mix of sustainable development and eco-entrepreneurship. It seeks to strike a balance between economic advancement and the preservation of cultural identity, environmental protection, and social justice. This tactic addresses the unique challenges presented by the steep terrain while capitalizing on the inherent capacity to effect positive change. By doing this, it seeks to create a future for Mizoram that is resilient, sustainable, and carefully balances expansion with environmental preservation.

4.6 Review of Literature

This section examines a range of theoretical perspectives and models related to eco-entrepreneurship and sustainable development. The Triple Bottom Line concept, popularized by John Elkington, is a framework that emphasizes evaluating companies and initiatives based on economic, social, and environmental dimensions (profit, people, and planet, respectively) in order to promote sustainable development (Santini, 2017). Paul Hawken coined the term ecopreneurship, which combines elements of social, environmental, and green entrepreneurship with a strong emphasis on environmental sustainability (Majid, 2015; Wagner, 2015). It highlights how companies may create profits and also drive environmental progress. According to Domanska et al. (2018) and Nikolaou et al. (2011), green entrepreneurship entails integrating the economic, environmental, and social pillars for sustainable development.

According to Rajasekaran (2013), sustainable entrepreneurship considers social and environmental factors, stressing the long-term effects of new business prospects and the significance of natural capital. Production and exchange of renewable energy are essential for fostering green economic growth (Wei et al., 2023). By reusing, recycling, and repurposing goods and materials, the circular economy model promotes sustainability by reducing waste and maximizing resource efficiency (Jayasinghe et al., 2021). Amartya Sen and Martha Nussbaum created the Capability Approach, which is in line with sustainable development and places an emphasis on people's well-being and access to health care, education, and clean environments (Masjud, 2020; Hummels & Argyrou, 2021).

The study of sustainable business models looks at how concepts like sharing economies and performance-based contracting can create value while lessening negative effects on the environment and society. For traditional entrepreneurship, this move towards sustainability-driven entrepreneurship is a major direction (Rajasekaran, 2013; Borkowski et al., 2020). Businesses with combined environmental and social missions for holistic sustainability are highlighted at the junction of eco-entrepreneurship and social entrepreneurship (Tabares et al., 2022). The goal of social enterprises is to balance financial viability with a long-term social benefit (Jayawardhana et al., 2022).

A thorough framework for worldwide sustainable development is provided by the Sustainable Development Goals (SDGs) of the United Nations (Mio et al., 2020). Research into sustainable solutions, such as the GOOD Eco-Entrepreneurship Model (GEEM), is stimulated by these SDGs, which act as a road map for coordinating eco-entrepreneurship and sustainable development initiatives (Doe, 2022). In addition, Moon (2018) emphasizes the necessity of developing novel metrics related to environmental and social concerns in order to address the UN-SDGs.

Acknowledging the value of indigenous knowledge, research indicates that traditional ecological wisdom from indigenous groups can guide sustainable development, which includes eco-entrepreneurship projects (Tien et al., 2020). In order to address resources and capacities to fight rural poverty, the sustainable livelihood viewpoint is also highlighted as a pertinent framework for rural entrepreneurship (Tabares et al., 2022). According to Marsden & Smith (2005), specialized networks play a critical role in forming local and regional responses to problems in conventional agriculture, which could lead to the creation of a more sustainable framework for rural development.

To sum up, the literatures offer a thorough comprehension of the complex interplay between eco-entrepreneurship and sustainable development. These theories and models prioritize social justice, economic success, and environmental stewardship. They provide frameworks for strategic planning, policy creation, and decision-making. The emphasis is on interdisciplinary collaboration among entrepreneurs, politicians, scholars, and local people to tackle complex difficulties that arise in mountainous regions such as Mizoram. The study seeks to provide a conceptual framework that is specific to Mizoram's environment, culture, and society, encouraging positive change through the synthesis of these theories.

4.7 Gaps and Areas in Existing Literature

The extant body of literature on eco-entrepreneurship and sustainable development is largely generic in nature, frequently lacking context-specific insights. This includes a notable lacuna in the body of research that thoroughly examines the opportunities and challenges posed by Mizoram's distinct mountainous landscape. Initiatives pertaining to eco-entrepreneurship and sustainable development must respect regional customs, values, and cultures. Nevertheless, a lot of research frequently ignore how crucial it is to include cultural factors in their frameworks. The practical difficulties in putting eco-entrepreneurship projects for sustainable development into practise are frequently overlooked, despite the abundance of ideas and models. Furthermore, Mizoram's unique regulatory obstacles and policy environment require greater focus in the literature. It is imperative to recognize that, being a conceptual article lacking primary data, this research acknowledges the deficiency of quantitative evidence and Mizoram-specific case studies. Although quantitative data is frequently used extensively in existing literature to support claims, theoretical contributions and qualitative insights are also useful. In mountainous areas, successful eco-entrepreneurship and sustainable development necessitate cooperation across a range of players, including enterprises, NGOs, local communities, and government agencies. The intricacy of these multi-stakeholder collaborations is sometimes underestimated in research. The literature on eco-entrepreneurship and sustainable development has to go further into methods and policies geared at climate adaptation and mitigation, particularly in susceptible mountainous places like Mizoram, given the growing threat posed by climate change.

Furthermore, a lot of research tends to ignore long-term sustainability in favour of short-term rewards. There aren't many talks in the literature about how eco-entrepreneurship projects can continue to have a good influence for long. In order to fill up these gaps and resolve these points of disagreement, this conceptual research study offers a framework tailored to the unique circumstances of eco-entrepreneurship and sustainable development in the hilly state of Mizoram. According to Rodríguez-García et al. (2019), the ecological sector is a system that supports ecopreneurship in its quest for environmental friendliness. Ecopreneurship is defined as a new way to sustainably produce economic activity. The aim of this paper is to provide policymakers, entrepreneurs, and stakeholders with a viewpoint, practical recommendations, and theoretical insights to facilitate the effective utilization of eco-entrepreneurship for long-term beneficial change in the region. Even though it might not supply numerical facts, it adds to the conversation by providing a thorough conceptual framework and synthesizing the body of previous information.

4.8 Conceptual Framework

The conceptual framework proposed here provides a structured method to investigating the potential of eco-entrepreneurship as a driver for sustainable development, environmental protection, and socioeconomic well-being in the hilly terrain of Mizoram. This framework provides an interdisciplinary and comprehensive approach to navigating the complexity of this relationship within Mizoram's distinct geographical and cultural setting, drawing on critical theoretical approaches in eco-entrepreneurship and sustainable development.

The framework, at its core, recognizes Mizoram's environmental capital, which includes its rich biodiversity, pure water sources, and fertile soil. The appropriate management of these resources, which is based on natural capital theory, is emphasized as critical to the success of eco-entrepreneurship programmes. The framework also recognizes the value of social entrepreneurship and indigenous knowledge in connecting sustainable development with local cultures, community empowerment, and traditional ecological understanding.

It also emphasizes the transformative power of eco-entrepreneurs who are committed to environmental sustainability and use new products, services, and technologies. It is emphasized that they use circular economy principles into their company structures. Within the framework, the policy landscape is also analysed, recognizing its influence on fostering or impeding eco-entrepreneurial activities, in line with the goal of promoting inclusive institutions for sustainable development.

Given Mizoram's vulnerability to climate change and natural catastrophes, the framework incorporates resilience theory, calling for eco-entrepreneurship activities that boost regional resilience through climate adaptation and mitigation. The outcome evaluation adheres to the Triple Bottom Line concept, which takes into account not only economic profitability but also contributions to social equity and environmental preservation.

Furthermore, the framework emphasizes the need of long-term sustainability, emphasizing the importance of programmes that are adaptable, scalable, and capable of continuing to have a beneficial impact over time. Finally, this conceptual framework offers a thorough, context-sensitive roadmap for exploring positive transformation in Mizoram's mountainous setting. It acknowledges eco-entrepreneurship as a critical engine of long-term development, addressing the particular difficulties and opportunities given by these various terrains. In pursuit of economic progress, cultural preservation, environmental protection, and community resilience, the framework emphasizes the necessity of responsible resource management, environmental conservation, and community empowerment.

4.9 Benefits of Eco-Entrepreneurship in Mizoram

Mountain communities frequently face economic issues as a result of limited access to markets and resources. Eco-entrepreneurship addresses these concerns by encouraging community-based firms such as cooperatives for sustainable farming and agro-tourism, as well as empowering local inhabitants, eliminating poverty, and fostering social equity. In the face of climate change consequences such as irregular rainfall and landslides, eco-entrepreneurs can implement climate-resilient agricultural practises, improving food security, protecting livelihoods, and strengthening communities. Furthermore, eco-entrepreneurs serve as environmental educators, spreading knowledge about conservation and sustainable practises and inspiring communities to take an active role in environmental protection.

Entrepreneurs introduce modern, sustainable agriculture techniques, water management solutions, and eco-friendly construction practises customized to the specific needs of hilly terrains, and eco-entrepreneurship thrives on creativity and technology. Collaboration among various stakeholders, such as government agencies, non-governmental organizations (NGOs), and local communities, is critical to the success of eco-entrepreneurship programmes in mountainous areas. Long-term sustainability is central to this notion, driving entrepreneurs to ensure that their projects are adaptable, scalable, and capable of retaining their good effects over time, aligning with the goal of sustainable development.

Thus, eco-entrepreneurship acts as a dynamic catalyst for long-term development in Mizoram, promoting economic prosperity, social fairness, and environmental resilience. Eco-entrepreneurs have the ability to effect significant positive change in this unique region by mixing creative economic practises with cultural preservation, environmental protection, community empowerment, and resilience development. Thus, including eco-entrepreneurship into Mizoram's development strategy has the promise of ensuring the state's long-term viability.

4.10 Challenges and Limitations

While eco-entrepreneurship is promising, it is not free from difficulties and constraints due to Mizoram's distinct geographical, cultural, and developmental qualities. Mountainous topography, steep slopes, and limited arable land can be barriers to agriculture and infrastructural development. Creating eco-friendly enterprises under these conditions is logistically difficult and expensive. Furthermore, Mizoram's distant mountainous position may limit market access due to costly and difficult transportation due to inadequate road connectivity. Inadequate infrastructure, such as a consistent power supply and clean water access, can standoff eco-entrepreneurship ventures, particularly those that rely on modern technologies or energy-intensive processes.

Obtaining finance and investment for eco-entrepreneurship projects is difficult. Financial institutions and seed capital may be difficult to come by. It might take time to stimulate market demand for eco-friendly products and services, as well as raise consumer knowledge and readiness to pay for sustainable items.

Inadequate comprehensive data and study on eco-entrepreneurship and its potential impact might stymie evidence-based decision-making and policy development. Despite these significant hurdles and restrictions, eco-entrepreneurship continues to be a viable option for encouraging sustainable development in Mizoram. To address these difficulties, a multi-stakeholder strategy involving governments, NGOs, communities, and entrepreneurs is required to develop context-specific solutions that cater to the region's particular requirements.

4.11 Policy Recommendations

1. **Create a Regional Eco-Entrepreneurship Strategy**: The Mizoram government should create a thorough regional eco-entrepreneurship strategy adapted to the mountainous landscape's specific traits and demands as it is completely lacking in the present Industrial Policy of the state.
2. **Provide Financial Incentives and Support**: The government should provide financial incentives to eco-entrepreneurs such as grants, subsidies, and low-interest loans. Creating a separate eco-entrepreneurship fund can help startups and small enterprises gain access to funding.
3. **Improve Infrastructure and Connectivity**: To eliminate logistical barriers for eco-entrepreneurs, improve infrastructure such as road networks, power supply, and digital connectivity. Infrastructure investments in transport and communication can open new markets and promote economic growth. The state has tremendous avenues for eco-tourism, it can receive magnificent response if tourism infrastructure would be created.
4. **Designate Specific Locations as Eco-Industrial Zones**: Designate specific locations as eco-industrial zones where enterprises adhere to sustainable practises. To stimulate eco-friendly entrepreneurship, provide tax breaks and simplified regulatory processes for enterprises operating in these zones.
5. **Encourage Eco-Tourism Initiatives**: Rivers, lakes, groundwater, and mountain spring water all play important roles in Mizoram (Lashkar et al., 2022). Supportive eco-tourism initiatives that highlight Mizoram's biodiversity and cultural history. Support eco-tourism startups through training programmes, marketing help, and infrastructure development in high-traffic locations.
6. **Promote Skill Development and Training**: Collaborate with educational institutions and non-governmental organizations to provide eco-entrepreneurship training and capacity-building programmes. Give local communities and potential entrepreneurs the skills and information they need to run sustainable businesses.

7. **Encourage the Integration of Indigenous Knowledge**: Recognize and incorporate indigenous knowledge and traditional ecological wisdom into eco-entrepreneurship efforts. Ensure that policies and programmes respect local cultures and traditions, and that culturally sensitive techniques are used to promote sustainability.
8. **Make Market Access Easier**: Market circumstances are motivators for ecopreneurship (Santini, 2017). Creating platforms or markets that connect state's eco-entrepreneurs with customers and enterprises both in and out of the region. Digital platforms can play an important role in increasing market access. Encourage eco-entrepreneurs to obtain eco-certifications and labelling for their products and services. These certifications can increase consumer trust and provide a business advantage.
9. **Create Business Incubators and Support Centres**: Create eco-entrepreneurship incubators and support centres that provide startups with mentoring, technical assistance, and pooled resources. These amenities can promote entrepreneurial growth and innovation.
10. **Raise Public Awareness**: Conduct public education programmes to educate citizens on the necessity of supporting eco-entrepreneurship and sustainable consumerism. Participate in environmental conservation and sustainable practises with local populations.

4.12 Collaborative Approach

Policymakers play a critical role in fostering eco-entrepreneurship by enacting and implementing policies that promote and support this innovative approach. Existing institutional arrangements frequently favour unsustainable enterprises, impeding progress towards sustainable entrepreneurship (Raimundo & Cruz, 2022). To counteract this, eco-entrepreneurship policies and initiatives should be region-specific, focussing on sustainability, natural resource conservation, and responsible business practises. Allocating resources to develop infrastructure, such as roads, energy, and digital access, improves connectivity and facilitates the flow of goods and services for eco-entrepreneurs. Furthermore, policymakers should encourage collaboration among government agencies, non-governmental organizations (NGOs), enterprises, and communities in order to identify concerns and develop solutions.

Businesses have an important role in promoting sustainability in the state. Prioritizing sustainability in their operations entails using eco-friendly manufacturing techniques, reducing waste, and lowering their carbon footprint. Encouraging innovation in line with sustainable development goals allows businesses to create products, services, and technology that address environmental and social issues. Building solid relationships with local communities via engagement and understanding their needs and concerns leads to mutually beneficial collaborations and long-term corporate growth.

A vital component of Mizoram's sustainable development is its communities. People can incorporate sustainable habits into their daily life, like cutting back on trash, using less water and energy, and buying locally produced eco-friendly goods. Engaging in eco-tourism activities that support the preservation of natural and cultural assets also brings in money. Traditional ecological knowledge is essential for biodiversity conservation and sustainable resource management, and communities should keep it alive and share it with future generations. Buying the goods and services of local eco-entrepreneurs not only helps these businesses flourish but also generates revenue for the community.

Eco-entrepreneurship has the capacity to bring about significant change in the difficult yet promising environment of Mizoram's hilly terrain, but this is a shared duty. Together, entrepreneurs, communities, and legislators can build an ecosystem that promotes environmentally friendly growth, conserves cultural assets, and fosters sustainable development.

4.13 Potential Avenues for Future Research

The information on eco-entrepreneurship and sustainability provided here will be especially helpful to areas with a high degree of cultural variety and environmental fragility. Assessing how policy changes and incentives affect eco-entrepreneurship and sustainable development in Mizoram can help determine how effective government assistance programmes are. The social and cultural aspects of eco-entrepreneurship can also be studied, taking into account how cultural values, gender roles, and community dynamics influence sustainable business practises. It is worthwhile to look into how technical advancements such as climate-resilient agriculture, renewable energy sources, and sustainable infrastructure development fit into hilly environments. Empirical research on the connection between ecopreneurship and sustainable development is also necessary. The effect of environmental education and awareness campaigns on community involvement in eco-entrepreneurship and sustainability efforts can also be studied by future researcher.

4.14 Conclusion

The idea gained from this research has broad ramifications for other hilly and difficult-to-access areas across the globe. The conclusions and suggestions highlight the significance of community involvement, policy support, and corporate practises that put sustainability first.

It has been demonstrated how indigenous knowledge, customs, and cultural values influence the development of sustainable business models, opening the door for more study and useful applications in these fields. Future projects will also need to comprehend the technological advancements' effects as well as the social and cultural aspects of eco-entrepreneurship.

Additionally, the study highlights the importance of inclusive entrepreneurship and government support, as noted by Lalhunthara (2019), and stresses the necessity for larger scale surveys and empirical research to corroborate the empirical data in this field. Through additional investigation into these pathways and expansion of the knowledge gained from Mizoram, it can promote sustainable development strategies in hilly areas of the world. It emphasizes that stakeholders must work together to drive good change in order to create an ecosystem that promotes sustainable development, safeguards the environment, and preserves cultural heritage, all of which will have long-term benefits for Mizoram and beyond.

References

Akku Y, Çalıyurt K (2022) The role of sustainable entrepreneurship in UN sustainable development goals: the case of TED talks. Sustain 14:2–17

Borkowski M, Moroz O, Paslavskyi M (2020) Analytical and information support of the structural development of eco-entrepreneurship in the system of sustainable development. Sustain Development–State Prospect 37–50

Demir Y, Dqfõr D, Demir E (2015) Applicability to green entrepreneurship in Turkey: a situation analysis. 195:1238−1245. https://doi.org/10.1016/j.sbspro.2015.06.266

Doe SSP (2022) Connecting innovative eco-entrepreneurship model and sustainable nature-based solutions to advance climate action biodiversity and SDGs. In: Environmental sciences proceedings. pp 15–53. https://doi.org/10.3390/environsciproc2022015053

Domańska A, Żukowska B, Zajkowski R (2018) Green entrepreneurship as a connector among social, environmental and economic pillars of sustainable development. Why some countries are more agile? Problemy Ekorozwoju 13(2):67–76

Hummels H, Argyrou A (2021) Planetary demands: redefining sustainable development and sustainable entrepreneurship. J Clean Prod 278:123804. https://doi.org/10.1016/j.jclepro.2020.123804

Jayasinghe R, Liyanage N, Baillie C (2021) Sustainable waste management through eco-entrepreneurship: an empirical study of waste upcycling eco-enterprises in Sri Lanka. J Mater Cycles Waste Manage 23(2):557–565. https://doi.org/10.1007/s10163-020-01140-0

Jayawardhana K, Fernando I, Siyambalapitiya J (2022) Sustainability in social enterprise research: a systematic literature review. SAGE Open 12(3). https://doi.org/10.1177/21582440221123200

Karagülle AÖ (2012) Green business for sustainable development and competitiveness: an overview of Turkish logistics industry. Procedia Soc Behav Sci 41:456–460. https://doi.org/10.1016/j.sbspro.2012.04.055

Kumar A, Thangjam NM, Rani P, Lalmuankima HT (2023) The potential of medicinal and aromatic plants cultivation to develop the agri-business in Mizoram. Environ Ecol 41(3A):1651–1661. https://doi.org/10.60151/envec/yock6929

Lalhunthara (2019) Factors inducing entrepreneurship: a study of MSMEs in Aizawl district, Mizoram. SEDME (Small Enterp Dev, Manag & Ext J): Worldw Wind MSME Stud 46(1):8–17. https://doi.org/10.1177/0970846419829970

Lalrokhawma TH, Singh EN (2020) Education as an entrepreneurial trait: evidence from Rural Mizoram. 9(1):7−15

Laskar N, Singh U, Kumar R, Meena SK (2022) Spring water quality and assessment of associated health risks around the urban Tuirial landfill site in Aizawl, Mizoram, India. Groundw Sustain Dev 17(May). https://doi.org/10.1016/j.gsd.2022.100726

Majid IA (2015) Sustainable entrepreneurship (SE): a revised model based on triple bottom line (TBL) Wei-Loon Koe. Int J Acad Res Bus Soc Sci 2(6):293–310

Marsden T, Smith E (2005) Ecological entrepreneurship: sustainable development in local communities through quality food production and local branding. Geoforum 36(4):440–451. https://doi.org/10.1016/j.geoforum.2004.07.008

Masjud YI (2020) Ecopreneurship as a solution to environmental problems: implications for university enterpreneurship education. J Environ Sci Sustain Dev 3(1):97–113

Mio C, Panfilo S, Blundo B (2020) Sustainable development goals and the strategic role of business: a systematic literature review. Bus Strateg Environ 29(8):3220–3245. https://doi.org/10.1002/bse.2568

Mishra VN, Rai PK, Singh P (2021) Geo-information technology in earth resources monitoring and management (edit. Book). Nova Science Publishers, USA, pp 1–395. ISBN: 978–1–53619–669–6

Moon CJ (2018) Contributions to the sdgs through social and eco entrepreneurship: new mindsets for sustainable solutions. Contemp Issues Entrep Res 8:47–68. https://doi.org/10.1108/S2040-724620180000008008

Nikolaou EI, Ierapetritis D, Tsagarakis KP (2011) An evaluation of the prospects of green entrepreneurship development using a SWOT analysis. Int J Sustain Dev & World 18(1):1–16. https://doi.org/10.1080/13504509.2011.543565

Rai PK, Mishra VN, Singh P (2021) Recent technologies for disaster management & risk reduction-sustainable community resilience & responses (edit. Book). Springer Nature, Switzerland, pp 1–471. ISBN: 978–3–030–76116–5. https://doi.org/10.1007/978-3-030-76116-5

Rai PK, Mishra VN, Singh P (2022) Geospatial technology for landscape and environment management: sustainable assessment & planning (edit. Book). Springer Nature, Singapore, pp 1–431. ISBN: 978–981–16–7373–3. https://doi.org/10.1007/978-981-16-7373-3

Rai PK (2023a) River conservation and water resource management. Springer Nature, Singapore, pp 1–352. ISBN No.: 978–981–99–2604–6. eBook ISBN: 978–981–99–2605–3

Rai PK (2023b) Advances in water resource planning & sustainability. Springer Nature, Singapore. pp 1–322. ISBN No. 978–981–99–3659–5

Raimundo RJ, Cruz SP (2022) Sustainable entrepreneurship: a literature review. pp 1–25

Rajasekaran B (2013) Sustainable entrepreneurship: past researches and future directions. J Entrep Manag 2(1):20–27

Rao RK, Gautam N, Chowdhury IR (2012) S d r o m. State development report of Mizoram

Rodríguez-García M, Guijarro-García M, Carrilero-Castillo A (2019) An overview of ecopreneurship, eco-innovation, and the ecological sector. Sustain (Switzerland) 11(10). https://doi.org/10.3390/su11102909

Şanal M, Zare L (2017) Sustainable development through eco-entrepreneurship. 6(4):109−119

Santini C (2017) Ecopreneurship and ecopreneurs: limits, trends and characteristics. Sustain (Switzerland) 9(4). https://doi.org/10.3390/su9040492

Sati VP (2022) Assessment of natural capital for sustainable rural livelihoods: an empirical study of Mizoram, India. Asia-Pac J Rural Dev 32(1):7–25. https://doi.org/10.1177/10185291221114684

Singh A, Rai PK, Deka G, Biswas B, Prasad D, Rai VK (2021) Management of natural resources through integrated watershed management in Nana Kosi micro watershed, district Almora, India. Ecol Environ & Conserv 27(February Suppl. Issue):S260−S268

Stål HI, Bonnedahl K (2016) Conceptualizing strong sustainable entrepreneurship. Small Enterp Res 23(1):73–84. https://doi.org/10.1080/13215906.2016.1188718

Tabares A, Londoño-Pineda A, Cano JA, Gómez-Montoya R (2022) Rural entrepreneurship: an analysis of current and emerging issues from the sustainable livelihood framework. Economies 10(6). https://doi.org/10.3390/economies10060142

Tien NH, Hiep PM, Dai NQ, Duc NM, Hong TTK (2020) Green entrepreneurship understanding in Vietnam. Int J Entrep 24(2):1–14

Wagner M (2015) Frontiers in eco-entrepreneurship research article information: In: Advances in the study of entrepreneurship, innovation and economic growth, vol 20. Elsevier. https://doi.org/10.1108/S1048-4736(2009)0000020009

Wei X, Ren H, Ullah S, Bozkurt C (2023) Does environmental entrepreneurship play a role in sustainable green development? Evidence from emerging Asian economies. Economic Research-Ekonomska Istrazivanja 36(1):73–85. https://doi.org/10.1080/1331677X.2022.2067887

Chapter 5
Micro-Credit for Micro-Entrepreneurs: A Study on PM Street Vendor's Loan with Special Reference to Mizoram State

Daniel Lalawmpuia

Abstract Urban street vendors hold a very important part of the urban economy which plays a pivotal role in delivering availability of essential commodities at affordable price to the city dwellers at their door step. The COVID-19 nationwide lockdowns have seriously impacted the livelihoods of street vendors across the countries. In connection with this, on 1 June 2020, the Ministry of Housing and Urban Affairs, Government of India launched a Central Sector Scheme called the Prime Minister Street Vendor's AtmaNirbhar Nidhi (PM SVANidhi) to facilitate provision of micro-credit for working capital to street vendors, through commercial banks in India for facilitating to recover their livelihoods. The present study tries to examine the impact of micro-credit on their livelihoods, their access to formal credit and financial inclusion as well as linkages to social security schemes among urban street vendors. An in-depth analysis has been made to identify the challenges and prospects of PM SVANidhi-street vendors' loan with special reference to Mizoram state. Forty street vendors have been randomly selected for purposive sampling in the study from six district capital, viz. Aizawl, Lunglei, Mamit, Serchhip, Champhai and Kolasib (except newly created district and two backward districts Lawngtlai and Siaha) where the scheme is being implemented by the nodal department. Primary and secondary date have been collected which was analysed with relevant statistical tools.

Keywords Micro-credit · Street vendors · Entrepreneurs · Urban livelihoods

5.1 Introduction

Street vendor in urban areas plays an important part at the bottom of the pyramid of the unorganized economic activities in the cities. Vending activities give a source of employment and act as a measure to reduce urban poverty. Street vending also

D. Lalawmpuia (✉)
Department of Economics, Pachhunga University College, Aizawl, Mizoram, India
e-mail: danielvarte12@gmail.com

P. K. Rai et al. (eds.), *Natural Resources Management and Sustainable Livelihoods in the Mountainous Region*, Advances in Geographical and Environmental Sciences,
https://doi.org/10.1007/978-981-97-2100-9_5

has a prominent place in the urban supply chain that gives reasonable and easy access to essential commodities for all segments of the population including the poor. Therefore, street vending is an integral part of the economic growth process in urban areas.[1]

Since street vendors are normally unorganized and temporarily self-employed, they often lack any linkage to social and economic security, welfare and assistance schemes, and initiatives of the Government. This makes street vendors and their families economically and socially vulnerable especially in economic slowdown or crisis, or when they may require assistance for unforeseen expenses.[2]

5.1.1 Background of the Scheme "PM SVANidhi"

Urban street vendors represent a very important part of the urban informal economy and play a major role in ensuring delivery of the goods and services at affordable price for the city dwellers at their door step. They are known as hawkers, vendors, rehriwala, thelewala and theliphadwala in different areas/contexts in India. The goods and services supplied by the vendors include fruits, vegetables, ready-to-eat street food, tea, breads, pakodas, textile, footwear, artisan products, apparel, eggs, books/stationary, etc. The services also include daily basic needs such as pan shops, cobblers, laundry services, and barber shops. The COVID-19 pandemic nationwide lockdowns have negatively impacted the life and livelihoods of street vendors in India. They usually work with a small financial capital and might have consumed the same during the lockdown. Therefore, it was felt that there is an urgent need to provide credit facility for working capital to street vendors to recover their business.[3]

In connection with this, on 1 June 2020, the Ministry of Housing and Urban Affairs, Government of India launched a Central Sector Scheme called the Prime Minister Street Vendor's AtmaNirbhar Nidhi (PM SVANidhi) to facilitate provision of micro-credit for working capital to street vendors, through commercial banks in India for facilitating to recover their livelihoods.

The scheme-PM SVANidhi is a Central Sector Scheme which means fully funded without state matching share by Ministry of Housing and Urban Affairs, Government of India with the following objectives; to facilitate/provide working capital loan up to Rs. 10,000; to reward regular repayment; and incentivize digital mode of payment and transactions. Only those States/UTs who have notified Rules and Scheme under Street Vendors (Protection of Livelihood and Regulation of Street Vending) Act, 2014 can avail the Scheme as per guidelines from the Ministry which will be available to

[1] Gov. of India (2013).

[2] *Ibid.*

[3] Gov. of India (2020a, b).

all street vendors those engaged in vending activities in urban areas as on or before 24 March 2020.

5.2 Objectives of the Study

The present study tries to analyse the following main objectives from the identified street vendors:

(i) To find out the impact of micro-credit on their livelihoods.
(ii) To examine their access to formal credit and financial inclusion.
(iii) To ascertain their linkages to social security schemes.

5.3 Research Methodology

An in-depth analysis has been made to identify the problems and prospects of PM SVANidhi-street vendor's loan with special reference to Mizoram state. Forty street vendors have been randomly selected for purposive sampling in the study from six district capital, viz. Aizawl, Lunglei, Mamit, Serchhip, Champhai and Koalsib (excepting newly created district and two district Lawngtlai and Siaha) where the scheme is being implemented by the nodal department at the initial.

The study also mainly focusses the effectiveness and efficacy of micro-credit facility and their level of financial inclusion including financial literacy and digital inclusion. The present study is mainly based on primary data by using a canvassing questionnaire collected from the main stakeholder, i.e. street vendors. Further, secondary data have been collected from official website of the Ministry of Housing and Urban Affairs, State Level Bankers Committee, published and unpublished sources, Magazines, Journals, e-books, websites and other online resources, etc. The collected data have been analysed with relevant statistical tools and systematically interpreted.

5.4 Overview and Implementation of PM SVANidhi in India

Since inception of the scheme of a Special Micro-Credit Facility for Street Vendors on 1 June, 2020 there has been a tremendous progress in terms of number of loan applications sanctioned and disbursed among all ULBs across the country. This section will highlight vending activities of the street vendors, top lending institutions of the micro-credit and best performing state in the country as on date.

Table 5.1 shows the vending activities of the street vendors in India which has been classified into 10 categories comprising different activities. It has clearly revealed

Table 5.1 Vending activities of the street vendors

Sl. no	Vending activities	Percentage (%)
1	Fruits and vegetables	46
2	Fast food and food items	22
3	Cloth and handloom items	13
4	Beauty and fashion accessories	5
5	Services	4
6	Flower and puja items	3
7	Home decoration and handicraft	2
8	Plastic items	2
9	Kitchen items	2
10	Electrical and electronic goods	2

Source https://pmsvanidhi.mohua.gov.in/Home/PMSDashboard

from the data that fruits and vegetables selling is the most popular activities in India, i.e. 46% followed by fast food and food items (22%); cloth and handloom items (13%); beauty and fashion accessories (5%); services (4%); flower and puja items (3%); home decoration and handicraft, plastic items, kitchen items and electrical and electronic goods were 2%, respectively.

It is argued that affordable and timely access to finance is prerequisite for small and marginalized micro-entrepreneurs including street vendors since they were adversely affected by pandemic in their livelihoods. Meanwhile, financing of the underprivileged and urban street vendors has challenges with bankers in many issues like their bankability and repayment capacity as well as their livelihood security. Graph 5.1 has highlighted the top 10 lending institutions across the country in the implementation of PM SVANidhi.

Graph 5.2 has shown performance of Hilly and Northeast state in the implementation of PM SVANidhi. As shown in the graph, Assam state is the best performing

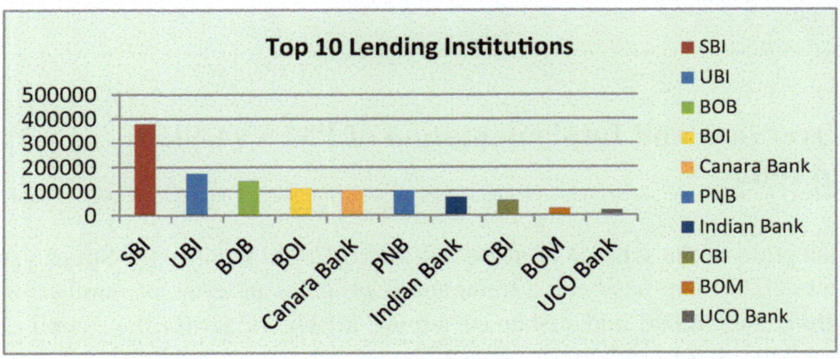

Graph 5.1 Top 10 lending institutions. *Source* https://pmsvanidhi.mohua.gov.in/Home/PMSDashboard

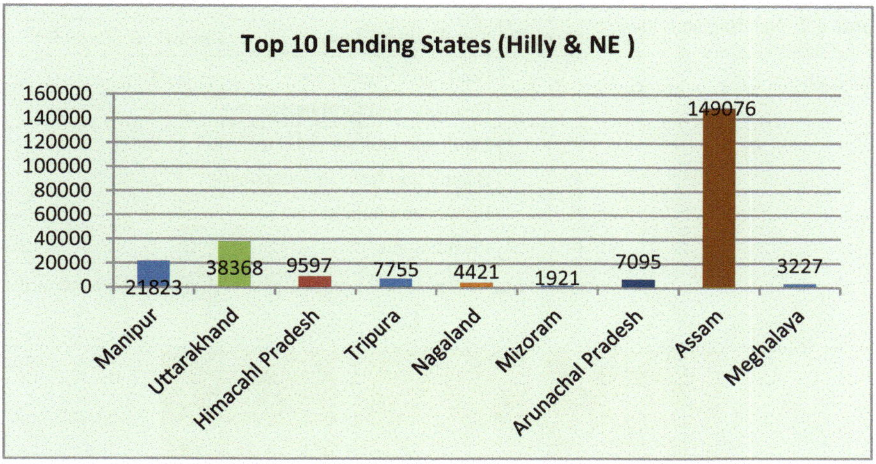

Graph 5.2 Top 10 lending states (Hilly and NE). *Source* https://pmsvanidhi.mohua.gov.in/Home/PMSDashboard

state in Hilly and Northeast state while Mizoram is the worst performing state in terms of loan sanction and disbursement under the scheme.

5.5 Overview of Mizoram State in the Implementation of PM SVANidhi

According to Ministry of Housing and Urban Affairs, as on 12 October 2023, in Mizoram there are 1024 loan applications in which 873 were sanctioned and 841 applications were disbursed by the banks. It is also remarkable that 92% of the street vendors in Mizoram are female while 8% are male. Besides, it is also important to note that all street vendors have accepted digital mode of payment which is one of the important initiatives by the government to ensure digital financial inclusion.

Table 5.2 Vending activities of the street vendors in Mizoram

Sl. no	Vending activities	Percentage (%)
1	Fruits and vegetables	64
2	Fast food and food items	24
3	Cloth and handloom items	8
4	Beauty and fashion accessories	2
5	Flower and puja items	1
6	Plastic items	1

Source https://pmsvanidhi.mohua.gov.in/Home/PMSDashboard

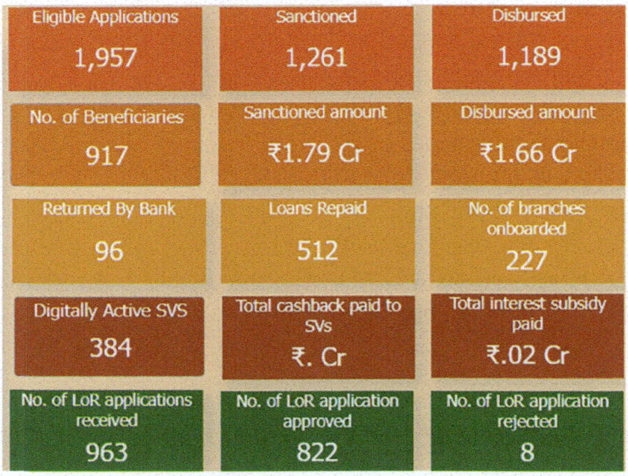

Source https://www.pmsvanidhi.mohua.gov.in/Home/PMSDashboard

Table 5.2 has shown vending activities in Mizoram which is almost similar with the country's vending activities as shown in Table 5.1.

Graph 5.3 has shown bank-wise performance in loan disbursement in the state of Mizoram. It reveals that Mizoram Rural Bank (which is under Regional Rural Bank) has disbursed the highest number of street vendors loan which is followed by SBI and other banks, viz. MCAB, Canara bank, Punjab & Sind Bank, BOM, PNB, HDFC and IDBI.

5.6 Empirical Analysis

In this section, we will highlight the progress and implementation of PM SVANidhi-Street vendor's loan in Mizoram. The present study has covered all the district capitals where the Nodal department has implemented the scheme. Primary data have been

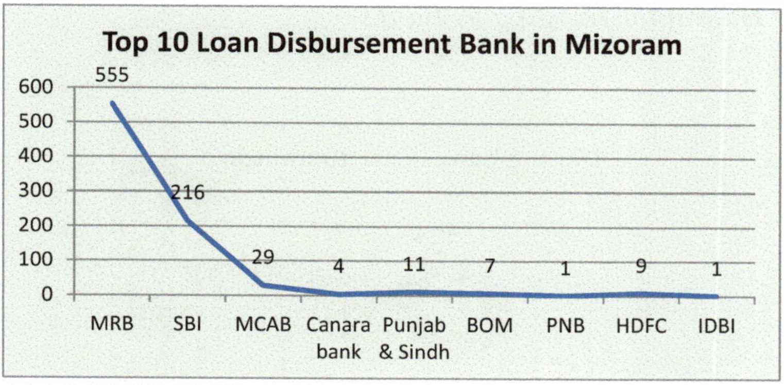

Graph 5.3 Bank-wise loan disbursements in Mizoram

collected from 40 beneficiaries (loanee) who have taken micro-loan from commercial banks.

5.6.1 General Information About the Loan

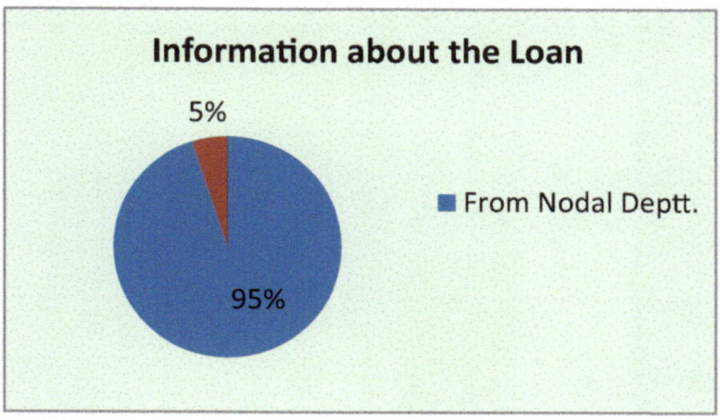

Table 2.1 has shown knowledge and information regarding the street vendor's loan among the street vendors. It was clearly revealed that majority of the vendors received their information from the nodal department while only 5% have got their information from their friends and relatives. This implies the activeness of the department in dissemination of awareness about the loan among the street vendors in all the districts of Mizoram.

5.6.2 Guarantor Insisting by Banks

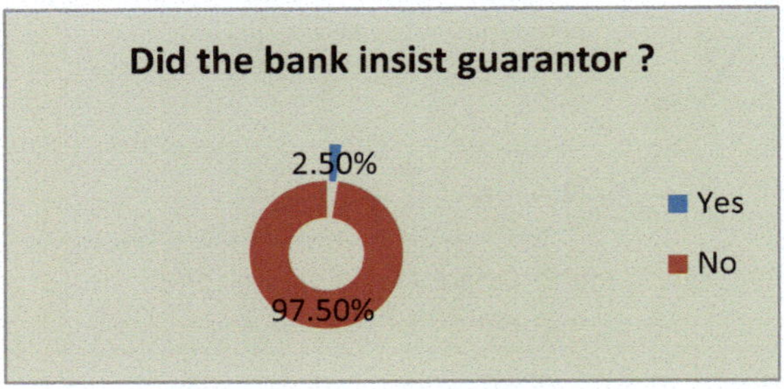

The present study has examined the problem faced by street vendor while applying for loan. Therefore, questionnaire was prepared accordingly in which they were asked about guarantor since it is one of the main stumbling blocks by the urban poor and street vendors to access credit even though the RBI has clearly instructed banks not to insist guarantors in this loan. The study shows that most of the street vendors (i.e. 97.5%) were sanctioned and disbursed loan without insisting guarantor in Mizoram which is a significant landmark in financing the micro-entrepreneurs.

5.6.3 Sufficiency of Loan Amount

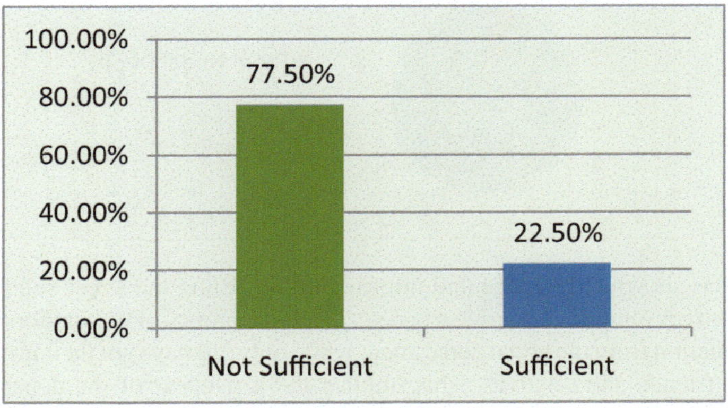

In this analysis, the respondents were asked about sufficiency of loan amount in order to recover for their livelihood activities due to lockdowns. The study shows that more than 77% of the street vendors responded that loan amount is not sufficient to meet their financial requirement while around 22% has observed that the loan amount is sufficient to meet their financial need during the pandemic. It is obvious from the analysis that due to transport cost and other variable factors like geographical constraints, hilly areas, the Centre may give special consideration in hilly areas not only in street vendor's loan but also in other programmes and schemes.

5.6.4 Whether the Scheme Will Help to Recover Their Livelihoods?

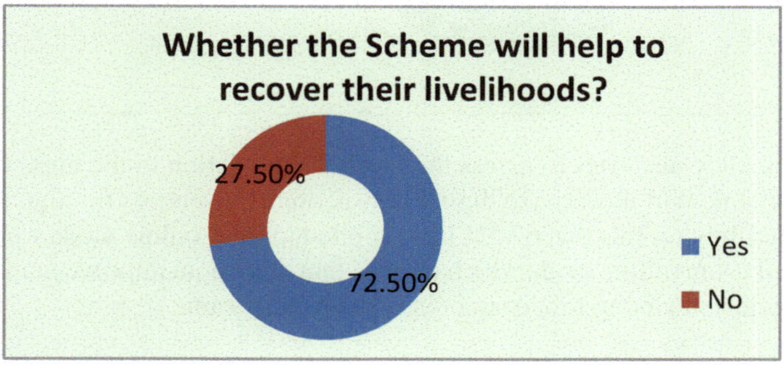

Different stages of lockdowns have been imposed throughout the year in 2020 because of COVID-19 pandemic which has severely impacted the economic activities of the vendors especially in urban areas. The present study attempts to identity whether the scheme will help the beneficiary in recovery of their occupation. It was found that 72.5% of the vendors expected that the micro-credit facility will help them to recover their livelihoods while 27.5% did not believe that the scheme will help them to recover from their livelihoods.

5.6.5 Digitization of PM SVANidhi

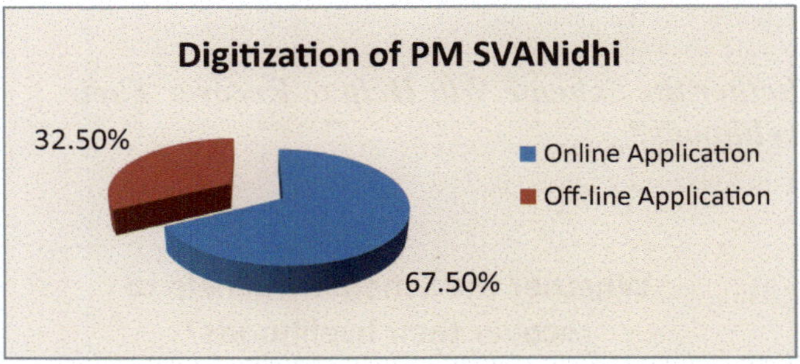

The present study also tries to assess the level of digitization in the implementation since the scheme will incentivize digital transactions by vendors through cash back facility. The study reveals that 67.5% have applied loan by online mode while 32.5% have applied using offline mode which implies that there is an improvement of digital financial inclusion among urban street vendors in Mizoram.

5.6.6 Digital Payment of Street Vendors

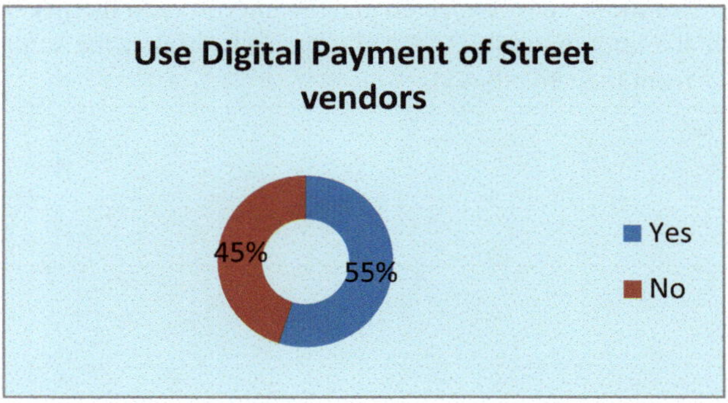

The study also tries to examine the level of digital payment among the street vendors as it is one of the initiatives taken by the government to promote digital transactions as part of Digital India as well as digital financial inclusion. It was found that 55% of the vendors have been using online payment, viz. G-pay, BHIM, Bharat Pay, Amazon Pay, PhonePe, etc. for transaction while 45% did not use any online payment.

5.6.7 Status of Street Vendors Loan Repayment

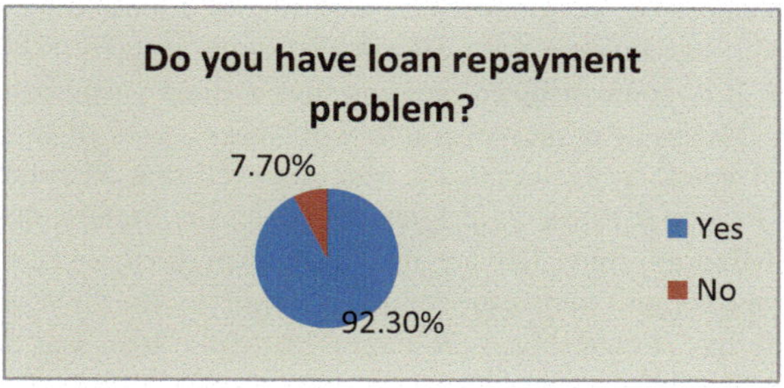

The loan repayment capacity is very significant as it shows their bankability and creditworthiness in the future from the supply side. The study found that 92.3% of the street vendors can repay their loan in time while the remaining (7.7%) has faced some problems in loan repayment. As of now, the loan recovery rate is very high in Mizoram.

5.6.8 Impact of Interest Subsidy on the Street Vendors

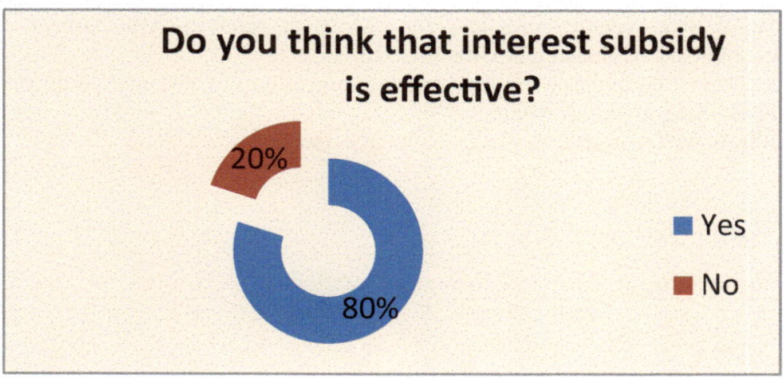

It is important to examine the impact of interest subsidy provision among the beneficiary which is the need of the hour in the country since many schemes have interest subsidy provision not only in SVANidhi but also in other schemes like Self-employment Programme, Women SHG loan and PMAY (housing for all) loan, etc. The present study has also observed that 80% of the beneficiaries have applied loan because of its interest subsidy provision (facilities) which also signifies effectiveness of the interest subsidy @ 7% per annum.

5.7 Major Findings and Conclusion

It is important for the state and the central governments that try to address the issues of urban street vendors by facilitating access to market for vending, formal credit, skill enhancement, and linkage with affordable social security scheme to ensure inclusive growth. It is noteworthy to mention that women participation among the street vendors in Mizoram is incomparable with other states of the country since 92% of the street vendors are female. It was observed that Mizoram Rural Bank (which is under Regional Rural Bank) has disbursed the highest number of street vendor's loan in Mizoram since they are also the most performing commercial banks in implementation of other interest subvention schemes by the Government of India. Further, the study has shown that more than 97% of the loan was sanctioned and disbursed by lending institution without insisting collateral/guarantor which is a significant landmark in financing the micro-entrepreneurs. It was also found that 72.5% of the vendors believe that the micro-credit facility will help them to recover their livelihoods that will mitigate the risk and vulnerability. Besides, the study also reveals that 67.5% street vendors have applied loan by online mode while 55% of the vendors have been using online payment that indicated the level of digital payment.

References

Gov. of India (2013) NULM operational guidelines for support to urban street vendors. Ministry of Housing & Urban Affairs

Gov. of India (2020a) Scheme guidelines for PM street vendor's AtmaNirbhar Nidhi (PM SVANidhi). Ministry of Housing & Urban Affairs

Gov. of India (2020b) Guidelines for socio-economic profiling of PM SVANidhi beneficiaries. Ministry of Housing & Urban Affairs

https://pib.gov.in/PressReleaseIframePage.aspx?PRID=1963799

Chapter 6
Global and Indian Perspectives on Sustainable Development, Remote Sensing, and Climate Change: An Overview

Satyavati Shukla, Deeksha, Sasmita Chand, Praveen Kumar Rai, Anoop Kumar Shukla, Rahul Kumar Misra, and Udham Singh

Abstract The intersection of sustainability, remote sensing, and climate change creates a dynamic trio of disciplines that provides novel answers to one of humanity's most important problems: the developing climate catastrophe. The overall foundation for a resilient and equitable future is sustainability, which is founded on the values of environmental stewardship, social equality, and economic success. Remote sensing, aided by cutting-edge technology, is a powerful toolkit that enables us to monitor, evaluate, and respond to the diverse effects of climate change with unprecedented accuracy. This work investigates how these interrelated factors interact to pave the way for a more sustainable and climate-resilient future. It looks into remote sensing's critical role in climate change research, as demonstrated by its ability to track temperature changes, sea level rise, deforestation, and urbanization. Remote sensing, viewed through this lens, enables evidence-based decision-making, stimulates cross-disciplinary partnerships, and improves our knowledge of the complex

---

S. Shukla
Key Laboratory of Geospatial Informatics, Guilin University of Technology, Guilin, PR China
e-mail: satyageomatics@gmail.com

Deeksha · S. Chand · A. K. Shukla (✉)
Manipal School of Architecture and Planning, Manipal Academy of Higher Education, Manipal, Karnataka, India
e-mail: anoop.shukla@manipal.edu

Deeksha
e-mail: deeksha.msapma2@learner.manipal.edu

S. Chand
e-mail: chand.sasmita@yahoo.in

P. K. Rai
Department of Geography, Khwaja Moinuddin Chishti Language University, Lucknow, India
e-mail: rai.vns82@gmail.com

R. K. Misra · U. Singh
Department of Economics, Khwaja Moinuddin Chishti Language University, Lucknow, India

P. K. Rai et al. (eds.), *Natural Resources Management and Sustainable Livelihoods in the Mountainous Region*, Advances in Geographical and Environmental Sciences,
https://doi.org/10.1007/978-981-97-2100-9_6

interactions between human activities and the environment. Our analysis suggests critical future measures to develop this synergy. These include the creation of integrated policy frameworks that prioritize sustainability, investments in remote sensing innovations, capacity building to fully utilize the potential of remote sensing technologies, interdisciplinary collaboration, and initiatives to promote sustainable lifestyles and responsible consumption. In conclusion, the convergence of sustainability, remote sensing, and climate change mitigation offers a game-changing path ahead. It promises to not only mitigate the effects of climate change but also to create peaceful cohabitation between humans and the world. This paper emphasizes the need of embracing this convergence as a key driver of global sustainable and climate-resilient development.

Keywords Sustainable development goals · Remote sensing · Geospatial technology · GIS · Climate change

6.1 Introduction

Sustainability, the technique of fulfilling current demands without jeopardizing future generations' capacity to meet their own, has emerged as a vital idea in our quickly changing society. Sustainability, at its foundation, strives to balance environmental, social, and economic factors in order to promote a peaceful coexistence of humans and the natural world. Climate change, which refers to long-term changes in Earth's climatic patterns caused mostly by human activities that emit greenhouse gases into the atmosphere, is one of the most significant concerns mankind has in this respect. Human activities, notably the use of fossil fuels, deforestation, and industrial operations, have raised the concentration of greenhouse gases such as carbon dioxide (CO_2) in the atmosphere dramatically. This increased greenhouse effect traps heat, resulting in global warming and a slew of other climate-related consequences. A leading scientific group, the Intergovernmental Panel on Climate Change (IPCC), has repeatedly warned about the increased hazards of extreme weather events, rising sea levels, damaged ecosystems, and challenges to food and water security caused by uncontrolled climate change (IPCC 2021). Climate change mitigation and sustainability need revolutionary activities across all sectors. Experts advocate methods such as switching to renewable energy sources, improving energy efficiency, implementing circular economic models, and supporting sustainable agriculture. Furthermore, global cooperation and policy frameworks, such as the Paris Agreement, play an important role in unifying states to combat climate change collectively (UNFCC 2015). Finally, sustainability is a critical pillar for ensuring the well-being of both present and future generations. Climate change, being the most pressing concern in the domain of sustainability, needs immediate and concerted measures to mitigate its effects. Humanity may aim for a more sustainable and resilient future by comprehending the interconnection of ecological, social, and economic systems.

6.2 Sustainability and Climate Change: Global Scenario

The current situation of sustainability and climate change is a complicated and critical global dilemma. Human activity continues to exacerbate environmental deterioration, with immediate and long-term effects for ecosystems, economies, and society. Climate change, a fundamental aspect of sustainability, is causing weather patterns to shift, sea levels to rise, and the frequency of extreme weather events to increase. The worldwide rise in average temperatures of around 1.2 degrees Celsius over pre-industrial levels, with grim estimates for the future, demonstrates the severity of the problem (IPCC 2021).

Climate change has far-reaching consequences that are unevenly distributed. Climate-related catastrophes frequently disproportionately affect developing nations, which have inadequate resources and infrastructure, compounding existing disparities. Rising sea levels endanger coastal populations' homes and livelihoods, while droughts and heatwaves reduce agricultural production and water availability. These difficulties have sparked worldwide debates and efforts such as the Sustainable Development Goals (SDGs) and the Paris Agreement, which attempt to inspire global collaboration to address both climate change and broader sustainability issues (UN 2015; UNFCC 2015).

To address these concerns, a comprehensive approach that incorporates environmental, social, and economic aspects is required. Renewable energy sources, such as solar and wind power, have grown in popularity as viable alternatives to fossil fuels, helping to reduce greenhouse gas emissions. Transitioning to circular economies, in which resources are reused and waste is reduced, can also help to relieve pressure on ecosystems. Furthermore, supporting sustainable urban development, responsible consumption, and biodiversity preservation are critical components of overall sustainability initiatives (UNEP 2019).

Global corporations and manufacturing sectors have come to understand their critical role in driving change, with many implementing environmentally friendly business procedures, establishing lower emission goals, and investing in cutting-edge technologies. Authorities are implementing policies that encourage sustainable practices, such as carbon pricing and emissions regulations, in an effort to reconcile economic development with ecological conservation (The World Bank 2022).

Finally, the global status of sustainability and climate change need immediate and coordinated action. The link between environmental health and human well-being emphasizes the importance of moving toward a more sustainable route. Humanity has the chance to limit the impact of climate change while building a more egalitarian and sustainable future by adhering to international accords, embracing renewable energy, and changing consumption standards.

6.3 Remote Sensing and Climate Change: Global Scenario

Remote sensing techniques have developed as critical instruments for monitoring and comprehending global climate change. These methods entail collecting data on the Earth's environment via satellites, airplanes, and ground-based sensors. Scientists may employ remote sensing to detect and evaluate climate-related data such as temperature, sea level, ice cover, and greenhouse gas concentrations, offering crucial insights into the dynamics of climate change (Richter-Menge et al. 2018).

Monitoring the Earth's polar regions is one of the most important uses of remote sensing in climate change research. Satellites with specialized sensors collect information on the extent and thickness of polar ice sheets and sea ice. This data is critical for determining the pace of glacier melt and sea level rise, as well as their contributions to global climate change (Zhang et al. 2022). Remote sensing tools also aid in tracking changes in land cover and vegetation, allowing scientists to better understand how deforestation, urbanization, and land-use changes affect regional and global climate trends (Turner et al. 2015; Mishra et al. 2014, 2021; Rai et al. 2021 and Rai et al. 2022).

Remote sensing has also been crucial in climate modeling and effect assessments. Remote sensing platform data is incorporated into climate models to increase their accuracy and prediction capabilities (Mishra et al. 2014; Stephens et al. 2015; Rai et al. 2014; Rai et al. 2023a and 2023b). These models are useful in estimating future climatic scenarios, which are critical for policymakers and academics seeking to minimize and adapt to the global effects of climate change.

If we have to look into the statistics of publication trends using remote sensing techniques to assess the climate change, we find China and the US with more than two thousand publications and Germany followed by India and the UK with more than eight hundred publications using remote sensing technique effectively (Fig. 6.1). In the next figure (Fig. 6.2), we find that there is a steady progress in the publication trends, annually, on climate change using remote sensing techniques as it is one of the most studied topics on the academic front.

Finally, remote sensing techniques have shown to be useful in the research of climate change. They give a plethora of data that helps us understand climate processes, monitor changes in Earth's polar regions, and influence climate modeling efforts. Remote sensing will continue to be an important component of climate study and response as the global climate issue worsens.

6.4 Sustainability and Climate Change: Indian Scenario

India, a fast-expanding country with a population of over 1.3 billion people, is confronted with a complex terrain of sustainability concerns and is severely impacted by climate change. As a party to international treaties such as the Paris Agreement,

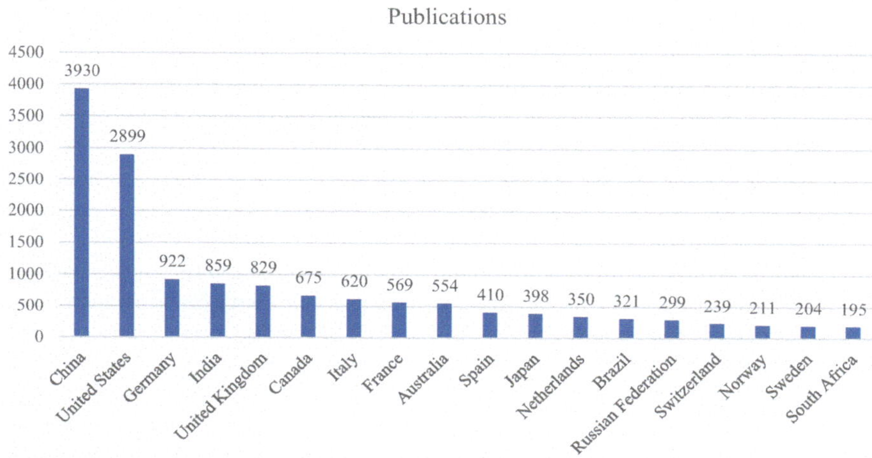

Fig. 6.1 World publication trends on remote sensing techniques used to assess climate change

Fig. 6.2 Annual global publication trends on remote sensing techniques used to assess climate change

India recognizes the need of addressing climate change as part of its larger sustainability goal. Environmental conservation, social development, and economic growth are all part of the country's sustainability initiatives.

Climate change is a serious danger to India's agriculture, water supplies, and infrastructure. Droughts and floods have become more common and severe as temperatures have risen and precipitation patterns have changed (NDMA 2019). These climate-related occurrences have disastrous effects on millions of people's livelihoods, especially in rural regions where agriculture is the major source of income.

India has taken significant initiatives to mitigate climate change, including the use of renewable energy. By 2022, the National Solar Mission hopes to have generated 100 GW of solar electricity (MNRE 2021). Wind and hydroelectric projects have also gained traction, helping to create a more sustainable energy mix. Furthermore, in order to counteract deforestation and increase biodiversity, India is investing in reforestation initiatives and sustainable land management (MoEFCC 2019).

Despite these efforts, India's economic growth faces long-term hurdles. Rapid urbanization has put a strain on resources, causing problems with waste management, air pollution, and water shortages (World Bank 2019). Industry and transportation expansion add to greenhouse gas emissions, demanding tougher environmental laws.

India understands the significance of balancing growth and sustainability and has incorporated SDGs into its national policy (NITI Aayog 2019). Initiatives such as the Swachh Bharat Abhiyan, which focuses on sanitation and cleanliness, illustrate the government's dedication to social and environmental reform.

Despite progress, India's path to sustainability and climate resilience remains complicated. In its pursuit of a more sustainable future, the country is actively working to achieve its climate and sustainability goals through policy measures, technological innovation, and international cooperation, emphasizing the interconnectedness of environmental health, social equity, and economic development.

The pursuit of sustainability and the fight against climate change in India is a diverse activity that includes a wide range of difficulties and possibilities. India faces a delicate balance between economic development and environmental preservation as one of the world's fastest-growing economies.

Impacts of Climate Change: India is extremely sensitive to climate change. The monsoon-dependent agricultural industry, which employs a sizable section of the population, is extremely vulnerable to climate change. Crop losses have resulted from erratic rainfall patterns and rising temperatures, threatening food security. Furthermore, sea level rise poses a threat to densely populated coastal areas, raising worries about displacement and loss of livelihood (NDMA 2019). These consequences highlight the importance of climate-resilient agriculture methods and coastal zone management techniques.

Renewable Energy: India has achieved significant progress in the utilization of renewable energy. One of the world's highest renewable energy ambitions is the country's ambitious aim of generating 100 GW of solar electricity by 2022 (MNRE 2021). Wind energy and hydropower projects also help to diversify the energy mix and reduce greenhouse gas emissions. These initiatives not only address climate change but also provide economic advantages, such as job development.

Issues of increasing Urbanization: India's increasing urbanization has posed a plethora of sustainability issues. Cities are dealing with air pollution, insufficient waste management, and overcrowded infrastructure. Initiatives such as the Swachh Bharat Abhiyan, or Clean India Mission, focus on sanitation, cleanliness, and waste management to solve these challenges. These projects attempt to improve urban residents' living standards while also encouraging sustainable urban growth (MoHUA 2019).

Policy Framework: A thorough policy framework guides India's sustainability initiatives. The National Action Plan on Climate Change (NAPCC) addresses mitigation and adaptation initiatives (MoEFCC 2019). Promoting energy efficiency, afforestation, and sustainable agriculture methods are among them. The incorporation of SDGs into national policy underscores India's commitment to combining economic growth with social and environmental equality (NITI Aayog 2019).

Challenges and Opportunities: India's path to sustainability is not without difficulties. The economic urge to raise millions out of poverty frequently collides with environmental aspirations. Balancing industrial expansion with emission reductions is an ongoing problem. Furthermore, shortage of resources, notably water, poses a substantial risk to long-term growth (World Bank 2019). These difficulties, however, create potential for innovation, technological transfer, and international collaboration.

To summarize, India's journey toward sustainability and resilience in the face of climate change is a dynamic and ever-changing process. The country's commitment to solving these difficulties is shown in regulatory measures, renewable energy investments, and attempts to improve urban living standards. As India develops, its attitude to sustainability and climate change will define not just its own destiny, but also global efforts to mitigate climate change and achieve a more sustainable world.

6.5 Remote Sensing and Climate Change: Indian Scenario

Remote sensing methods have played a critical role in determining the impact of climate change, assessing its repercussions, and developing mitigation solutions in India. This corpus of knowledge has been enriched by several publications. If we see the publication trends (Fig. 6.3), we find a steady increase in the graph annually, showing the use of remote sensing technique to assess climate changes.

In this part, we highlight a few notable articles that demonstrate the convergence of remote sensing methods with climate change in the Indian setting.

The paper "Climate Change and Its Impact on Himalayan Glaciers in India" (Bhambri et al. 2012) provides a thorough examination of glacier dynamics in the Indian Himalayas using remote sensing data. The work demonstrates the importance of remote sensing in monitoring glacier retreat and its implications for downstream water resources.

"Monitoring Land Use/Cover Changes Using Remote Sensing and GIS Techniques: A Case Study, Uttarakhand, India" (Pandey et al. 2019; Rawat and Kumar 2015) analyzes land-use changes in Uttarakhand using remote sensing and GIS techniques. It gives insight into how land-use changes contribute to climate change and its local consequences.

Fig. 6.3 Annual Indian publication trends on remote sensing techniques used to assess climate change

"Monitoring Forest Cover Change" (Reddy et al. 2013; Sudhakar Reddy et al. 2018) emphasizes remote sensing's vital role in tracking forest cover changes in Odisha. This study emphasizes the link between deforestation and climate change, underlining the importance of sustainable forest management.

The paper "Remote Sensing and GIS-Based Assessment of Land Use/Land Cover Change and Its Impact on Urban Heat Island" (Das et al. 2022; Mohan et al. 2012; Sahoo et al. 2022) shows how remote sensing and GIS can be used to better understand the urban heat island effect in rapidly urbanizing areas of India, contributing to climate resilience planning.

"Soil Erosion Vulnerability Assessment Using Remote Sensing and GIS" (Dash and Maity 2023; Jain et al. 2001) assesses soil erosion risk, at region prone to climate-related calamities. Remote sensing techniques enable the identification of vulnerable locations, which aids in climate adaptation planning.

"Assessing Agriculture's Vulnerability to Climate Change in India Using GIS-Based Approaches" (Kantamaneni et al. 2020; Singh and Kumar 2021) uses remote sensing and GIS to analyze Indian agriculture's susceptibility to climate change, emphasizing the necessity of precision agriculture and climate-resilient farming practices.

6.6 Conclusions

The convergence of sustainability, remote sensing, and climate change creates a powerful synergy that is crucial in our worldwide efforts to solve the issues of climate change. With its emphasis on balancing environmental, social, and economic factors, sustainability serves as a guiding concept for creating a more resilient and equitable world. Remote sensing, with its improved technology, provides a vital tool for monitoring, analyzing, and responding to the effects of climate change. These parts, when combined, constitute a formidable framework for comprehending, alleviating, and responding to the changing climate issue.

Sustainability requires a multifaceted strategy that includes environmental conservation, prudent resource management, and fair development. Remote sensing makes a substantial contribution by providing fast and precise data on a variety of climate-related indicators, such as temperature changes and sea level rise, as well as deforestation and urbanization. These data help politicians, scientists, and communities make evidence-based decisions and get a better understanding of the intricate interplay between human activity and the environment.

Governments and international organizations must develop integrated policy frameworks that promote sustainability while exploiting the possibilities of remote sensing technology. These frameworks should encourage climate-resilient development, the use of renewable energy, and the protection of vital ecosystems.

Innovation and Data Accessibility: Priority should be given to investment in innovation, particularly in remote sensing technology and data analytics. Governments and institutions should collaborate to enable open access to remote sensing data, establishing a culture of collaboration and information sharing.

Capacity Building: It is critical to build the capacity of communities, researchers, and institutions in remote sensing methods. Education and training programs should be created to provide individuals with the skills necessary to effectively use remote sensing data in climate change research and sustainable practices.

Interdisciplinary Collaboration: It is critical to encourage interdisciplinary collaboration among scientists, politicians, and local communities. It has the potential to lead to more comprehensive solutions that address both climate change and other environmental issues.

Public awareness campaigns and educational programs should stress the necessity of sustainable lifestyles and responsible consumption habits. Individuals and communities may lower their environmental footprints through these actions.

Climate financing: Using climate financing to assist long-term initiatives, such as those that rely on remote sensing data, can help to speed climate change mitigation and adaptation efforts.

Finally, the integration of sustainability principles with the capabilities of remote sensing technology gives a once-in-a-lifetime chance to address climate change holistically. To go forward effectively, we must include these features in our policies, provide information and resources to individuals and communities, and stimulate collaboration across sectors and disciplines. We can work together to limit the effects of climate change, promote sustainability, and secure a more resilient and fair future for all by doing so.

References

Bhambri R, Bolch T, Chaujar RK (2012) Frontal recession of Gangotri Glacier, Garhwal Himalayas, from 1965 to 2006, measured through high-resolution remote sensing data. J Geol Soc India 102(3):489–494. https://doi.org/10.5167/uzh-59630

Das T, Jana A, Mandal B, Sutradhar A (2022) S patio-temporal pattern of land use and land cover and its effects on land surface temperature using remote sensing and GIS techniques: a case study of Bhubaneswar city, Eastern India (1991–2021). GeoJournal 87:765–795. https://doi.org/10.1007/s10708-021-10541-z

Dash SS, Maity R (2023) Effect of climate change on soil erosion indicates a dominance of rainfall over LULC changes. J Hydrol: Reg Stud 47:101373. https://doi.org/10.1016/j.ejrh.2023.101373

IPCC (2021) Climate change 2021: the physical science basis Intergovernemental panel on climate change. Working group 1

Jain SK, Kumar S, Varghese J (2001) Estimation of soil erosion for a himalayan watershed using GIS technique. In: Water resources management, vol 15

Kantamaneni K, Rice L, Yenneti K, Campos LC (2020) Assessing the vulnerability of agriculture systems to climate change in coastal areas: a novel index. Sustainability (Switzerland) 12(11). https://doi.org/10.3390/su12114771

Mishra V, Rai PK (2016) A remote sensing aided multi-layer perceptron-marcove chain analysis for land use and land cover change prediction in Patna district (Bihar), India. Arabian J Geosci 9(1):1–18. https://doi.org/10.1007/s12517-015-2138-3

Mishra VN, Rai PK, Mohan K (2014) Prediction of land use changes based on land change modeler (LCM) using remote sensing: a case study of Muzaffarpur (Bihar), India. J Geograph Inst, Jovan Cvijić" SASA (Serbia) 64(1):111–127.https://doi.org/10.2298/IJGI1401111M

Mishra VN, Rai PK, Singh P (2021) Geo-information technology in earth resources monitoring and management (edit. Book). Nova Science Publishers, USA, pp. 1–395. ISBN: 978-1-53619-669-6

MNRE (2021) National Solar Mission. Ministry of New and Renewable Energy

MoEFCC (2019) India State of forest report 2019. Ministry of Environment, Forest and Climate Change, Government of India

Mohan M, Kikegawa Y, Gurjar BR, Bhati S, Kandya A, Ogawa K (2012) Urban heat island assessment for a tropical urban airshed in India. Atmos Clim Sci 02(02):127–138. https://doi.org/10.4236/acs.2012.22014

MoHUA (2019) Swachh Bharat Mission (Urban). Ministry of Housing and Urban Affairs, Government of India. http://mohua.gov.in

NDMA (2019) National Disaster Management Plan 2019. National Disaster Management Authority, Government of India

NITI Aayog (2019) Sustainable Development Goals. National Institution for Transforming India, Government of India

Pandey A, Palni S, Rawat JS, Pundir C, Mohite S, Kumar Bhatt P (2019) Monitoring land use/cover change using remote sensing and GIS techniques: a case study of hill station pithoragarh town, Uttarakhand, India. Int J Emerg Sci Engin 6(4). http://earthexplorer.usgs.gov/

Rai PK, Mishra S, Ahmad A, Mohan K (2014) A GIS-based approach in drainage morphometric analysis of Kanhar River Basin, India. Appl Water Sci (springer) 7:217–232. https://doi.org/10.1007/s13201-014-0238-y

Rai PK, Nathawat MS (2017) Geoinformatics in health facility analysis. Springer International Publishing, Switzerland, pp 1–231. ISBN (Hard copy): 978-3-319-44623-3., eBook ISBN: 978-3-319-44624-0. https://doi.org/10.1007/978-3-319-44624-0

Rai PK, Mishra VN, Singh P (2021) Recent technologies for disaster management and risk reduction-sustainable community resilience and responses (edit. Book). Springer Nature, Switzerland, pp. 1–471. ISBN: 978–3–030–76116–5. https://doi.org/10.1007/978-3-030-76116-5
Rai PK, Mishra VN, Singh P (2022) Geospatial technology for landscape and environment management: sustainable assessment and planning (edit. Book),. Springer Nature, Singapore, pp 1–431. ISBN: 978–981–16–7373–3. https://doi.org/10.1007/978-981-16-7373-3
Rai PK (2023a) River conservation and water resource management. Springer Nature, Singapore, pp 1–352. ISBN No.: 978–981–99–2604–6. eBook ISBN: 978–981–99–2605–3
Rai PK (2023b) Advances in water resource planning and sustainability. Springer Nature, Singapore, pp 1–322. ISBN No. 978–981–99–3659–5
Rawat JS, Kumar M (2015) Monitoring land use/cover change using remote sensing and GIS techniques: a case study of Hawalbagh block, district Almora, Uttarakhand, India. Egyptian J Remote Sens Space Sci 18(1):77–84. https://doi.org/10.1016/j.ejrs.2015.02.002
Reddy CS, Jha CS, Dadhwal VK (2013) Assessment and monitoring of long-term forest cover changes in Odisha, India using remote sensing and GIS. Environ Monit Assess 185(5):4399–4415. https://doi.org/10.1007/s10661-012-2877-5
Richter-Menge J, Osborne E, Jeffries M (2018) Arctic report card 2018. www.arctic.noaa.gov/Report-Card
Sahoo S, Majumder A, Swain S, Gareema PB, Al-Ansari N (2022) Analysis of decadal land use changes and its impacts on Urban Heat Island (UHI) using remote sensing-based approach: a smart city perspective. Sustainability (Switzerland) 14(19). https://doi.org/10.3390/su141911892
Singh RK, Kumar M (2021) Assessing vulnerability of agriculture system to climate change in the SAARC region. Environ Chall 5:100398. https://doi.org/10.1016/j.envc.2021.100398
Stephens GL, O'Brien D, Webster PJ, Pilewski P, Kato S, Li JL (2015) The albedo of earth. Rev Geophys 53(1):141–163. Blackwell Publishing Ltd. https://doi.org/10.1002/2014RG000449
Sudhakar Reddy C, Saranya KRL, Vazeed Pasha S, Satish KV, Jha CS, Diwakar PG, Dadhwal VK, Rao PVN, Krishna Murthy YVN (2018) Assessment and monitoring of deforestation and forest fragmentation in South Asia since the 1930s. Global Planet Change 161:132–148. https://doi.org/10.1016/j.gloplacha.2017.10.007
The World Bank (2022) State and trends of carbon pricing. https://documents1.worldbank.org/curated/en/099045006072224607/pdf/P1780300092e910590acb201757ecd54322.pdf
Turner W, Rondinini C, Pettorelli N, Mora B, Leidner AK, Szantoi Z, Buchanan G, Dech S, Dwyer J, Herold M, Koh LP, Leimgruber P, Taubenboeck H, Wegmann M, Wikelski M, Woodcock C (2015) Free and open-access satellite data are key to biodiversity conservation. In: Biological conservation, vol. 182, pp. 173–176. Elsevier Ltd. https://doi.org/10.1016/j.biocon.2014.11.048
UN (2015) Transforming our world: the 2030 agenda for sustainable development. United Nations
UNEP (2019) Global environment outlook: summary for policymakers 2019
UNFCC (2015) The Paris agreement - united nations framework convention on climate change
World Bank (2019) India - systematic country diagnostic. World Bank Group.
Zhang H, Wang FT, Zhou P (2022) Changes in climate extremes in a typical glacierized region in central Eastern Tianshan Mountains and their relationship with observed glacier mass balance. Adv Clim Change Res 13(6):909–922

Chapter 7
Status and Challenges for Sustainable Rural Livelihood Faced by the Small Tea Growers (STG) in the Darjeeling Hills, India

Bivek Subba, Sajal Mandal, and Lakpa Tamang

Abstract Small Tea Growers (STGs) play a vital role in the tea production landscape of the Darjeeling Himalayan region, and comprehending their livelihoods is essential for informed decision-making in both policy and industry circles. This study delves deep into the lives of STGs, drawing upon an extensive dataset (n = 927) collected from various blocks in the region using a purposeful random sampling technique. The primary data revealed the existence of five primary livelihood categories, viz. Tea Growers (78%), Agriculture and Horticulture (12%), Service (8%), Tourism (1%), and the Unorganized sector (1%). The average farm size in Pulbazar-Bijanbari is 0.18 ha, resulting in high land productivity, translating to an impressive annual tea production of 922.14 kg. In contrast, regions like Gorubathan, Mirik, and Algarah, with larger farm sizes ranging from 0.52 to 0.86 ha. Demographically, the age distribution of STGs reveals that those in Kurseong are relatively younger, with an average age of 49.25 yr, potentially contributing to the better adoption of modern agricultural techniques. Furthermore, Kurseong boasts relatively higher educational levels, with an average of 7 years of formal education. Income levels vary across blocks, with Pulbazar-Bijanbari generating the highest annual income of Rs. 208,604.44, while Algarah reports a comparatively lower income of approximately Rs. 110,448.98. To gauge overall well-being, a Livelihood Index was formulated, with Kurseong registering the highest index at 0.33, signifying relatively better overall living conditions. The Tea sector exhibited a Livelihood Index of 0.32, indicating a somewhat better quality of life compared to the Tourism sector, which displayed a lower index of 0.30. Additionally, the study pinpointed a strong correlation (r = 0.98) between land productivity and income, underscoring the pivotal role of land productivity in determining STGs' income across different sectors. This outcome suggests that tea production significantly influence the income of the growers in this region which controls their livelihood pattern. Therefore, it can be concluded on the basis of annual income

B. Subba · S. Mandal · L. Tamang (✉)
Department of Geography, University of Calcutta, Calcutta-700019, India
e-mail: ltgeog@caluniv.ac.in

P. K. Rai et al. (eds.), *Natural Resources Management and Sustainable Livelihoods in the Mountainous Region*, Advances in Geographical and Environmental Sciences,
https://doi.org/10.1007/978-981-97-2100-9_7

that tea production-based livelihood is the most sustainable livelihood which needs to be improved through modern farming mechanisms and financial opportunities.

Keywords Small Tea farming · Sustainable rural livelihood · Occupation structure · Pearson correlation coefficient

7.1 Introduction

The livelihoods that depend on agriculture in rural areas are a fundamental aspect of human existence (Chambers and Conway 1992; Mosse et al. 2002; Singh et al. 2021, Brototi et al. 2022). It represents the means by which individuals and communities sustain themselves in agrarian and non-urban settings (Scott et al. 2007; Ilbery 2014). Rural livelihood encompasses various strategies, activities, and resources that rural populations rely on to meet their basic needs, generate income, and improve their quality of life (Chambers & Conway 1992; Kamanga et al. 2009). In contrast to urban areas, where diverse employment opportunities and services are readily available, rural livelihoods are often deeply rooted in traditional practices and closely connected to the natural environment (Ellis 2000; Van der Ploeg 2017; Mishra et al. 2021; Rai et al. 2021; 2022). In rural areas, livelihoods are typically characterized by a combination of agricultural and non-agricultural activities, including farming, livestock rearing, forestry, fishing, and small-scale enterprises (Smith et al. 2001; Ruben and Pender 2004; Martin and Lorenzen 2016). These activities are influenced by factors such as climate, geography, culture, and local resources. Since rural livelihoods are mainly dependent on agriculture and climate and the natural environment determine the type of agriculture in that region (Osbahr et al. 2008; Kangalawe and Lyimo 2013; Shah et al. 2013). This is why most growers in Darjeeling, Assam Valley, and Nilgiris of India are engaged in tea cultivation to earn their living (Yadav 1950; Rai 2019; Mondal and Samaddar 2021). One of these tea growers is small-scale tea growers who cultivate tea on their own small plots of land.

Small Tea Growers (STG) hold particular significance in tea-producing nations such as India, Sri Lanka, China, and Africa (Qiao 2016). These small tea cultivators engage in the growth of their tea leaves, constituting a vital part of the country's overall tea production (Graham 1992). In the latter half of the twentieth century, STGs emerged as a distinct group due to shifts in land ownership and governmental policies (Little 1992; Wiggins et al. 2010). Generally, these small tea growers possess or lease relatively diminutive plots of land, often measuring < 10 ha (Nyakweba 2019; Hannan and Mondal 2021; Mallik 2022). These modest land holdings restrict the scale of their tea cultivation when compared to vast tea estates or plantations. STGs work independently, functioning outside the purview of larger tea estates or plantations (Mansingh and Johnson 2021). They may either own the land they cultivate or lease it, which grants them a greater degree of autonomy in decision-making. In addition to the customary tea varieties, STGs often cultivate local or indigenous tea plant variants (Besky 2014). These native cultivars may exhibit distinctive flavors and characteristics, contributing to the diversification of tea products available in the market.

STGs chiefly rely on family labor or employ local workers for the maintenance of their tea bushes (Tao and Wall 2009). The labor-intensive nature of tea cultivation frequently entails tasks such as plucking tea leaves, pruning, and caring for the bushes (Ahmed and Stepp 2013; Bitama et al. 2020). Certain small tea growers adhere to traditional and manual tea processing techniques, encompassing activities like hand-rolling, withering, and sun-drying the tea leaves (Hinsch 2008; Cowan-Gore and Sein 2020). Others may opt for more contemporary machinery to boost efficiency and consistency in the final tea product. Small tea growers often sell their tea directly to local markets, tea factories, or through cooperatives (Raynolds and Ngcwangu 2010; Wenner 2020; Manoraj 2021; Saha 2021; Mansingh and Johnson 2021). This direct sales approach enables them to forge connections with buyers and potentially secure better prices for their tea, circumventing the conventional auction system prevalent in certain tea-producing regions.

STGs wield considerable influence over local and regional economies (Carr and Chen 2002; McNeely and Scherr 2012; Majumdar 2017). They not only generate income for themselves but also create employment opportunities in rural areas, which can aid in poverty alleviation and support the livelihoods of individuals in tea-producing regions (Borah and Das 2015; Qiao et al. 2016; Arjunan 2018). However, STGs frequently encounter limitations concerning access to resources, including capital for investment, modern machinery, and technical knowledge. These challenges can impede their capacity to enhance the quality and quantity of their tea production (Alam et al. 2009; Salami 2010; Rais et al. 2013). Furthermore, STGs find themselves susceptible to fluctuations in global tea prices. These price variations can impact their income and profitability, making long-term planning challenging (Ross 1999; Mohan 2016; Saha 2021). Maintaining consistent quality proves to be an ongoing hurdle for small tea growers, as fluctuations in quality can affect the marketability of their tea (Lakshmi 2015; Das 2019; Hannan 2021; Khandelwal et al. 2022). They may encounter obstacles related to resources and expertise, which are essential for rigorous quality control processes. Additionally, STGs face challenges in expanding their market reach due to their limited access to marketing and distribution networks, unlike the comprehensive networks possessed by larger tea companies (Renard 2003; Blowfield and Dolan 2010).

As STGs are facing various problems, they engage in other activities along with cultivation and earn their living from it (Galt 2013; Besky 2014). In India's Darjeeling district, 70% of the total tea growers are involved in small tea farming and many of them are involved in tea farming as well as other activities, such as services, tourism, etc. (Sen 2009; Datta 2010; Lama 2022). Darjeeling tea is famous all over the world and its price is very high for the worldwide demand for this tea (Arya 2013; Besky 2014). So, it is very important to know the economic status of those who are involved in this cultivation. Therefore, the primary objectives of the current study are to examine the livelihoods of STGs in different blocks of Darjeeling Hills, including their farm characteristics, social status, and income, subjects that have not been previously investigated.

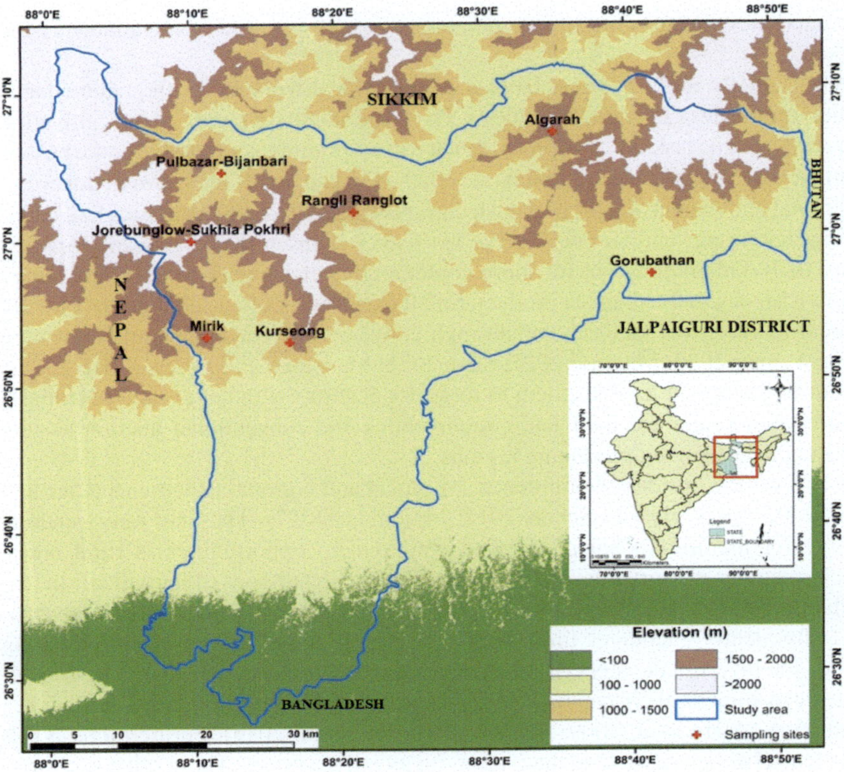

Fig. 7.1 Elevation map of Darjeeling district, prepared from 30-m Shuttle Radar Topography Mission data where the blue line is the administrative boundary of Darjeeling District and sampling sites are represented by (+) Red Cross mark

7.2 Study Area

Darjeeling Hills, located in the Indian state of West Bengal (Fig. 7.1), is renowned for its stunning natural beauty, diverse landscapes, and rich cultural heritage (Bhutia 2014; Kannegieser 2015; Gogoi 2016; Mondal et al. 2022). Situated in the eastern Himalayas, the district is famous for its lush tea gardens, which produce some of the world's finest teas (Heiss and Heiss 2012; Bajaj 2021). It boasts an average elevation of around 2200 m above sea level and features rugged terrain with steep slopes, deep valleys, and sharp ridges (Rai 2002; Raveendranathan 2017). The topography consists of the Lesser Himalayas, numerous river valleys, and dense forests. Several significant rivers, such as the Teesta, Rangeet, and Balason, flow through the district (Lahiri 1992; Cajee 2018; Rudra et al. 2018). Due to its high elevation, Darjeeling District experiences a temperate climate with distinct seasons. The summer months, from March to June, are relatively mild, with temperatures ranging from 10 °C to

25 °C The monsoon season, occurring from June to September, brings vital heavy rainfall for the tea gardens and agriculture. The period from October to November is pleasant with clear skies and mild temperatures. From December to February, temperatures drop significantly, bringing cold weather, and higher elevations occasionally witness snowfall (Ghosh et al. 2017; Thapa 2021). Situated at the boundary between the Indian Plate and the Eurasian Plate, tectonic activity has led to the complex geology of the region, featuring a mix of sedimentary rocks, schist, granite, and gneiss (Hutchison 1989; Kellett et al. 2014; Dhital 2015; Kumar et al. 2020). This geology has significantly influenced the landscape, including the formation of hills and valleys. Soil types in the district are primarily influenced by physiographic and climatic conditions. Alluvial soil suitable for agriculture is found in the lower valleys, while forest soils predominate at higher elevations (Sandhu and Singh 1981; Pal et al. 1985; Majumdar et al. 2014; Rai 2021). The region's tea gardens possess a unique type of soil known as "terroir," which contributes to the distinctive flavor of Darjeeling tea (Datta 2010; Dewan 2019; Kumpf 2020). This terroir results from a combination of well-drained, loamy soil and the region's climatic conditions.

Tea farming plays a significant role in the economy of Darjeeling Hills, with the region's tea gardens known for producing highly sought-after Darjeeling tea (Besky 2014; Sahu et al. 2023). In addition to tea, other crops like maize, paddy, ginger, and cardamom are grown in the lower valleys (Mukherjee 2012; Samajdar 2015). The steep terrain limits large-scale agriculture, but terraced farming practices are employed to maximize land utilization. The agricultural backbone of the region is tea cultivation, conducted in extensive tea gardens (Khawas 2011; Sarkar 2018). The exceptional quality of Darjeeling tea is a testament to the exceptional climate, soil, and altitude of the region (Christison 1895; Datta 2010). The livelihoods of the people in Darjeeling are closely tied to agriculture, with the tea industry serving as a primary source of employment and revenue, making a significant contribution to the local economy (Crucefix 1998). A key element in the sustainability and inclusivity of the tea industry in Darjeeling is the presence of small tea growers (Majumdar 2017; Deka and Goswami 2022). These individuals, often working with limited resources, bring diversity to the region's tea production landscape.

7.3 Database and Methods

A total of 927 small tea growers were selected as the study's participants and to ensure a representative sample, a random sampling technique was employed, guaranteeing that each grower had an equal chance of being included in the study (Kruskal and Mosteller 1979; Bellon 2001; Mugo 2002; Akudugu 2012; Lohr 2021). A structured questionnaire was designed and administered to the selected participants. This survey instrument aimed to gather essential information regarding the livelihoods of these small tea farmers.

To facilitate the meaningful comparison of diverse data points, a max–min normalization method was applied to the collected data (Jain et al. 2005; Ardil 2019; Vafaei et al. 2020, 2022; Mhlanga and Lall 2022). This process transformed the raw data into unitless values, which is crucial when dealing with variables with different units and scales. The normalization allowed for a fair and consistent assessment of each participant's contribution to the livelihood index.

The STG livelihood index was computed based on the normalized data. The key step involved determining the arithmetic mean of the normalized data points for each participant. This index served as an aggregate measure of their livelihood status. To explore the relationship between tea production and income, the Pearson correlation coefficient was employed (Zou 2003; Sarfaraz et al. 2014; Oke et al. 2016; Schober et al. 2018; Akoglu 2018). This statistical method was used to quantify the degree and direction of the linear relationship between these two key variables. It provided valuable inputs into how changes in tea production influenced income among small tea farmers in the Darjeeling region.

7.4 Results and Discussions

7.4.1 Characteristics of Tea Farms

Farm characteristics encompass a diverse set of attributes and features related to agricultural operations and farms (Brodt et al. 2006; Therond et al. 2017; Meuwissenet et al. 2019). These characteristics, which include factors like farm size, cropping intensity, and the types of crops cultivated, play a crucial role in evaluating the efficiency, productivity, and sustainability of farming operations (Karlı et al. 2006; Playan and Mateos 2006; Murshed-E-Jahan and Pemsl 2011; Ren et al. 2019). In the Darjeeling Hills, a significant portion of Small Tea Growers (STGs) relies solely on tea cultivation for their livelihood (Majumdar 2017; Hannan 2019). At the same time, traditional agricultural farmers are gradually transitioning to tea cultivation, although not exclusively (Chatterjee et al. 2013; Rasaily 2021). Approximately 12% of traditional farmers are also involved in tea cultivation, while 78% of STGs depend on tea cultivation as their primary source of livelihood, supplemented by government or private services (8%). There is a growing trend among STGs to diversify their income by combining tea cultivation with the tourism sector (1%) (Fig. 7.2). Even workers from the unorganized sector, who have limited land holdings, show interest in tea cultivation. The appeal of substantial long-term profits and reduced annual maintenance costs likely contributes to the popularity of this activity in the studied region.

7.4.1.1 Size of Tea Plantations

The size of the tea plantation area plays a pivotal role in influencing various aspects of tea cultivation and production (Costa et al. 2007; Karki et al. 2011). It is typically

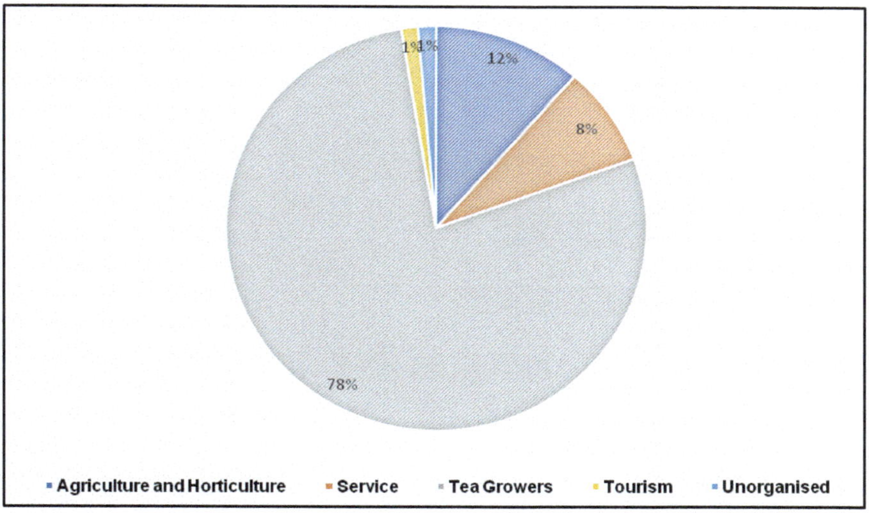

Fig. 7.2 Distribution of the Livelihood practices of the STGs (*Source* Questionnaire Survey, 2022)

measured by the total land area dedicated to tea cultivation, often owned and operated by families (Eastwood et al. 2010; Besky 2014). Different blocks in the region have varying plantation sizes, for instance, Pulbazar-Bijanbari boasts relatively smaller farms, with an average size of 0.18 ha, while Gorubathan has slightly larger farms, averaging 0.86 ha. Algarah (0.56 ha) and Mirik (0.52 ha) have moderately sized plantations, whereas Jorebunglow-Sukhiapokhri, Kurseong, and Rangli-Rangliot have relatively larger farms (Fig. 7.3). The size of the farm significantly influences tea production and the livelihoods of growers (Kanyua 2015). Smaller farms come with unique challenges and opportunities distinct from larger ones.

In this region, the occupational structure is diverse, with varying plantation areas across different sectors. Tea cultivation dominates, with the largest plantation area averaging 0.42 ha (Fig. 7.4). This underscores the significance of tea cultivation in the local economy. In contrast, agricultural and horticultural activities have a relatively smaller plantation area of 0.23 ha.

7.4.1.2 Tea Production

In Pulbazar-Bijanbari, with an average tea plantation area of 0.18 ha, tea production reaches 922.14 kg/yr, showcasing high land productivity (Fig. 7.5). This suggests the potential for high-value or specialized tea production in this block. Similarly, Jorebunglow-Sukhiapokhri (0.39 ha) and Rangli-Rangliot (0.31 ha) achieve 770.55 kg/yr and 767.55 kg/yr, respectively, with relatively high land productivity. Kurseong (0.33 ha) produces 639.32 kg/yr. Conversely, Gorubathan (0.86 ha), Mirik

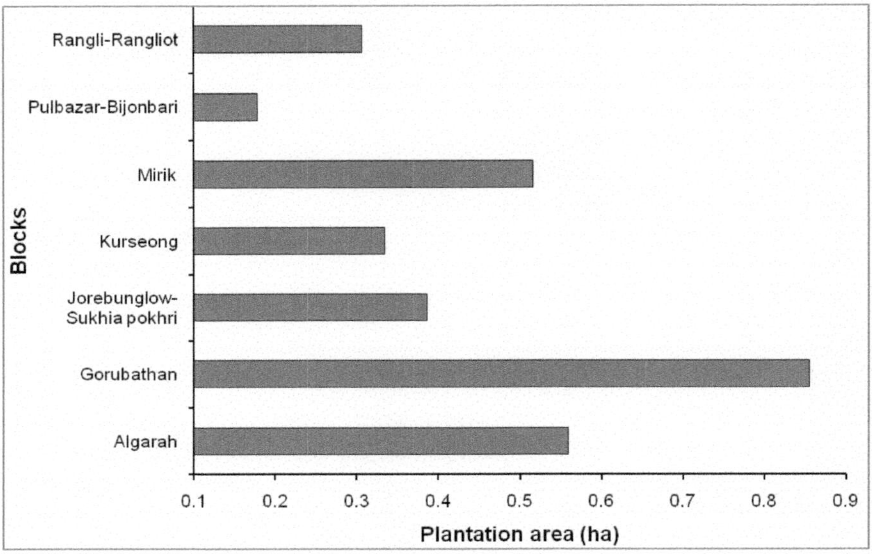

Fig. 7.3 Block-wise distribution of average tea plantation area (in ha) owned by STGs in Darjeeling Hills as per the data collected by questionnaire survey, 2022

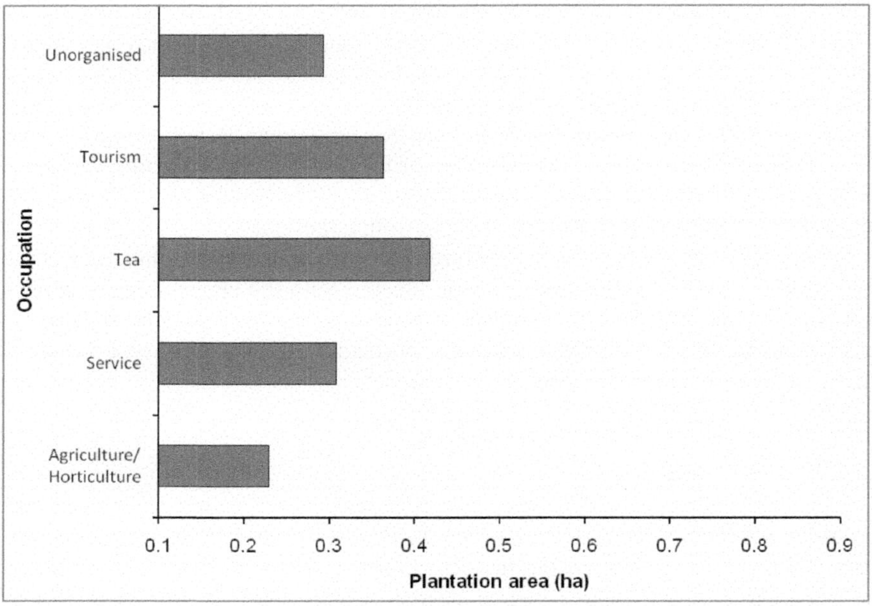

Fig. 7.4 The association of STGs Occupational Structure with plantation area (in ha.) in the selected tea growing blocks based on the questionnaire survey, 2022

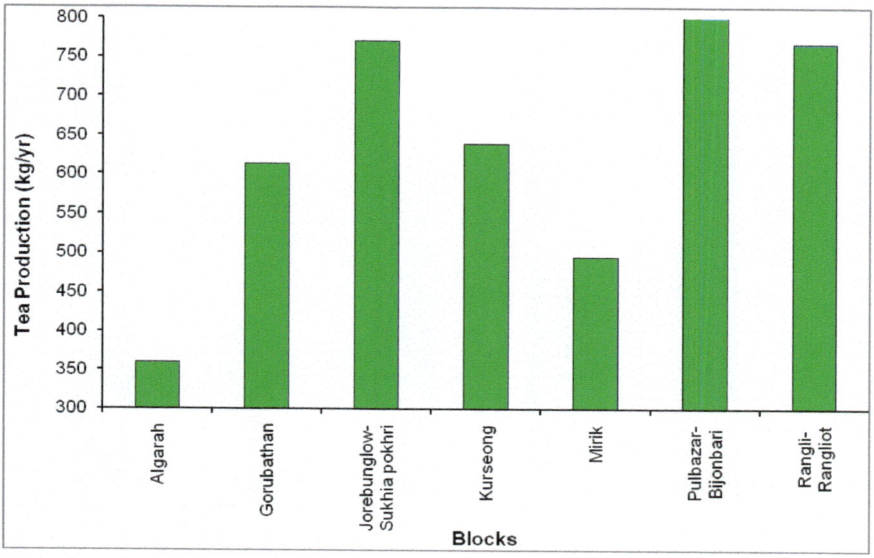

Fig. 7.5 Block-wise distribution of STGs average tea production (in Kg/Yr) as obtained through the questionnaire survey, 2022

(0.52 ha), and Algarah (0.56 ha) exhibit comparatively lower production, indicating room for improvement.

Each sector has a unique approach to tea cultivation. The Tea sector, with the largest plantation area (0.42 ha), has a slightly lower land productivity of 720.54 kg/yr, suggesting a balance between area and productivity (Fig. 7.6). The Unorganized sector, encompassing various informal labor activities, excels in land productivity, achieving 926.72 kg/yr. The Service sector focuses on non-agricultural services, with a moderate plantation area (0.31 ha) and high land productivity (894.24 kg/yr). The Tourism sector, related to tourism activities, has a plantation area of 0.36 ha, complementing tourism with a land productivity of 536.36 kg/yr.

7.4.1.3 Cropping Intensity in Tea Cultivation

Cropping intensity measures the extent to which agricultural land is used for crop cultivation and reflects the intensity of land use (Yu et al. 2021; Li et al. 2023). In Pulbazar-Bijanbari, the cropping intensity is exceptionally high, with 11,196.32 kg/ha/yr, the highest among the listed blocks (Fig. 7.7). Kurseong exhibits a considerable cropping intensity of 7,938.38 kg/ha/yr, and Jorebunglow-Sukhiapokhri shows a very high cropping intensity of 5,237.83 kg/ha/yr. Higher cropping intensity suggests efficient farming practices, better crop management, or the cultivation of high-value or high-yielding crops (Oerke et al. 1997; Pooniya et al. 2017).

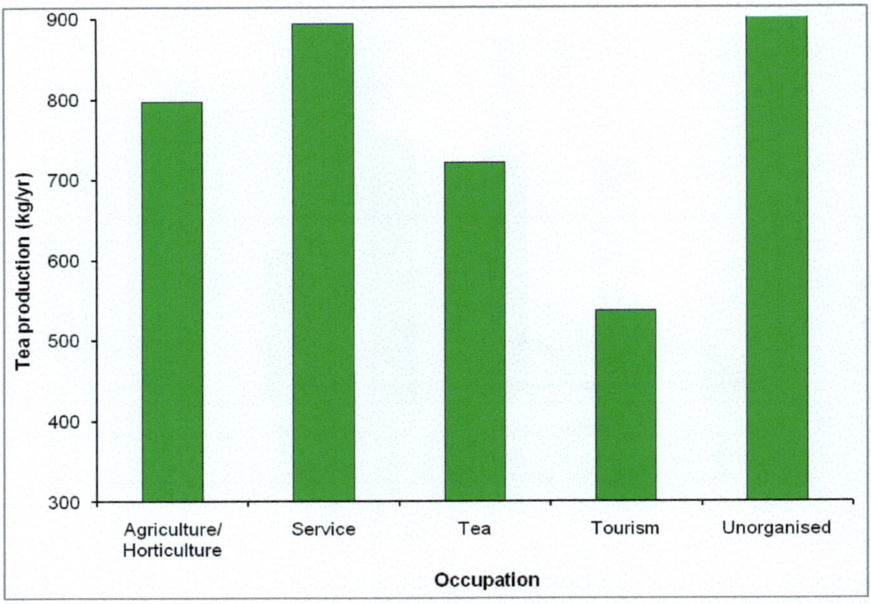

Fig. 7.6 The association between Tea production with occupation of STGs in Darjeeling Himalayan region based on the questionnaire survey, 2022

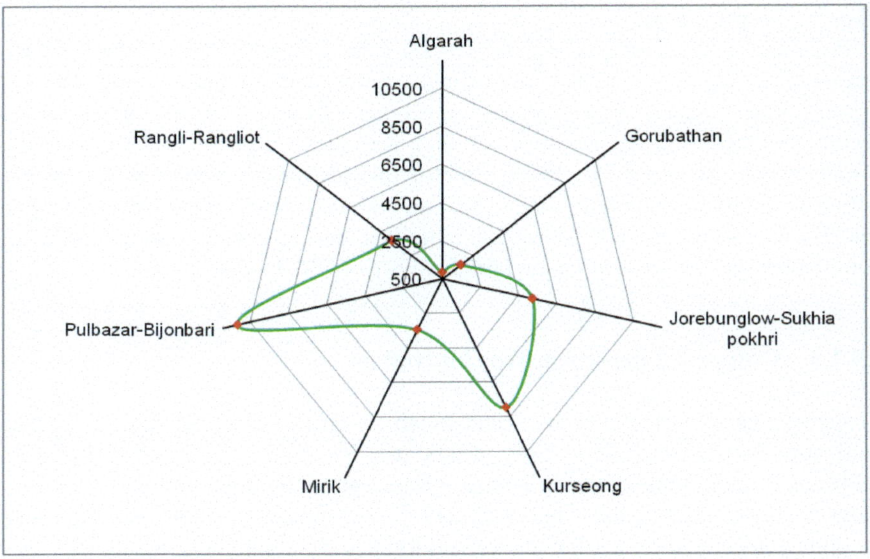

Fig. 7.7 Block-wise distribution of average cropping Intensity (in Kg/ha) by STGs calculated from data collected through the questionnaire survey, 2022

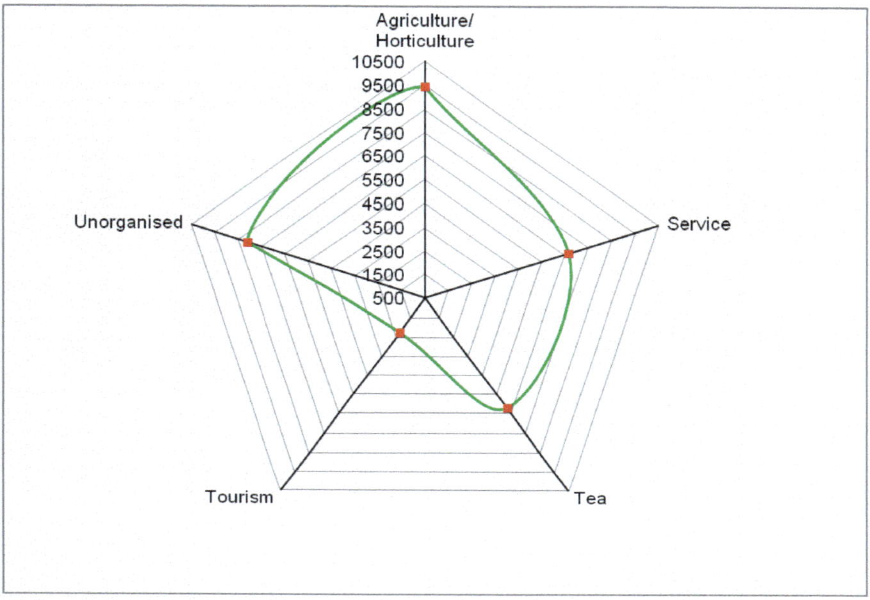

Fig. 7.8 The association between STGs Occupational Structure with Cropping Intensity in the study area based on the questionnaire survey, 2022

The Agriculture/Horticulture sector, despite having a smaller land area, achieves a very high cropping intensity of 9,428.79 kg/ha/yr (Fig. 7.8), indicating efficient land utilization for crop production. The Tourism sector, although not primarily agricultural, incorporates some crop production with a lower cropping intensity of 2,233.59, reflecting less efficient land use for agriculture within tourism activities.

7.4.2 Farmers Characteristics

7.4.2.1 Age of Tea Growers

Small Tea Growers (STGs) were categorized into three age groups for analysis: up to 30 yr (Youth), 30–50 yr (Moderate Age), and 50 yr and above (Old Age). In Kurseong, the average age of STGs is approximately 49.25 yr, which is relatively lower compared to other regions (Fig. 7.9). Youth and those in the moderate age group are mentally sharp and physically strong, which enables them to maintain small tea gardens. In Mirik, STGs have an average age of 53.40 yr, which is relatively higher than in some other areas. Most of them have converted their agricultural land into small tea gardens. Field data indicates that the majority of STGs are male, and female workers assist during the plucking season or when their husbands work

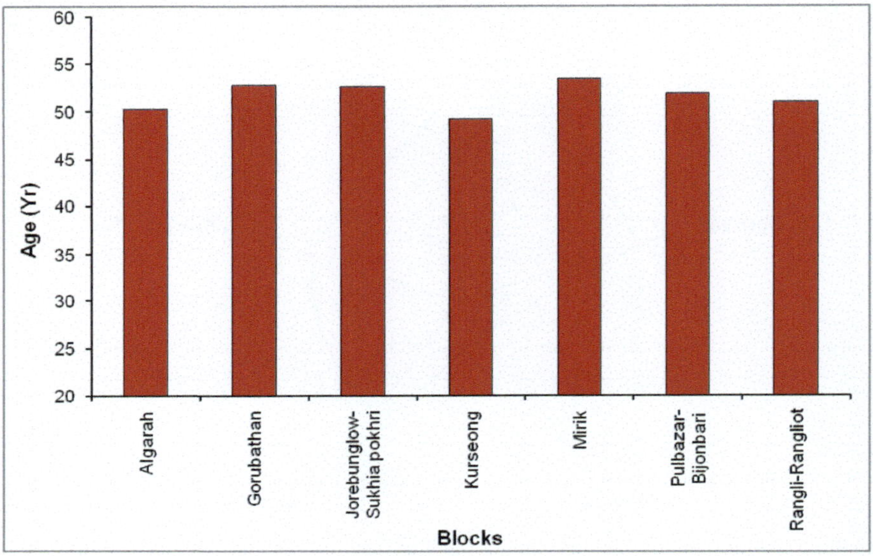

Fig. 7.9 Block-wise distribution of STGs age (in years) as collected through the questionnaire survey, 2022

far from their residence. Therefore, female workers often take on the responsibility of tea farming. The age of growers has implications for succession planning and the adoption of modern agricultural practices, especially by the younger generation (Potter and Lobley 1992; Inwood and Sharp 2012).

The age distribution within the occupational structure suggests varying workforce demographics in different sectors. In the Agriculture/Horticulture sector, the average age is 51.99 yr (Fig. 7.10). This sector combines a relatively high age with high cropping intensity, indicating that older individuals actively engage in highly productive agricultural activities. In the Tea sector, the average age is 51.58 yr, and this sector focuses on tea cultivation with a slightly younger workforce compared to Agriculture/Horticulture. The Unorganized sector, with an average age of 54.06 yr, includes informal labor activities and has the oldest workforce, yet it achieves high cropping intensity.

7.4.2.2 Education Level

The level of education plays a significant role in the adoption of new agricultural technology (Weir and Knight 2000). The educational status of STGs was evaluated, and it was observed that, on average, respondents had completed up to the sixth grade. In Kurseong, the education level is about 7 yr, which is relatively higher than

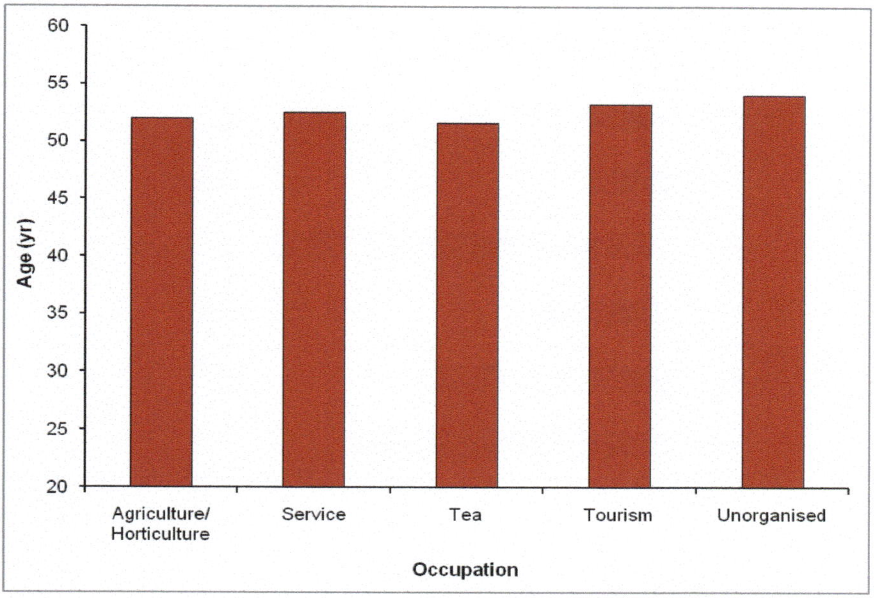

Fig. 7.10 The association between Occupation structure with Age of the STGs in the tea growing areas of the study area based on the questionnaire survey, 2022

in other regions (Fig. 7.11). This suggests a better capacity to understand and implement advanced agricultural techniques and participate in decision-making processes. Algarah, Gorubathan, Pulbazar-Bijanbari, and Rangli-Rangliot have, on average, received almost 6 yr of formal education, indicating a moderate level of education that may enable them to better understand and implement modern agricultural practices. In Mirik, STGs have an average of 5.44 yr of formal education, indicating a relatively lower level of education, which may present challenges in adopting advanced farming techniques and effectively engaging in tea-related activities. It also suggests that due to a lack of employment opportunities, STGs engage themselves in tea cultivation.

Educational levels vary among sectors, with the Service sector having the highest education level and the Tourism sector having the lowest. The data suggests a mix of education levels in different occupational sectors, with some sectors associated with more traditional or specialized activities having lower education levels. As indicated in Fig. 7.12, the service sector has a more educated workforce with an average education level of 6.73 yr. The sector has slightly older individuals with moderate cropping intensity, suggesting that higher education levels may be associated with the provision of non-agricultural services. The Tourism sector has a relatively lower education level of 4.40 yr, which affects income levels.

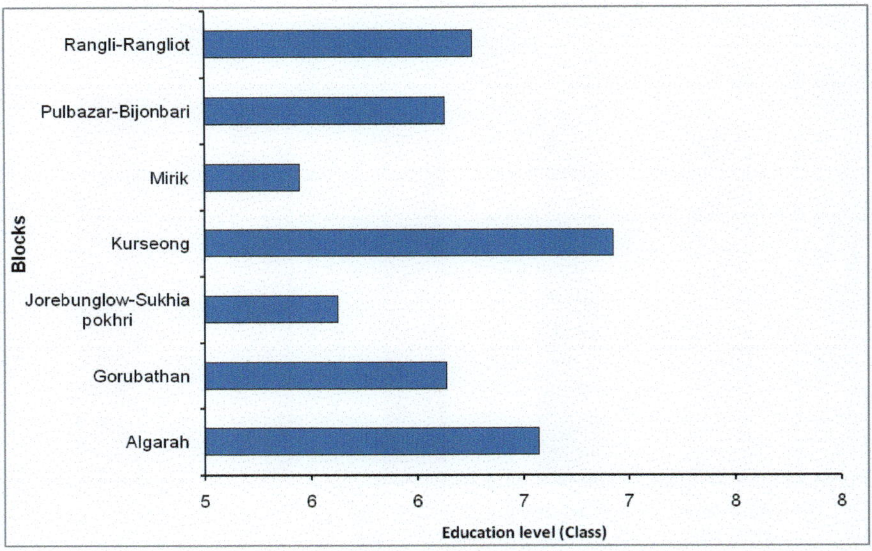

Fig. 7.11 Block-wise distribution of STGs education level as collected during the questionnaire survey, 2022

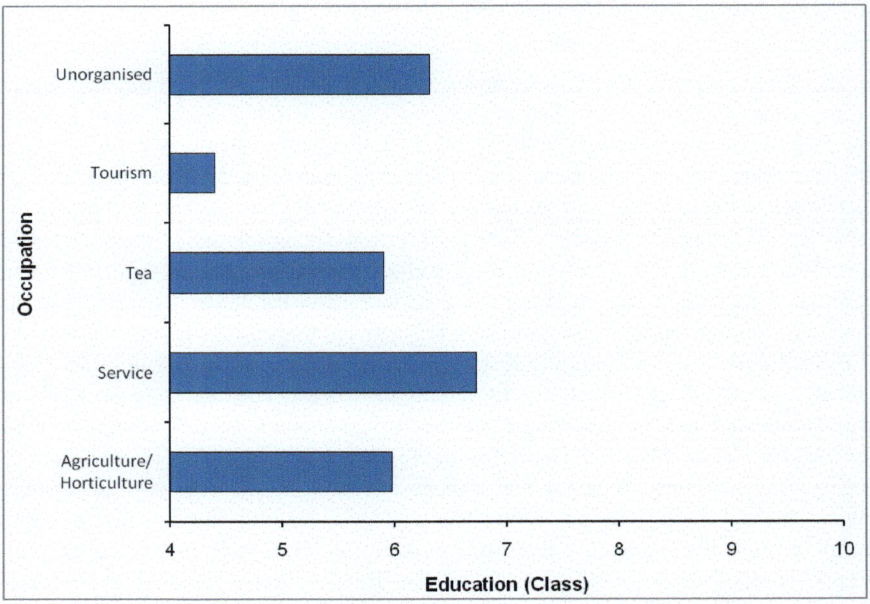

Fig. 7.12 The association between the Occupation and Education level of STGs in the study area based on the questionnaire survey, 2022

7.4.2.3 Tea Growing Experience

Small Tea Growers in this region have significant experience in tea farming, which has helped them understand the variations in tea cultivation, improve productivity, and adapt to changes over the years (Tirkey 2005; Singh et al. 2016; Gurung 2022). STGs in Mirik have an average experience of 22.15 yr (Fig. 7.13), indicating that they have been engaged in tea cultivation for a relatively long period, potentially leading to a wealth of experience and knowledge in farming. Growers in other remaining regions have an average of 20 yr of farming experience, suggesting a deep understanding of tea farming and management practices.

Figure 7.14 illustrates the tea cultivation experiences of tea growers from various occupations, with those engaged in the tourism sector possessing the most extensive experience (avg.22 yr). Conversely, farmers involved in horticulture exhibit the least experience (avg.19) in tea cultivation. This discrepancy can be attributed to the initial limited interest in tea farming among agricultural practitioners when small-scale tea cultivation began to gain traction in Darjeeling. These individuals eventually embraced tea cultivation, recognizing its superior profitability in comparison to other agricultural pursuits. On average, it can be approximated that small-scale tea farmers in Darjeeling possess approximately 20 yr of experience.

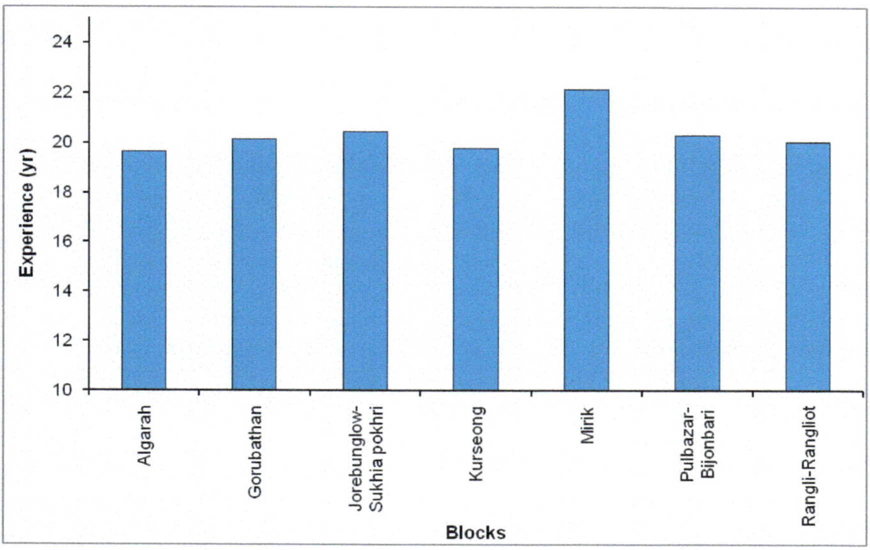

Fig. 7.13 Blockwisedistribution of STGs tea growing experience (in years) in Darjeeling Hills based on the questionnaire survey, 2022

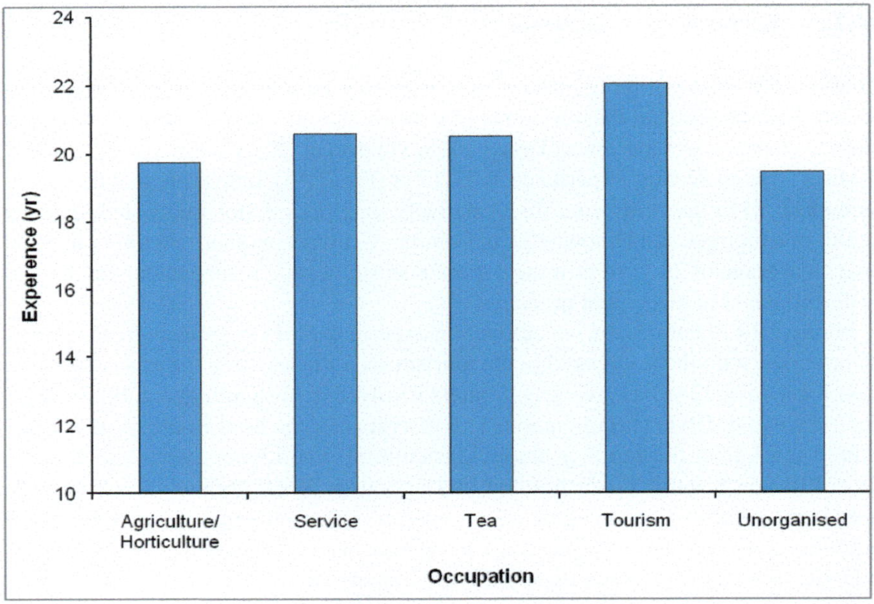

Fig. 7.14 The association between the tea growing Experience and Occupation of STGs in the study area based on the questionnaire survey, 2022

7.4.2.4 Income Level

The income of STGs from all sources directly reflects their economic status and the standard of living of their households. The income data provided in Fig. 7.15 represents the annual income generated by these growers from their tea cultivation and allied activities. Growers in Pulbazar-Bijanbari achieve a relatively high annual income of about Rs. 208,604.44, indicating highly productive tea plantations and substantial income. STGs in Algarah generate an annual income of Rs. 110,448.98 from their tea cultivation activities. This income level reflects the financial outcome of their tea farming, influenced by factors such as land productivity, cropping intensity, and market conditions. Higher-income levels generally suggest more successful tea cultivation practices, while lower income levels may indicate areas for improvement in tea farming operations.

The relationship between education and income varies based on the specific sector and the nature of the economic activities. Sectors like Service and Unorganized, which have individuals with higher education levels, tend to generate higher incomes, while sectors like Tourism with lower education levels may have lower incomes. The Agriculture/Horticulture sector generates an income of Rs. 185,808.19/yr, reflecting the efficiency and productivity of agricultural activities (Fig. 7.16). The Unorganized sector generates the highest income of Rs. 214,412.50/yr, despite the older workforce.

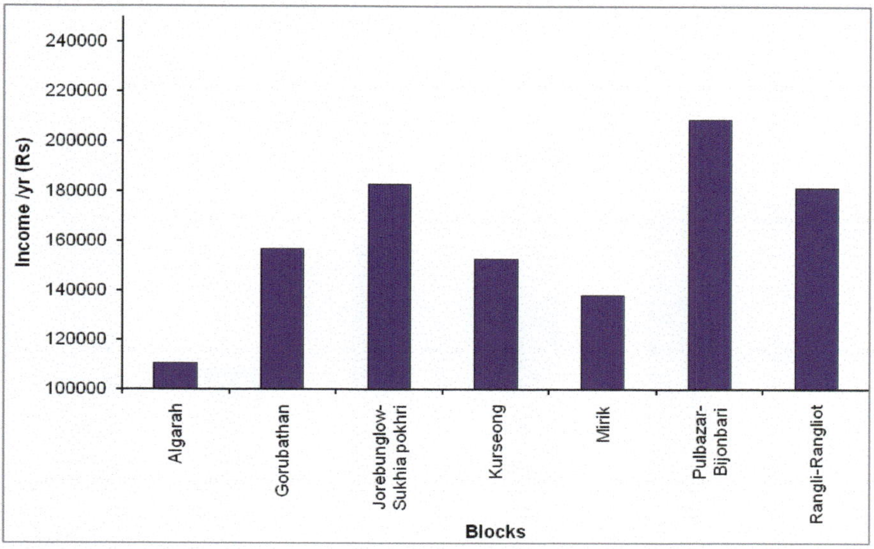

Fig. 7.15 Block-wise distribution of annual Income (in Rs.) of the STGs from all sources based on the questionnaire survey, 2022

The Tourism sector generates an income of Rs. 136,368.00/yr and has an older workforce with lower education levels, affecting income levels.

7.4.2.5 Housing Condition

Shelter is a basic necessity of life, and housing conditions reflect the socio-economic profile of small tea growers in the study area. Analyzing the distribution of sample respondents across different blocks highlights that most growers in Pulbazar-Bijanbari have a housing condition index of 2.37, which is lower than in other tea-growing areas. Growers in Gorubathan have a housing condition index of 2.68, slightly better than Algarah, indicating relatively more satisfactory housing conditions in this region (Fig. 7.17). This suggests that there is room for improvement in housing quality in the area. The data reflects the quality of housing for STGs in these regions, ranging from moderate to relatively satisfactory, with variations in housing quality across different areas. Improving housing conditions can contribute to the overall well-being and comfort of STGs and their families.

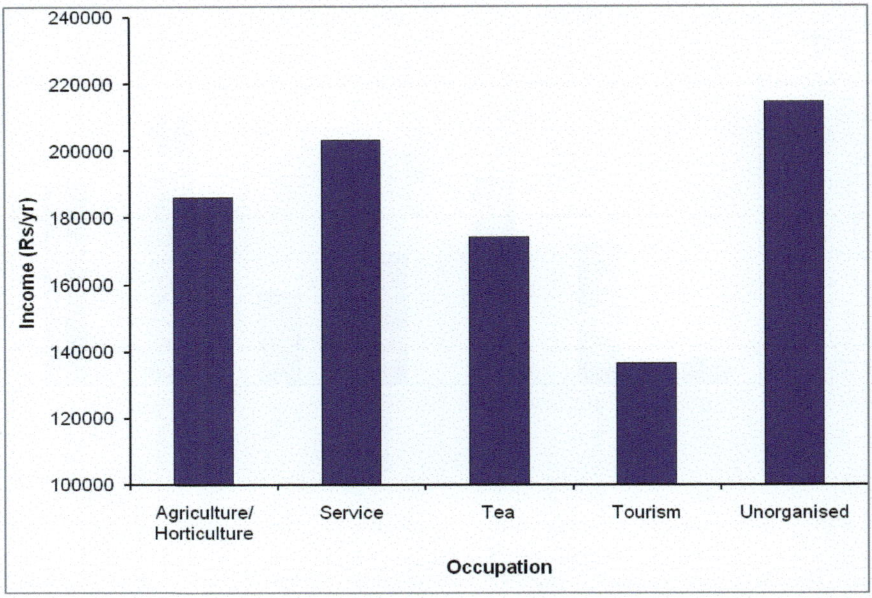

Fig. 7.16 The association between the STGs Income and Occupational Structure in the study area based on the questionnaire survey, 2022

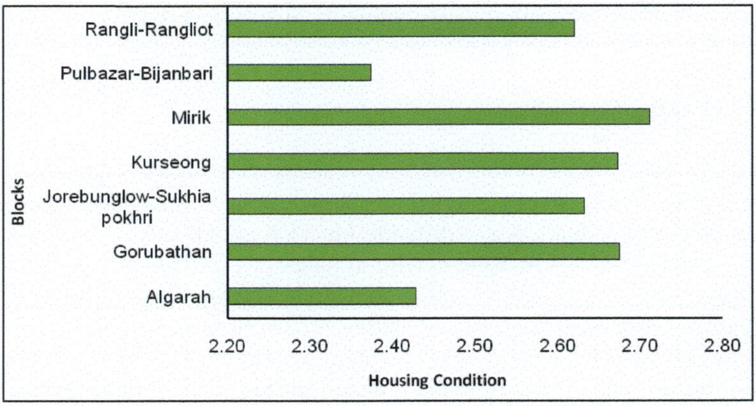

Fig. 7.17 Housing Condition of STGsin different blocks of Darjeeling Hills based on the questionnaire survey, 2022

7.4.3 Livelihood Index of STGs in Tea Growing Areas of Different Blocks in the Darjeeling Himalayan Region

The "Livelihood Index" is a composite measure (Paul et al. 2020) that evaluates the overall well-being or quality of life of STGs in the Darjeeling Himalayan region (Lindenberg 2002; Rai et al. 2008; Khan et al. 2015; Kayal et al. 2018). It is calculated based on eight indicators, including land area, land productivity, cropping intensity, age, education level, income, housing conditions, and the year they began their tea cultivation activities. The Livelihood Index for the Kurseong block stands at 0.33, indicating a relatively higher overall well-being for STGs in this area (Fig. 7.18). Their livelihood conditions are somewhat favorable compared to other blocks, such as Mirik and Gorubathan, which have a Livelihood Index of 0.32, suggesting slightly better livelihood conditions. Pul Bazar-Bijanbari, Rangli-Rangliot, and Jorebunglow Sukhiapokhri have a Livelihood Index of 0.30, indicating a moderate level of well-being for the STGs in this region. Algarah, with a Livelihood Index of 0.30, shows a relatively lower overall well-being or quality of life for STGs in the region.

The Livelihood Index for the STGs also varies by occupational sector, reflecting the overall quality of life in each sector. The tea sector has a Livelihood Index of 0.32, indicating a somewhat better quality of life (Fig. 7.19). This sector maintains reasonable income levels, cropping intensity, and housing conditions, which contribute to the well-being of tea growers. The Service sector has a slightly higher Livelihood Index of 0.31, indicating a relatively better quality of life. This is likely

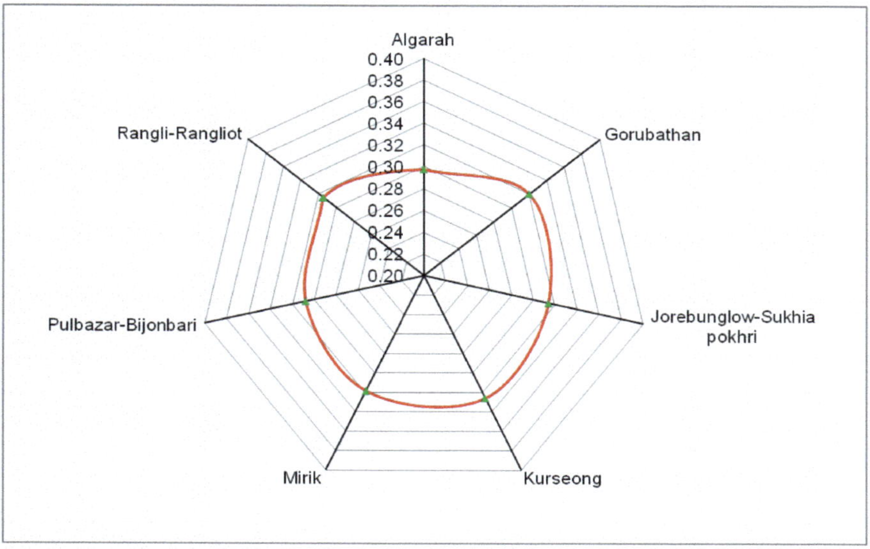

Fig. 7.18 The distribution of STGs livelihood index in different blocks of Darjeeling based on the questionnaire survey, 2022

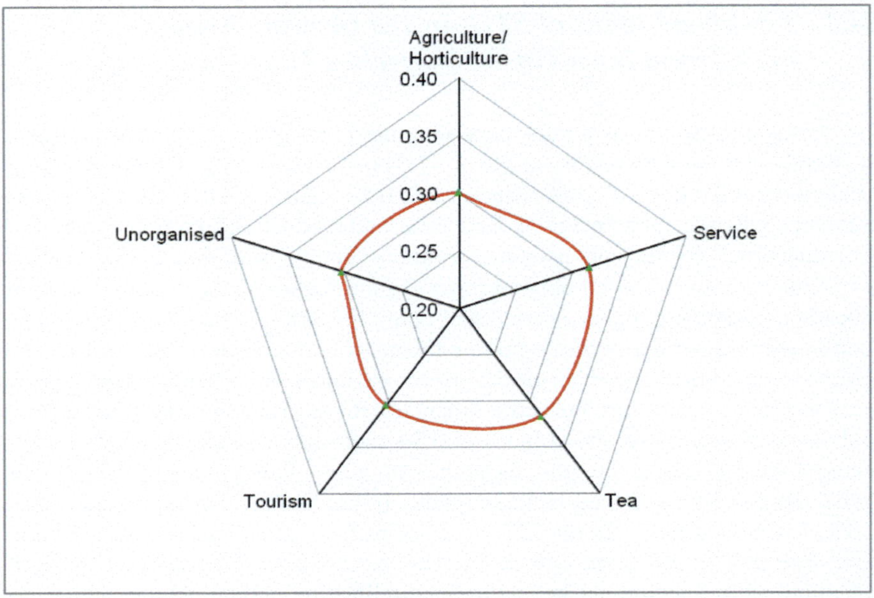

Fig. 7.19 The association between the STGs livelihood index and Occupational Structure in the study area based on the questionnaire survey, 2022

influenced by higher income levels, increased education attainment, and moderate housing conditions. The Agriculture/Horticulture, Tourism, and Unorganized sectors all have a similar, moderate level of well-being. The Livelihood Index can guide policies and interventions aimed at improving the livelihoods of individuals in these sectors (Hussein 2002).

Figure 7.20 depicts the livelihood index of small tea growers in the Darjeeling region. Over 85% of the small tea growers in this area have a livelihood index that falls below 0.4, while only 0.4% of them have a livelihood index exceeding 0.6. This indicates that the majority of small tea growers in this region experience a low standard of living. Several factors contribute to the subpar quality of life among these small tea growers, including limited access to education, inadequate housing conditions, and the challenge of securing fair prices for their tea production.

7.4.4 Correlations Between Land Productivity and Income in Various Sectors

The relationship between tea production and income has been assessed using the Pearson correlation coefficient. The average "r" value for this relationship is 0.98 (Fig. 7.21), indicating that 98% of the total income of Small Tea Growers (STGs)

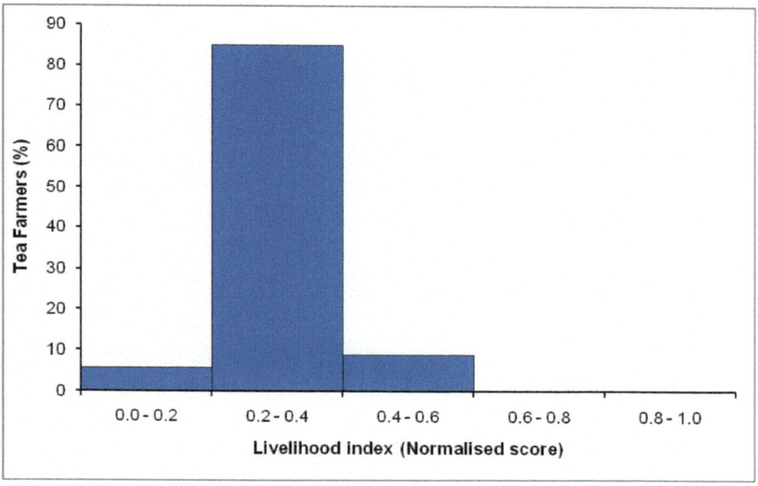

Fig. 7.20 Distribution livelihood index of STGs is shown by histogram. In the X-axis, the livelihood index is represented and the percentage of STGs is represented in the Y-axis

in this region is derived from tea production. Despite engaging in other professions alongside tea cultivation, the annual income of tea farmers involved in various occupations has also been evaluated using the Pearson correlation Coefficient. The "r" value for tea growers involved in Horticulture is 0.99, for STGs engaged in services, it is 0.97, for those exclusively dedicated to tea cultivation, it is 0.99, for STGs involved in Tourism, it is 0.97, and for those working in the Unorganized sector, it is 0.99 (Fig. 7.21). It is evident that tea production significantly influences the income of small tea growers in this region and the influential role of land productivity as a key determinant of income across various sectors.

7.5 Role of STG Organizations to Support Their Livelihood Condition

The presence of various organizations and institutions plays an important role in supporting and empowering the Small Tea Growers (STGs) in Darjeeling Hills. These organizations aid, and resources, and act as a platform for STGs to address their common challenges and achieve their collective and individual goals. There are various STG organizations or welfare societies in the Darjeeling region.

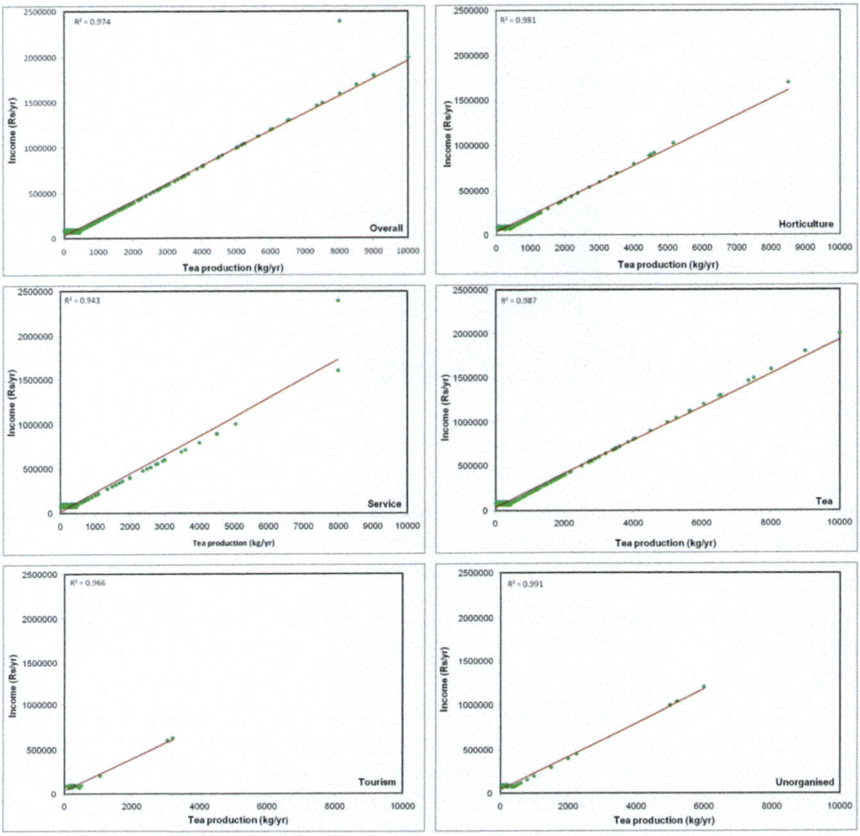

Fig. 7.21 The relationship between Tea production (X) and Income of the STGs in the Darjeeling Hill region where Tea production is plotted in the X-axis and the income of the farmers is plotted in Y-axis and red line represents the linear regression line (*Data Source* Questionnaire survey, 2022)

7.5.1 Mineral Spring Sanjukta Vikash Sangh (MSSVS)

This organization is located at Mineral Spring, Lebong in Darjeeling Hills. It was established in the year 1996. At present, it has 341 active members working in different agricultural activities. Tea plantation is the major livelihood of the region followed by various other activities. This organization started the STG registration process from the year 2002 with the Tea Board of India. It provides training and capacity-building programs for STGs, including workshops on sustainable farming practices, crop management, pest control, and financial literacy. Promoting environmentally friendly and sustainable farming practices is a key role of this organization.

7.5.2 *Darjeeling Hills Small Tea Growers Welfare Society (DHSTGWS)*

This society was formed in the year 2014 and currently has a total number of 178 registered households. Besides more members have been added recently but the registration process is incomplete. This society covers the larger area like Pulungdung, Rangbhang, Pokhriabong, Pubung, and Tumsong. The main aim of society is to act as a unified voice to negotiate better prices, policies, and working conditions on behalf of their members.

7.5.3 *Shittong Organic Small Tea Growers Welfare Society*

It was recently formed in the year 2020. The society has around 23 members. The society is mostly involved in organizing training and workshops for the growers and often promoting sustainable and environmentally friendly farming practices to ensure the long-term viability of the tea industry.

7.5.4 *ShelpuThamdara Small Tea Growers Welfare Society (STSTGWS)*

This society was established in the year 2002 with 30 members. The main objective of the society is to encourage the local people to plantation for the betterment of their livelihood. They act as a bridge between the growers and the company-owned Bought leaf factories.

7.5.5 *Teesta Small Tea Farmers Welfare Society (TSTFWS)*

This society was formed in the year 2005 when the Small Tea farmers of the village started plantation of tea crops in replacement of the traditional agricultural crops. The main aims and objectives of the society are to represent their members in negotiations with tea processing companies, ensuring that growers get a fair share of the profits. They facilitate networking among members, allowing them to learn from each other's experiences and share information.

7.5.6 Teesta Lamahatta Organic Small Tea Growers Welfare Society (TLOSTGWS)

The society was formed in the small village of Lamahatta with 22 members. It was established in the year 2015. The society's main function is to encourage the local community to settle in their respective homes and engage themselves in the tea farming culture recently established.

7.5.7 SeeyokKhasmal Busty Organic Tea Growers Welfare Society

These organizations serve as a collective voice and platform for individual tea growers to address common challenges, advocate for their rights, and promote sustainable practices. It was established in the year 2015 and has a total of 80 members at present. The society engages in collective bargaining on behalf of its members, negotiating better prices, contracts, and working conditions with tea processors and buyers.

7.5.8 Yanki Tea—Darjeeling Orthodox Small Tea Growers Welfare Society

They are a network of over 60 small growers in the society to "Promote Small Farmers, Socially and Economically." It started from the year 2014. They have built up their own tea processing small factory in Bunkulung Valley in Mirik. They are now producing their own standard Darjeeling teas for the market, and handcrafting quality batches.

7.5.9 Nim Busty Small Tea Farmers Producers Society

The society was started in the year 2014 but the registration processes were completed in the year April 2023. It has 59 members in the society. The main aim of the society that it can provide STGs with access to resources, knowledge, and financial support to the growers. It is engaged on behalf of its members, negotiating better prices and other supports from the BLFs and Tea Board of India.

7.5.9.1 Subarna Small Tea Farmers Welfare Society

The society was established in the year 2012. It has a total number of 109 registered members covering different villages in Algarah and Gorubathan block. The Subarna Small Tea Farmers Welfare Society likely plays a significant role in supporting the welfare and interests of small tea growers in its area. It serves as their collective voice in interactions with government bodies, tea industry associations, and Bought leaf factories or companies. They also organize training programs and workshops to help tea growers improve their cultivation techniques, pest management, and quality control. The society promotes sustainable and responsible farming practices among its members, encouraging environmentally friendly methods and social responsibility standards.

7.6 Conclusion

In conclusion, the analysis of STG in the Darjeeling Hills provides valuable insights into various aspects of their farming practices and livelihood conditions. The study covers a wide range of parameters, shedding light on the intricacies of tea cultivation in the region. The tea plantation area varies significantly among different blocks, with Pulbazar-Bijanbari having smaller farms and Gorubathan having slightly larger ones. The size of the farm has a substantial impact on tea production and the economic well-being of growers. Smaller farms tend to exhibit higher land productivity, suggesting potential for high-value or specialized tea production. Cropping intensity, measuring how efficiently agricultural land is utilized, also varies across blocks. Pulbazar-Bijanbari stands out with an extremely high cropping intensity, indicating efficient farming practices. Higher cropping intensity is associated with better crop management and possibly the cultivation of high-value or high-yielding crops. Kurseong has a relatively younger population, while Mirik has a higher average age. The region generally has a higher proportion of older tea growers, which has implications for succession planning and the adoption of modern agricultural practices. Education levels among tea growers differ, with Kurseong having a relatively higher level of education compared to other regions. It is crucial for the adoption of modern agricultural techniques, and it can lead to more efficient farming practices. Experience in tea farming is an asset, Mirik tea growers having the most extensive experience. This experience contributes to a wealth of knowledge in tea cultivation and management. Pulbazar-Bijanbari generates a relatively high annual income. Housing conditions vary across blocks, with some regions having room for improvement in housing quality. Improving housing conditions can contribute to the overall well-being of STG and their families. The livelihood index provides an overall measure of well-being, with Kurseong having a relatively higher quality of life. This index considers various factors, including income, education, and housing conditions. The study also examines the correlation between tea production and income in different sectors, highlighting the strong positive relationship between land productivity and income. Higher land productivity is consistently associated with higher income across various sectors. The STG of Darjeeling Hills faces unique challenges and opportunities based on their farm characteristics, cropping intensity, education, and experience. The findings emphasize the importance of efficient farming practices, education, and modernization to ensure the sustainability and growth of the tea industry in the region. Additionally, policies and interventions aimed at improving housing conditions and overall well-being can contribute to the livelihood and prosperity of STG in the Darjeeling Hills.

Acknowledgements This study was supported by UGC financial assistance received by Bivek Subba [Ref ID: 3510 (NET-JULY 2018)] and would like to thank all those who provided the logistics and transportation support during the questionnaire survey.

References

Ahmed S, Stepp JR (2013) Green tea: The plants, processing, manufacturing and production. Tea Health Dis Prev, 19–31

Akoglu H (2018) User's guide to correlation coefficients. Turk J Emerg Med 18(3):91–93

Akudugu MA (2012) Estimation of the determinants of credit demand by farmers and supply by rural banks in Ghana's Upper East Region. Asian J Agric Rural Dev, 2(393–2016–23992), 189–200

Alam GM, Hoque KE, Khalifa MTB, Siraj SB, Ghani MFBA (2009) The role of agriculture education and training on agriculture economics and national development of Bangladesh. Afr J Agric Res 4(12):1334–1350

Ardil C (2019) Aircraft selection using multiple criteria decision making analysis method with different data normalization techniques. Int J Ind Syst Eng 13(12):744–756

Arjunan R (2018) Livelihood security of tea plantation workers

Arya N (2013) Indian tea scenario. Int J Sci Res Publ 3(7):1–10

Bajaj R (2021) Indian tea landscapes. In: International Council on Monuments and Sites 11, rue du Séminaire de Conflans 94220 Charenton-le-Pont France, 232

Bellon MR (2001) Participatory research methods for technology evaluation: A manual for scientists working with farmers. CIMMYT

Besky S (2014) The Darjeeling distinction: Labor and justice on fair-trade tea plantations in India (Vol. 47). Univ of California Press

Bhutia S (2014) The role of tourism for human resource development in darjeeling district of west Bengal, India. J Tour Hosp Manag 2(1):113–128

Bitama PC, Lebailly P, Ndimanya P, Burny P (2020) Socioeconomic constraints to tea productivity: a case of Small-Scale tea farmers in burundi. J Appl Econ Sci, 15(2)

Blowfield ME, Dolan C (2010) Fairtrade facts and fancies: What Kenyan Fairtrade tea tells us about business' role as development agent. J Bus Ethics 93:143–162

Borah K, Das AK (2015) Growth of small tea cultivation and economic independence of the indigenous people of Assam. Int J Res Soc Sci HumIties 5(1):82–93

Brodt S, Klonsky K, Tourte L (2006) Farmer goals and management styles: Implications for advancing biologically based agriculture. Agric Syst 89(1):90–105

Brototi B, Singh A, Rai PK, Kumar J, Walker S (2022) GIS based study of reclamation of degraded semi-arid soil: A case study of Rajasthan from India. Indian J Environ Prot 42(3):302–315

Cajee L (2018) Physical aspects of the Darjeeling Himalaya: understanding from a geographical perspective. IOSR J HumIties Soc Sci 23(3):66–79

Carr M, Chen MA (2002) Globalization and the informal economy: How global trade and investment impact on the working poor. Int Labour Off, Geneva, pp 92–102

Chambers R, Conway G (1992) Sustainable rural livelihoods: practical concepts for the 21st century. Inst Dev Stud (UK)

Chatterjee AK, De GC, Barik AK, Seal A, Bera R (2013) Relevance of organic approach in tea physiology as a C3 Plant & suggestive recommendations for effective organic tea cultivation in Assam & Darjeeling plantations. Publ Inhana Biosci, Bulletin-I, pp 1–72

Christison GW (1895) Darjeeling, tea planting in. RSA J 44:624

Cowan-Gore I, Sein TT (2020) Myanmar tea cultivation and processing guide (No. 995071489802676). Int Labour Organ

Crucefix D (1998) Organic agriculture and sustainable rural livelihoods in developing countries. Rep Nat Resour Ethical Trade Program, June

Das K (2019) (The) Small tea growers of Assam: A study of their monopsonistic exploitation and production (Doctoral dissertation)

Datta TK (2010) Darjeeling tea in India. Lecoent A, Vandecandelaere E, Cadilhon J (2010) Quality linked to geographical origin and geographical indications: Lessons learned from six case studies in Asia. Bangkok: Food and Agricultural Organization of the United Nations, Regional Office for Asia and the Pacific, 113–160

De Costa WA, Mohotti AJ, Wijeratne MA (2007) Ecophysiology of tea. Braz J Plant Physiol 19:299–332

Deka N, Goswami K (2022) Organic cultivation and sustainable value chain development for tea smallholders: Findings from Assam, India. Sustain Prod Consum 32:562–579

Dewan A (2019) 'T'(ea) is for terroir: an analysis of the branding and valuation of Darjeeling tea (Doctoral dissertation, University of Oregon)

Dhital MR (2015) Geology of the Nepal Himalaya: regional perspective of the classic collided orogen. Springer

Eastwood R, Lipton M, Newell A (2010) Farm size. Handb Agric Econ 4:3323–3397

Ellis F (2000) Rural livelihoods and diversity in developing countries. Oxf Univ Press

Galt RE (2013) The moral economy is a double-edged sword: Explaining farmers' earnings and self-exploitation in community-supported agriculture. Econ Geogr 89(4):341–365

Ghosh R, Bruch AA, Portmann F, Bera S, Paruya DK, Morthekai P, Ali SN (2017) A modern pollen–climate dataset from the Darjeeling area, eastern Himalaya: Assessing its potential for past climate reconstruction. Quatern Sci Rev 174:63–79

Gogoi J (2016) Biotourism potential in Northeast India. Bioprospecting Indig Bioresour North-East India, 339–347

Graham HN (1992) Green tea composition, consumption, and polyphenol chemistry. Prev Med 21(3):334–350

Gurung M (2022) Livelihood Diversification among the tea plantation workers in West Bengal: A Case Study of Darjeeling Hills (Doctoral dissertation, Mizoram University)

Hannan A (2019) Farm size and trade relations of small tea growers (STGs) in Assam and North Bengal. Soc Chang Dev 16(2):7

Hannan A (2021) Organizational innovations and small tea growers (STGs) in India

Hannan A, Mondal N (2021) Small-Scale tea economy and arsla organic tea cooperative: a grassroots initiative of Khasi Tribes of Meghalaya, North-East India. Antrocom: Online J Anthropol 17(2)

Heiss ML, Heiss RJ (2012) The tea enthusiast's handbook: a guide to the world's best teas. Ten Speed Press

Hinsch B (2008) The ultimate guide to Chinese tea. Bret Hinsch

Hussein K (2002) Livelihoods approaches compared. Dep Int Dev, London

Hutchison CS (1989) Geological evolution of South-east Asia. Oxford: Clarendon Press 13, p. 368

Ilbery B (2014) Dimensions of rural change. In The geography of rural change. Routledge pp 1–10)

Inwood SM, Sharp JS (2012) Farm persistence and adaptation at the rural–urban interface: Succession and farm adjustment. J Rural Stud 28(1):107–117

Jain A, Nandakumar K, Ross A (2005) Score normalization in multimodal biometric systems. Pattern Recogn 38(12):2270–2285

Kamanga P, Vedeld P, Sjaastad E (2009) Forest incomes and rural livelihoods in Chiradzulu District. Malawi. Ecological Economics 68(3):613–624

Kangalawe RY, Lyimo JG (2013) Climate change, adaptive strategies and rural livelihoods in semiarid Tanzania

Kannegieser I (2015) A Home In The Hills: Examining the socioeconomic benefits of homestay tourism on rural women and their communities in the Darjeeling District

Kanyua MJ, Waluse GKISK, Wairimu WR (2015) Factors influencing profitability of diversified cash crop farming among smallholder tea farmers in Gatanga District, Kenya. J Econ Sustain Dev, ISSN, 2222–1700

Karki L, Schleenbecker R, Hamm U (2011) Factors influencing a conversion to organic farming in Nepalese tea farms. J Agric Rural Dev Trop Subtrop (JARTS) 112(2):113–123

Karlı B, Bilgiç A, Çelik Y (2006) Factors affecting farmers' decision to enter agricultural cooperatives using random utility model in the South Eastern Anatolian region of Turkey. J Agric Rural Dev Trop Subtrop (JARTS) 107(2):115–127

Kayal P, Chowdhury IR (2018) Level of human development of Raiganj CD Block, Uttar Dinajpur District, West Bengal: A geographical analysis. Int J Res Anal Rev 5:619–628

Kellett D, Grujic D, Mot-Tram C, Mukul M, Larson KP (2014) Virtual field guide for the Darjeeling-Sik-kim Himalaya India. Geological field trips in the Himalaya Kar-akoram and Tibet. J Virtual Explor, Electron Ed, ISSN, 1441–8142

Khan A, Chatterjee S, Mistri K, Das A (2015) Level and pattern of human development in Paschim Medinipur District, West Bengal. India. J Hum Ecol 49(3):219–229

Khandelwal A, Agarwal N, Jain B, Gupta D, John AT, Farah Cox P (2022) Scaling up sustainability through new product categories and certification: Two cases from India. Front Sustain Food Syst 6:1014691

Khawas V (2011) Status of tea garden labourers in eastern Himalaya: a case of Darjeeling tea industry. Cloud, stone and the mind: the people and environment of Darjeeling hill area. KP Bagchi & Company, India

Kruskal W, Mosteller F (1979) Representative sampling, III: The current statistical literature. Int Stat Rev/Rev Int Stat, 245–265

Kumar KA, Thayalan S, Reddy RS, Lalitha M, Kalaiselvi B, Parvathy S, Mishra BB (2020) Geology and geomorphology. Soils India, 57–79

Kumpf D (2020) Organic taste and labour on Indian tea plantations. Soc Anthropol/Anthropol Soc 28(4):789–802

Lahiri S (1992) Landuse in the eastern himalayan region. New Dimens Agric Geogr 31(4):205

Lakshmi M (2015) An analysis of risk among the small tea growers In Nilgiris District, Tamilnadu (Doctoral Dissertation, Christ University)

Lama SD (2022) Tea plantation workers and the human cost of Darjeeling Tea. In Caste, COVID-19, and Inequalities of Care: Lessons from South Asia (pp 333–354). Singapore: Springer Nature Singapore

Li G, Wang J, Cao Y, Savage D, Chen H, Wu Q, Pijanowski BC (2023) Quantifying recent trends in multifaceted cropland intensification in China: Cropping intensity, land productivity, and their effects on farmer livelihoods. Land Degrad & Dev

Lindenberg M (2002) Measuring household livelihood security at the family and community level in the developing world. World Dev 30(2):301–318

Little PD (1992) The elusive granary: herder, farmer, and state in northern Kenya (No. 73). Camb Univ Press

Lohr SL (2021) Sampling: design and analysis. CRC Press

Majumdar K, Ray DP, Chakraborty S, Pandit T (2014) Change of nutrient status of hilly soil in darjeeling district within five years. Int J Bioresour Sci 1(1):25–30

Majumdar TR (2017) Economics of small tea plantations in West Bengal a study of production, efficiency and productivity performance using DEA and stochastic frontier regression (Doctoral dissertation, University of North Bengal)

Mallik C (2022) Smallholder tea farming in West Bengal, India: An exploratory insight. In Livelihood Enhancement Through Agriculture, Tourism and Health pp 137–148. Singapore: Springer Nature Singapore

Manoraj SJ (2021) Branding of green tea leaf: a disruptive innovation for building market competitiveness of small tea growers in North East India. J Agribus Dev Emerg Econ 11(2):88–104

Mansingh P, Johnson L (2021) Comparative analysis of existing models of small tea growers in tea value chain in the Nilgiris

Martin SM, Lorenzen KAI (2016) Livelihood diversification in rural Laos. World Dev 83:231–243

McNeely JA, Scherr SJ (2012) Ecoagriculture: strategies to feed the world and save wild biodiversity. Island Press

Meuwissen MP, Feindt PH, Spiegel A, Termeer CJ, Mathijs E, De Mey Y, Reidsma P (2019) A framework to assess the resilience of farming systems. Agric Syst 176:102656

Mhlanga ST, Lall M (2022) Influence of normalization techniques on Multi-criteria Decision-making Methods. In Journal of Physics: Conference Series 2224(1), p 012076. IOP Publishing

Mohan S (2016) Institutional change in value chains: Evidence from tea in Nepal. World Dev 78:52–65

Mondal BK, Acharya A, Nandan T (2022) Assessing the Geo-Ecotourism Potentiality of West Bengal with Special Reference to its Coastal Region Using Geospatial Technology. In: Morphology S, Welfare H (eds) And Sustainability. Springer International Publishing, Cham, pp 687–719

Mondal S, Samaddar K (2021) Exploring the current issues, challenges and opportunities in tea tourism: a morphological analysis. Int J Cult, Tour Hosp Res 15(3):312–327

Mishra VN, Rai PK, Singh P (2021) Geo-information technology in earth resources monitoring and management (edit. Book), Nova Science Publishers, U.S.A., pp 1–395. ISBN: 978–1–53619–669–6

Mosse, D., Gupta, S., Mehta, M., Shah, V., Rees, J. F., & Team, K. P (2002) Brokered livelihoods: Debt, labour migration and development in tribal western India. J Dev Stud 38(5):59–88

Mugo FW (2002) Sampling in research

Mukherjee D (2012) Resource conservation through indigenous farming system in hills of West Bengal. J Crop Weed 8(1):160–164

Murshed-E-Jahan K, Pemsl DE (2011) The impact of integrated aquaculture–agriculture on small-scale farm sustainability and farmers' livelihoods: experience from Bangladesh. Agric Syst 104(5):392–402

Nyakweba TO (2019). Financial service accessibility strategies for farmers 'economic empowerment in kenya: a survey of small-scale tea farmers in kisii county (doctoral dissertation, kabarak university)

Oerke EC, Dehne HW (1997) Global crop production and the efficacy of crop protection-current situation and future trends. Eur J Plant Pathol 103:203–215

Oke AO, Kamolshotiros P, Popoola OY, Ajagbe MA, Olujobi OJ (2016) Consumer behavior towards decision making and loyalty to particular brands. Int Rev Manag Mark 6(4):43–52

Osbahr H, Twyman C, Adger WN, Thomas DS (2008) Effective livelihood adaptation to climate change disturbance: scale dimensions of practice in Mozambique. Geoforum 39(6):1951–1964

Pal DK, Nath S, Banerjee SK, Sharma SK (1985) Characteristics of some forest soils of Darjeeling Himalayan region under pinus patula. J Indian Soc Soil Sci 33(1):84–91

Paul S, Das TK, Pharung R, Ray S, Mridha N, Kalita N, Singh AK (2020) Development of an indicator based composite measure to assess livelihood sustainability of shifting cultivation dependent ethnic minorities in the disadvantageous Northeastern region of India. Ecol Ind 110:105934

Playan E, Mateos L (2006) Modernization and optimization of irrigation systems to increase water productivity. Agric Water Manag 80(1–3):100–116

Pooniya V, Choudhary AK, Swarnalakshmi K (2017) High-value crops' imbedded intensive cropping systems for enhanced productivity, resource-use-efficiency, energetics and soil-health in Indo-Gangetic Plains. Proceedings of the National Academy of Sciences, India Section B: Biological Sciences 87:1073–1090

Potter C, Lobley M (1992) Ageing and succession on family farms: The Impact on Decision-making and Land Use. Sociologia ruralis, 32

Qiao Y, Halberg N, Vaheesan S, Scott S (2016) Assessing the social and economic benefits of organic and fair trade tea production for small-scale farmers in Asia: a comparative case study of China and Sri Lanka. Renewable Agric Food Syst 31(3):246–257

Rai A, Sharma SD, Sahoo PM, Malhotra PK (2008) Development of livelihood index for different agro-climatic zones of India. Agric Econ Res Rev, 21(347–2016–16693), 173–182

Rai PC (2019) An empirical study on The Socio-Economic conditions of the plantation workers in the tea industry of Darjeeling. Int J Appl Sci Eng 7(1):83–93

Rai SK (2002) Studies on the ethnobotany of Darjeeling Himalaya (Doctoral dissertation, University of North Bengal)

Rais M, Acharya S, Sharma N (2013) Food processing industry in India: S&T capability, skills and employment opportunities. J Rural Dev, 451–478

Rasaily R (2021) Changing land utilisation patterns in tea plantation sector in West Bengal: Some policy imperatives

Raveendranathan D (2017) Himalayas melting and peninsular India Wilting. Notion Press

Raynolds LT, Ngcwangu SU (2010) Fair trade rooibos tea: Connecting South African producers and American consumer markets. Geoforum 41(1):74–83

Ren C, Liu S, Van Grinsven H, Reis S, Jin S, Liu H, Gu B (2019) The impact of farm size on agricultural sustainability. J Clean Prod 220:357–367

Renard MC (2003) Fair trade: quality, market and conventions. J Rural Stud 19(1):87–96

Ross ML (1999) The political economy of the resource curse. World Politics 51(2):297–322

Ruben R, Pender J (2004) Rural diversity and heterogeneity in less-favoured areas: the quest for policy targeting. Food Policy 29(4):303–320

Rudra K, Rudra L, Leuchner (2018) Rivers of the Ganga-Brahmaputra-Meghna Delta. Springer International Publishing, Berlin, Germany

Rai PK, Mishra VN, Singh P (2021) Recent technologies for disaster management and risk reduction-sustainable community resilience and responses (edit. Book), Springer Nature, Switzerland, pp 1–471. ISBN: 978–3–030–76116–5. https://doi.org/10.1007/978-3-030-761 16-5

Rai PK, Mishra VN, Singh P (2022) Geospatial technology for landscape and environment management: sustainable assessment and planning (edit. Book), Springer Nature, Singapore. pp 1–431. ISBN: 978–981–16–7373–3. https://doi.org/10.1007/978-981-16-7373-3

Saha D (2021) Producer collectives through self-help: sustainability of small tea growers in India. In OwnShip GovAnce Co pp 241–260. Routledge

Sahu N, Das P, Saini A, Varun A, Mallick SK, Nayan R, Kumar A (2023) Analysis of tea plantation suitability using geostatistical and machine learning techniques: a case of Darjeeling Himalaya. India. Sustainability 15(13):10101

Salami A, Kamara AB, Brixiova Z (2010) Smallholder agriculture in East Africa: Trends, constraints and opportunities. Tunis, Tunisia: African Development Bank. p. 52

Samajdar A (2015) Gender and natural resource management among the Lepchas of Sikkim and Darjeeling. Dev, Resour Livelihoods, 365

Sandhu DK, Singh S (1981) Distribution of thermophilous microfungi in forest soils of Darjeeling (Eastern Himalayas). Mycopathologia 74(2):79–85

Sarfaraz L, Faghih N, Majd AA (2014) The relationship between women entrepreneurship and gender equality. J Glob Entrep Res 4:1–11

Sarkar Tahiti (2018) Commercial cash crop and the development of capitalist economy: A study of colonial tea plantations in Darjeeling hills. (2018)

Schober P, Boer C, Schwarte LA (2018) Correlation coefficients: appropriate use and interpretation. Anesth Analg 126(5):1763–1768

Scott A, Gilbert A, Gelan A (2007) The urban-rural divide: Myth or reality?. Aberdeen: Macaulay Institute

Sen D (2009) From illegal to organic: Fair trade-organic tea production and women's political futures in Darjeeling. Rutgers The State University of New Jersey-New Brunswick, India

Shah KU, Dulal HB, Johnson C, Baptiste A (2013) Understanding livelihood vulnerability to climate change: Applying the livelihood vulnerability index in Trinidad and Tobago. Geoforum 47:125–137

Singh AK, Bisen JS, Chauhan RK, Choubey M, Kumar R, Kumar N (2016) Tea research for Darjeeling tea Industry-Various aspects. Tea Technol Initiat, 195–239

Singh A, Rai PK, Deka G, Biswas B, Prasad D, Rai VK (2021) Management of natural resources through integrated watershed management in Nana Kosi Micro Watershed, District Almora, India, Ecol Environ & Conserv, 27 (February Suppl. Issue); pp S260–S268

Smith DR, Gordon A, Meadows K, Zwick K (2001) Livelihood diversification in Uganda: patterns and determinants of change across two rural districts. Food Policy 26(4):421–435

Tao TC, Wall G (2009) Tourism as a sustainable livelihood strategy. Tour Manage 30(1):90–98

Thapa S (2021) Revisiting thermal comfort in the cold climate of Darjeeling, India-Effect of assumptions in comfort scales. Build Environ 203:108095

Therond O, Duru M, Roger-Estrade J, Richard G (2017) A new analytical framework of farming system and agriculture model diversities. A Review Agron Sustain Dev 37:1–24

Tirkey LP (2005) Tea plantations in the Darjeeling district, India: Geo-ecological and socio-economic impacts in post-independence period (Master's thesis)

Vafaei N, Ribeiro RA, Camarinha-Matos LM (2020) Selecting normalization techniques for the analytical hierarchy process. In Technological Innovation for Life Improvement: 11th IFIP WG 5.5/SOCOLNET Advanced Doctoral Conference on Computing, Electrical and Industrial Systems, DoCEIS 2020, Costa de Caparica, Portugal, July 1–3, 2020, Proceedings 11 pp 43–52. Springer International Publishing

Vafaei N, Ribeiro RA, Camarinha-Matos LM (2022) Assessing normalization techniques for simple additive weighting method. Procedia Computer Science 199:1229–1236

Van der Ploeg JD, Renting H, Brunori G, Knickei K, Mannion J, Marsden T, Ventura F (2017) Rural development: from practices and policies towards theory. In The Rural (pp. 201–218). Routledge

Weir S, Knight J (2000) Adoption and diffusion of agricultural innovations in Ethiopia: the role of education

Wenner M (2020) Towards an alternative Indian tea economy. Econ Pol Wkly 55(45):53–60

Wiggins S, Kirsten J, Llambí L (2010) The future of small farms. World Dev 38(10):1341–1348

Yadav LS (1950) Entrepreneurship development in tea industries. A J Manag Mak, 21

Yu Q, Xiang M, Sun Z, Wu W (2021) The complexity of measuring cropland use intensity: An empirical study. Agric Syst 192:103180

Zou KH, Tuncali K, Silverman SG (2003) Correlation and simple linear regression. Radiology 227(3):617–628

Chapter 8
Marginal Uplands in the Philippines: Characteristics, Occurrence, and Management for Improved Crop Production and Environmental Quality

Victor B. Asio, Luz G. Asio, Cahyoadi Bowo, and Johannes Reiner G. Asio

Abstract Marginal uplands are defined in this paper as undulating and steep lands ranging in elevation from near sea level to about 1000 m elevation and having low crop productivity due to poor soil fertility. Deforestation, shifting cultivation, and unsustainable cultivation practices have resulted in the widespread occurrence of these problematic lands in the Philippines. The common plant species in marginal uplands which can be used as indicator plants are (A). Cogon (*Imperata cyclindrica*), (B). Bugang or Talahib (*Saccharum spontaneum*), (C). Carabao grass (*Paspalum conjugatum*), and (D). Hantutuknaw (*Melastoma malabathricum*). Native guava (*Psidium guajava*) is almost always present in marginal uplands. The soils in marginal uplands possess various constraints to crop production that vary with the soil genesis and land use history. The constraints include physical such as shallow depth, compaction, unfavorable slopes, and low infiltration rates as well as chemical constraints such as acidic or alkaline pH, low OM, and nutrient contents. The soil fertility constraints vary widely among marginal upland soils. Agroforestry is widely accepted as the sustainable form of management strategy for marginal upland soils. The use of native tree species to rehabilitate denuded lands called *Rainforestation Farming* has been proven effective at various sites throughout the country and has been adopted as part of the national strategy to rehabilitate marginal or degraded uplands. Lastly, contour farming is also a proven strategy to improve soil and environmental quality in marginal uplands.

V. B. Asio (✉)
Department of Soil Science, CAFS, Visayas State University, Baybay City, Philippines
e-mail: vbasio@vsu.edu.ph

L. G. Asio
Department of Agronomy, CAFS, Visayas State University, Baybay City, Philippines

C. Bowo
Department of Soil Science, University of Jember, Jember, Indonesia

J. R. G. Asio
Formerly at the School of Environmental Science and Management (SESAM), University of the Philippines Los Baños, Laguna, Philippines

P. K. Rai et al. (eds.), *Natural Resources Management and Sustainable Livelihoods in the Mountainous Region*, Advances in Geographical and Environmental Sciences,
https://doi.org/10.1007/978-981-97-2100-9_8

Keywords Degraded uplands · Problem soils · Soil degradation · Tropical soils

8.1 Introduction

The degradation and marginalization of upland areas is one of the most serious agricultural and ecological problems in the Philippines today. This has resulted not only in decreased crop productivity of these lands but also caused environmental problems like soil erosion, floods, siltation in lowlands and water bodies, and the drying up of drinking water sources. Marginal uplands, also called in various publications as degraded uplands, possess biophysical and socio-economic constraints that seriously affect their use for crop production (Lightfoot et al. 1988). Thus, efforts to rehabilitate and improve the productivity of these problematic lands will benefit both upland and lowland communities, especially in view of the increasing storm and typhoon occurrences due to climate change.

Marginal uplands are widespread in the Philippines as a result of soil erosion and other degradation processes which in turn were brought about by decades of deforestation and forest conversion into agricultural lands. The World Bank (1989) reported that commercial logging was a major contributing factor to land degradation and soil erosion in the country.

Regardless of migration pattern, migrants to the marginal uplands have common socioeconomic characteristics. These migrants are the "poorest of the poor," least educated, least hopeful, and considered the most neglected by the Philippine government in terms of exerting effort for their agricultural development. Labor in the farm is confined within the family (DENRD-ERDB 2010). Roa (2007) observed that the poorest households, who are also the most vulnerable and most food insecure households, are living and farming on these marginal lands. Cruz et al. (1989) stressed that the upland migrant must be viewed as a victim rather than the perpetrator of forest destruction and called for more government programs to address poverty in the uplands.

The purpose of this paper is to compile and synthesize the available published data on the characteristics, occurrence, and management of marginal uplands in the Philippines. We hope that this book chapter will contribute to the discussion on the topic and hopefully encourage more research and development interests in marginal uplands. The paper draws heavily from the published results of our recently concluded project on marginal uplands in Eastern Visayas funded by the Commission on Higher Education Philippine Higher Education Research Network (CHED-PHERNET) (Asio et al. 2014a, b).

8.2 Definition of Marginal Uplands

The term "marginal uplands" does not have a widely accepted definition until now. The term is sometimes used synonymously with degraded uplands. Although low soil fertility is a common characteristic of these two problematic lands, they differ however in terms of formation. Degraded uplands are less productive uplands resulting from land degradation processes such as soil erosion, nutrient depletion, contamination, and others. Marginal uplands, on the other hand, may be formed through the same land degradation processes, but they can also be formed by natural soil-forming processes acting on parent materials containing low nutrients (e.g., sandstone, limestone) or excessive amounts of certain elements (e.g., ultramafic rocks).

The term "marginal" is defined by The Free Dictionary (www.thefreedictionary.com) as "barely within a lower standard or limit of quality." A land that is marginal for crop production may be good for grazing (Shahid and Al-Shankiti 2013). Kang et al. (2013) describe marginal lands as characterized by low productivity and reduced economic return or by severe constraints for agricultural cultivation. These lands are generally fragile and their use is environmentally risky. The World Bank (1989) used the term marginal uplands to mean uplands unsuitable for crop cultivation or suitable only for forests or pasture.

The term "upland" has no consistent definition as well (Sajise and Ganapin 1991). It is defined differently in several Southeast Asian countries like Indonesia, Malaysia, the Philippines, and Vietnam (Sajise and Ganapin 1991). The official definition of upland in the Philippines is that it refers to all lands with a slope gradient exceeding 18% including the tableland and plateaus lying at higher elevations which are not generally suitable for wet rice culture (Sajise and Ganapin 1991; Garrity et al. 1993; World Bank 1989). This 18% slope rule dates back to the land classification system proposed by the Forestry Bureau in 1919 (Boado 1988). It is arbitrary and is an insufficient guide to desirable land use since it poorly reflects the geomorphological and physiognomic characteristics of upland areas. It is legal and bureaucratic but not biogeographic (World Bank 1989; Nelson 1994).

Despite the debate, it thus seems a big puzzle why until now agricultural and environmental scientists in the Philippines have continued to adhere to this non-scientific and antiquated definition of the term upland. The present paper uses the definition by Garrity (1993) as cited by Nelson (1994) that uplands are "*undulating and steep lands that range in elevation from near sea level to about 1000 m elevation.*" This definition is geomorphological and clearly separates the flat lowlands from the elevated uplands as illustrated in Fig. 8.1. Further, undulating and rolling uplands with slope of less than 18% which are used for upland agriculture easily fit to this definition. It also conforms to the definition of upland by Merriam-Webster Dictionary (http://www.merriam-webster.com) that it is "a ground elevated above the lowlands." It captures the basic idea behind the suggestion of Rambo and Sajise (1985) to divide the Southeast Asian landscape into two general areas for research and development based on topography, dominant land use, and human activities as

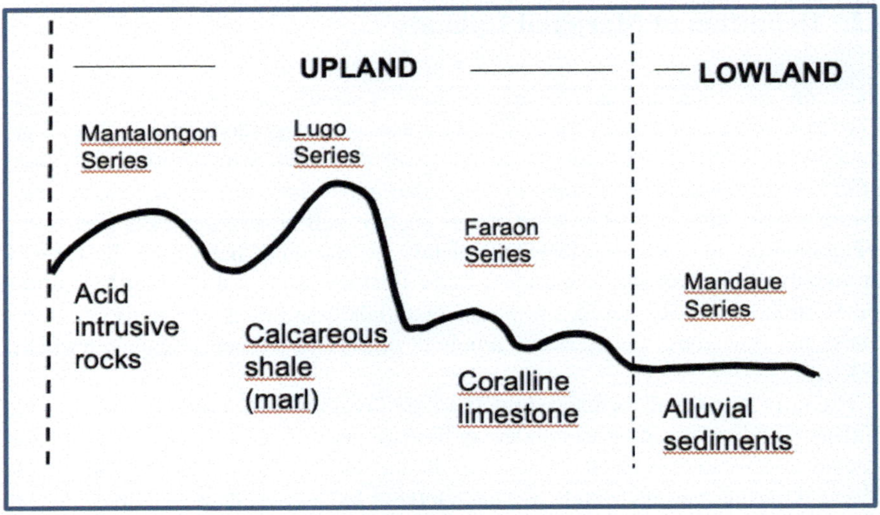

Fig. 8.1 Illustration of the proposed definition of "upland" using Argao, Cebu as an example. The lowland is the flat alluvial land dominated by the Mandaue Series, while the upland starts with the rolling topography of the Faraon Series much of which has less than 18% slope

follows: (1) the core areas where large-scale monoculture rice production predominates, and (2) the hinterlands which are generally sloping lands where small-scale mixed farming, grazing, and forestry occurs.

8.3 Extent and Importance of Marginal Uplands

Marginal uplands are largely the result of land degradation which is a severe global problem of modern times (Lal 1998). Every year 5–7 million hectares of agricultural land worldwide are degraded due to various physical and chemical alterations (Steiner 1996) accelerated by deforestation, overgrazing, and unsuitable agricultural practices on mountain slopes. Lapar and Pandey (1999) reported that soil degradation in the sloping lands of Asia is a serious problem that threatens the sustainability of agriculture. In the Philippines, the 2004–2010 National Action Plan identified land degradation as a major threat to food security because about 5.2 million hectares are severely degraded annually resulting in a 30–50% reduction in soil productivity. Thus, soil and land degradation is the main agricultural and environmental problems which are impacting the food security and rural development in the Philippines (Asio et al. 2009; Crambs et al. 2000). Agricultural practices have sternly degraded the agricultural resource base, associated with accelerated soil erosion, siltation of irrigation systems, flooding, and water pollution (Briones 2009; Mishra et al. 2021; Rai et al. 2021; Singh et al. 2021; and Rai et al. 2022).

Sajise and Ganapin (1991) reported that the uplands are of considerable importance because they contain the tropical rainforest ecosystem which is the oldest productive and protective ecosystem on earth. Likewise, Kummer (1992) mentioned that upland areas in the country are important since they contain the majority of the remaining Philippine forests, they have been the destination of migrants in the postwar period, and land degradation is severe and widespread in the uplands. Pailagao et al. (2010) noted that the most fragile ecosystem in the Philippines is located in the uplands. These consist of more than 40% of the country's total land area. Studies in other countries in Southeast Asia have shown that upstream cultivation activities, especially with intensive use of fertilizers, impact the water quality and in the long term can damage the hydrological conditions of the watershed (Alfarisy et al. 2020).

Soil erosion is the most important cause of land degradation in the Philippines (World Bank 1989; Asio et al. 2009; Brototi et al. 2022). Although soil erosion data are difficult to interpret due to high spatial, temporal, and methodological variability, they nevertheless reflect the extent of occurrence of marginal uplands in the country. The NAP (2004) estimated that 33, 21, and 46% of Luzon, Visayas, and Mindanao, respectively, are severely eroded. The data for the Visayas appear to be an underestimation considering the fact that many Visayas islands are severely degraded. It has been reported that one third of the total land area of the Philippines has been excessively eroded (David 1988) suggesting that marginal uplands are widespread throughout the country. Table 8.1 shows soil erosion rate estimates which vary considerably with land use systems and across different areas in the country. As can be seen, the highest erosion rate was measured in a banana–coffee–pineapple intercropping (421 t ha^{-1} $year^{-1}$) and the lowest was recorded under a primary forest (0.2 t ha^{-1} $yearr^{-1}$). These values are comparable to those reported by The World Bank (1989) which range from 1 t ha^{-1} $year^{-1}$ under undisturbed forest to around 300–400 t ha^{-1} $year^{-1}$ on kaingin (slash and burn) or shifting cultivation plots. Figure 8.2 indicates that among the 13 regions of the country, Region II has the lowest soil erosion rate of 56.41 t ha^{-1} $year^{-1}$, while Region I has the highest (128.5 t ha^{-1} $year^{-1}$) both regions are located in Luzon (Fig. 8.2). Only five regions (II, IV, VIII, X, and XI) are below the average soil erosion rate for the entire country which is 81 t ha^{-1} $year^{-1}$ (FAO 1998).

8.4 Biophysical Characteristics and Occurrence

A. Climate

The Philippines has a wet tropical climate with an annual rainfall ranging from about 1,000 mm in the south to more than 4,000 mm in mountainous areas, particularly along the eastern coast of the archipelago. This wet climate results in a net leaching environment where the rain that falls on the land percolates into the soil or occurs as runoff depending on the slope and soil cover. Runoff causes soil erosion, while excessive percolation results in nutrient leaching. Both processes cause soil degradation

Table 8.1 Soil erosion rates under various land use systems in the Philippines (*Source* Asio et al. 2009)

Location	Mean annual rainfall (mm)	Land use systems	Soil loss (Mg/ha/year)	References
Leyte	2,600	Bare plot	137	Presbitero et al. (1995)
Leyte	2,600	Planted to Zea mays	76	Presbitero et al. (1995)
Leyte	2,600	Leucaena leucocephala	37	Presbitero et al. (1995)
Leyte	2,600	Arachis hypogaea	5	Presbitero et al. (1995)
Mindanao	3,381	Tomato	23	Poudel et al. (1999)
Mindanao	3,381	Potato then corn	15	Poudel et al. (1999)
Mindanao	3,381	Cabbage	15	Poudel et al. (1999)
Mindanao	2,825	Cabbage–fallow–tomato	17	Poudel et al. (1999)
Mindanao	2,825	Cabbage–potato–cauliflower	1	Poudel et al. (2000)
Mindanao	2,825	Fallow–corn–cabbage	18	Poudel et al. (2000)
Mindanao	2,825	Fallow–cabbage–potato	34	Poudel et al. (2000)
Mindanao	2,825	Fallow–fallow–fallow	24	Poudel et al. (2000)
Mindanao	2,825	Fallow–fallow–potato	23	Poudel et al. (2000)
Mindanao	2,825	Sweet paper–fallow–cabbage	13	Poudel et al. (2000)
Mindanao	2,825	Potato–cauliflower–sweet pea	53	Poudel et al. (2000)
Leyte	3,300	Forest	0.2	Asio (1996)
Leyte	3,300	Grassland	11	Asio (1996)
Leyte	3,300	Pasture	14	Asio (1996)
Leyte	3,300	Bushland	2	Asio (1996)
Leyte	2,600	Reforestation	23	Asio (1996)
Leyte	2,600	Coconut and bushes	23	Asio (1996)
Leyte	2,600	Shifting cultivation	100	Asio (1996)
Leyte	2,600	Coconut with kudzu	6	Asio (1996)
nd	nd	Grassland/Pastureland	269	FAO (1998)
nd	nd	Paddy rice, irrigated	2	FAO (1998)
nd	nd	Fruit trees	22	FAO (1998)
nd	nd	Open grassland	80	FAO (1998)
nd	nd	Secondary forest	3	FAO (1998)
nd	nd	Upland agriculture	113	FAO (1998)
nd	nd	Gmelina, ipil-ipil, coffee, grass	1	FAO (1998)
nd	nd	Trees, shrubs, grasses	13	FAO (1998)

(continued)

Table 8.1 (continued)

Location	Mean annual rainfall (mm)	Land use systems	Soil loss (Mg/ha/year)	References
Luzon	nd	Pineapple	308	David (1988)
Luzon	nd	Coffee	318	David (1988)
Luzon	nd	Tiger grass	396	David (1988)
Luzon	nd	Castor bean	360	David (1988)
Luzon	nd	Banana	414	David (1988)
Luzon	nd	Banana–coffee–Pineapple intercropping	421	David (1988)
Luzon	nd	Undisturbed areas	251	David (1988)
Luzon	2,353	Farmers practice	140a	Paningbatan et al. (1995)
Luzon	2,353	Alley cropping	23a	Paningbatan et al. (1995)
Luzon	2,353	Alley cropping	3a	Paningbatan et al. (1995)
Luzon	2,353	Zero tillage	2a	Paningbatan et al. (1995)

a—average of 3 year sampling nd= no data

and thus contribute to the formation of marginal lands throughout the archipelago. The degradation of soil resources by erosion is aggravated by the occurrence of an average of 15–20 typhoons per year (PCARRD 2009).

There are four types of climate in the country according to the Coronas Climate Classification (Coronas 1920). Type 1 has two pronounced seasons, dry from November to April and wet during the rest of the year. Type II has no dry season with a very pronounced maximum rain period from December to February. Type III has no pronounced maximum rain period, with a short dry season of 1–3 months either from December to February or from March to May. Type IV is characterized by a more or less even distribution of rain throughout the year. Other factors like topography, land use, and soil type being comparable, areas under the Types II and IV climates in the eastern side of the archipelago are expected to experience higher erosion and faster land degradation rates, especially since these areas are exposed to more frequent typhoons. Super Typhoon Haiyan has demonstrated how strong typhoons can cause massive land degradation due to the complete destruction of the vegetation cover and the tremendous runoff which caused erosion and landslides (Figs. 8.3 and 8.4).

B. **Geology and Geomorphology**

Majority of the marginal uplands in the Philippines range in topography from undulating to steeply sloping lands. Based on the slope and landform classes by Young

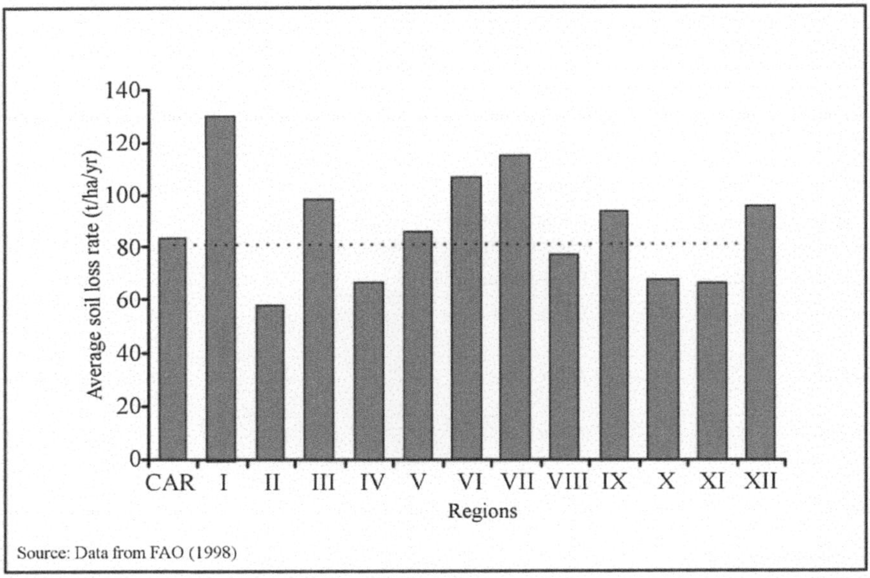

Fig. 8.2 Average soil erosion rate by region in the Philippines based on 1993 data. The dotted line represents the national average soil erosion rate (*Source* Asio et al. 2009)

(1989) given in Table 8.2, marginal uplands generally range from moderate to steep. They vary in geological characteristics although the most widespread marginal uplands are underlain by igneous rocks particularly andesitic and basaltic volcanics as well as sedimentary rocks such as limestone, mudstone, and sandstone.

Figure 5a–c show the typical marginal uplands in Leyte, Samar, Bohol, Cebu, Claveria and Davao. In Leyte, large areas of marginal uplands can be found on volcanic and limestone landscapes. Cebu, Samar, and Bohol are sedimentary islands dominated by mudstone and limestone, respectively, which is the reason why most marginal upland soils in these islands developed from these parent rocks. The marginal soils in Claveria have developed from volcanic parent materials. In Davao, the severely degraded landscape is underlain by mudstone rocks.

Our research has shown that sedimentary landscapes are more prone to degradation compared to volcanic landscapes probably because rock weathering in the former type of rocks cannot easily replenish the nutrients lost through erosion, leaching, and crop removal.

Metamorphic rocks are not widespread in the country, thus, very few degraded lands with metamorphic geology can be found. One example is the marginal upland in the schist landscape in Bukidnon.

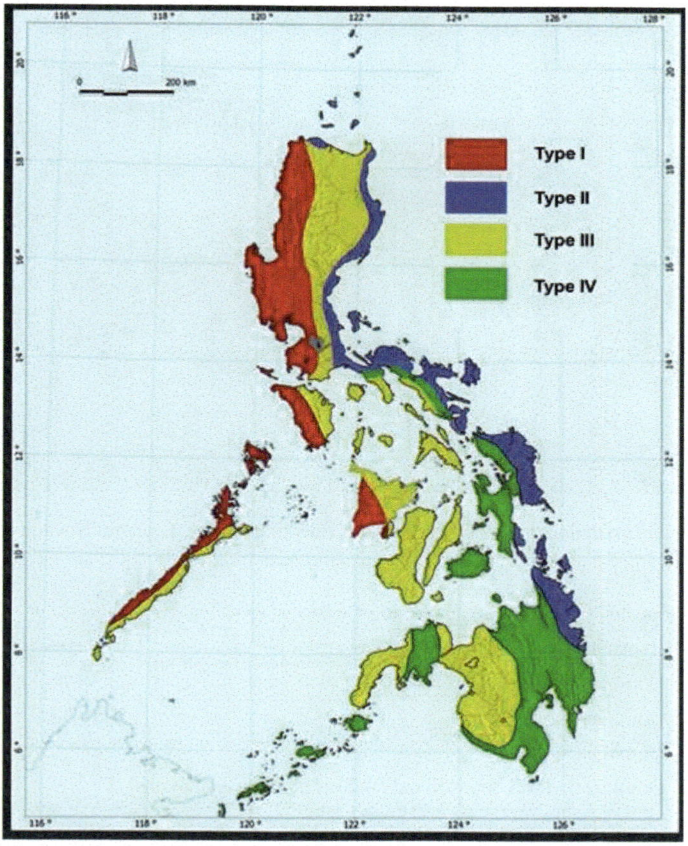

Fig. 8.3 Climate map of the philippines according to the coronas climate classification (*Source* kidlat.pagasa.dost.gov.ph)

C. Vegetation Cover

Marginal uplands can be easily identified through their vegetation cover which is generally dominated by grass and shrub species. In their study in the marginal uplands of Inopacan, Leyte, Belonias et al. (2014) documented 72 plant species in the valleys the most dominant of which were grasses and ferns. On the ridges, they found 33 plant species the most dominant among them were *Chrysopogon aciculatus*, *Desmodium triflorum*, *Imperata cylindrica*, and *Melastoma malabathricum*. They also found tree species such as *Antidesma ghaesembilla* and *Psidium guajava*.

Quimio in his vegetation analysis of the grasslands in Western Leyte found that geology and elevation define the occurrence of plant species in some areas. He found

Fig. 8.4 Land degradation due to Super Typhoon Haiyan in Dulag, Leyte on November 8, 2013

Table 8.2 Slope and landform classes (Young 1989)

Class	Explanation	Degrees	Percent
Steep	Dominant slopes	>17°	
Moderate	Dominant slopes	5°–17°	>30%
Gentle	Dominant slopes	<5°	8–30%
Flat	Dominantly level depositional landforms, e.g., flood plains		<8%

that grass species dominating open areas wide enough for the purpose of management planning were *Chrysopogon aciculatus*, *Imperata cylindrica*, *Saccharum spontaneum*, *Calopogonium mucunoides*, and *Chromolaena odorata*, and *Axonopus compressus* for moist sites and *Chromolaena odorata* for dry sites.

In the marginal uplands of Cagayan Valley, Snelder and Masipiquena identified a total of 31 trees, 12 shrubs, 5 herbaceous plants in woody patches on protected grassland and 18 trees, 10 shrubs, 8 herbaceous plants on grassland under intense utilization. The most common trees were *Antidesma ghaesembilla* and *Psidium guajava* and the most common shrubs were *Desmodium pulchellum* and *Psychotria luzoniensis*.

The most common plant species in marginal uplands in Leyte and Samar which can be used as indicator plants for such lands are *Imperata cylindrica*, *Saccharum spontaneum*, *Paspalum conjugatum*, *Axonopus compressus*, *Cyrtococcum accrescens*, *Melastoma malabathricum*, and the native *Psidium guajava* (Asio 1996; Asio et al. 2009) (Fig. 8.6).

D. **Soils**

a. Undulating to steeply sloping marginal uplands underlain by A) volcanics in Inopacan, Leyte; B) mudstone in Pinabacdao, Western Samar.

b. Degraded watershed underlain by limestones in A) Balamban, Cebu, and B) Carmen, Bohol.

c. A gently sloping marginal upland underlain by volcanics in Claveria, Misamis, Oriental (A) and the rugged marginal upland built up by mudstone in Marilog District, Davao City (B).

Fig. 8.5 **a** Undulating to steeply sloping marginal uplands underlain by (A) volcanics in Inopacan, Leyte; (B) mudstone in Pinabacdao, Western Samar. **b** Degraded watershed underlain by limestones in (A) Balamban, Cebu, and (B) Carmen, Bohol. **c** A gently sloping marginal upland underlain by volcanics in Claveria, Misamis, Oriental (A) and the rugged marginal upland built up by mudstone in Marilog District, Davao City (B)

A B C D

Fig. 8.6 The common plant species in marginal uplands. **A** Cogon (*Imperata cyclindrica*), **B** Bugang or Talahib (*Saccharum spontaneum*), **C** Carabao grass (*Paspalum conjugatum*), and D). Hantutuknaw (*Melastoma malabathricum*)

In Leyte, the degraded upland (or marginal upland) soils may be divided into three groups (Fig. 8.7) based on the landforms where they are found. The first group is comprised of the relatively fragile young soils from recent volcanic materials located on higher elevations, usually above 300 m above sea level, and characterized by rugged topography. These soils are widely used for upland crop production although some areas are still covered with primary and residual forests. Most of these soils belong to the soil order Andisols and Inceptisols.

The second group of degraded soils comprises the highly weathered upland soils (mostly Ultisols) on the older and more stable geomorphic surfaces located on the lower elevations and on footslope positions. Derived from older volcanic rocks and volcanic sediments, these soils are red, deep, heavy clay, strongly acidic (pH < 5.0), and have high iron oxide content. In addition, they have low effective cation exchange capacity and OM contents. The clay fraction of these soils is dominated by 1:1 clay minerals (halloysite and kaolinite) which explains their low nutrient-holding capacity and fertility status. Because their location is more accessible than that of the young Andisols, the Ultisols have been intensively used for agriculture despite their low inherent soil fertility status. They are, however, generally more stable and resilient compared to the Andisols.

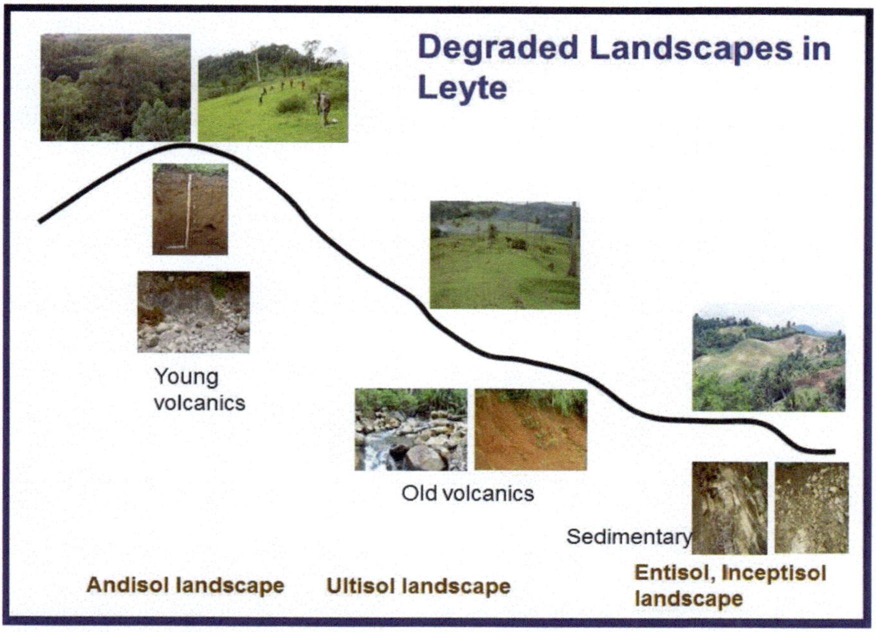

Fig. 8.7 The typical degraded landscapes in Leyte (Asio 2012)

The third group of degraded upland soils in Leyte are the poorly to moderately developed soils derived from mixed sediments as well as from sedimentary rocks, particularly limestone. They are widespread in the northwestern and southern parts of the island with elevations generally below 100 m above sea level. Because of their high initial productivity and good accessibility, these soils which belong to the soil order Mollisols and Inceptisols, have a very long history of intensive cultivation and thus have been degraded due to destructive practices employed by farmers.

In the marginal uplands of Inopacan, Leyte (Figs. 5a and 8.8), we found that the mature soils derived from volcanics are degraded by soil erosion as a result of decades of unsustainable cultivation practices. The soils are strongly acidic with soil pH values close to 5.0 and they contain low amounts of OM, total N, and available P. Most of the soils are compact and difficult to cultivate. For the marginal upland soils from limestone rocks in Leyte (Fig. 8.8), our research has shown that the soils on the upper slopes have neutral to alkaline pH values, thin solum due to erosion, clayey texture, granular structure, high contents of OM, N, Ca, and $CaCO_3$ but low contents of P, K, Fe, Mn, and B. In addition, the soils on the lower slopes have thicker solum, higher clay content, and subangular blocky structures that turn hard when dry and become plastic and sticky when wet. These soils have neutral to strongly alkaline pH values, high OM, N, Ca, and $CaCO_3$ contents, and low contents of P, K, Fe, Mn, and B. In various parts of Bohol island, the marginal upland soils derived from limestone mostly have shallow solum due to erosion and the poorly developed nature of the soils. They generally have low OM, total N, available P, and extractable Zn contents (Asio et al. 2006; Asio et al. 2015a).

On the island of Samar, we evaluated two sites, one in Sta. Rita, Western Samar (Fig. 8.8) has mudstone parent material and another in Salcedo, Eastern Samar from peridotite (ultramafic rock) parent material. Results revealed that the nutrient status

Fig. 8.8 Marginal upland soils developed from **A** mudstone in Sta. Rita, Western Samar; **B** volcanics in Inopacan, Leyte; **C** volcanics in Abuyog, Leyte; and **D** limestone in Baybay, Leyte. Except for the shallow limestone soil, all the other soils show a compact and poorly structured condition

of the marginal upland soils in both sites is low. In general, the Salcedo site has more severe soil fertility problems which are related to its more strongly weathered characteristic compared to the Sta. Rita site. Both sites, however, have strongly acidic soils and are low in OM, total N, available P, and exchangeable base contents. Considerable variations in the soil nutrient status occur under different vegetation covers or crops planted. Soil erosion appears to be the major soil degradation process in both degraded uplands studied. The study also revealed that bulk density, porosity, aggregate stability, soil pH, organic matter, and carbon dioxide evolution are good indicators of soil degradation (Asio et al. 2015b; Prajapati et al. 2021).

In Bukidnon, Poudel and West (1999) observed that the marginal upland soils derived from volcanic ash on the slope of Mt. Kitanglad, differ in chemical and fertility characteristics which vary with the degree of soil development.

In the marginal upland of Marilog District in Davao, soil samples we collected from vegetable gardens indicated low available P and exchangeable K contents. The soils have moderate OM and total N contents as well as sufficient amounts of Cu, Fe, Mn, and Zn.

In terms of soil constraints to crop production, limestone soils have more severe soil constraints than volcanic soils (Fig. 8.9 and Tables 8.3). While the marginal upland soils from volcanic rocks have limitations in terms of some nutrients (available P and K), the marginal upland soils from limestone have several physical and chemical constraints many of which are related to the nature of the parent material.

In summary, the marginal upland soils possess various constraints to their use for crop production. Depending on the genesis and land use history of the soil, the constraints may include physical such as shallow depth, compaction, unfavorable slopes, and low infiltration rates as well as chemical constraints such as acidic or alkaline pH, low OM, and nutrient contents.

8.5 Sustainable Management of Marginal Uplands

Sustainable land use is that which achieves production combined with conservation of the resources on which that production depends thereby permitting the maintenance of productivity (Young 1989). Professor Percy Sajise, a pioneer in grassland ecology and upland development in the Philippines, stressed that the technology for upland development should be soil, water, and nutrient conserving for the farming system to be sustainable. Moreover, the technology should provide immediate benefits and answer the basic needs of the farmers (Sajise and Rambo 1991).

Snelder (2001) who worked on the degraded grasslands of northeast Luzon, observed that the variable soil conditions and topography are inadequately considered in regional attempts to develop and rehabilitate these lands. Cramb (2001) observed that one cause for the failure of soil management technologies in the Philippines is the introduction of inappropriate technologies to farmers in degraded lands. According to him, many crop production technologies developed for the marginal upland areas

Soil Properties	Threshold pedon value*	Soil Profiles									
		1	2	3	4	5	6	7	8	9	10
1. Depth (cm)	>50	■	■	■	■	■	■	■	■	■	■
2. Texture	medium	●	●	●	●	●	●	●	●	●	●
3. Consistence	fr;np;ns	■	■	●	■	■	■	●	●	■	■
4. Bulk Density (g/cm^3)	<1.45	●	●	●	●	●	●	●	●	●	●
5. pH (water)	5.5-7.0	■	■	■	■	■	■	■	■	■	■
6. OM (%)	>3.0	●	●	■	■	■	■	■	●	●	●
7. Total N (%)	>0.2	●	●	■	●	■	■	■	●	●	●
8. Avail. P (mg/kg)	>8-15	■	■	■	■	■	■	■	●	■	■
9. Exch. K (cmol/kg)	>0.20	■	■	■	●	●	●	■	●	●	●
10. Exch Ca (cmol/kg)	>0.40#	●	●	●	●	●	●	●	●	●	●
11. Exch Mg (mg/kg)	>0.50	●	●	●	●	●	●	●	●	●	●

fr, friable; np, non-plastic; ns, non-sticky; *except for depth, which considers solum thickness; #excessive levels (Ca:Mg 7:1) is a constraint (Tisdale et al., 1985).

■ unfavorable property (constraint) ● favorable property

Fig. 8.9 Fertility constraints to crop production of marginal soils from limestone based on the properties of surface horizons* (Oraiz et al., 2021)

have not been adopted by farmers or have failed to increase crop yields in these stressed environments.

Bales et al. (2014) in their socio-ecological study in the marginal uplands of Inopacan, Leyte, observed that interventions would more likely succeed if these are designed considering the needs and resources of the target beneficiaries. They were able to identify the problems and needs which can be divided into seven categories: (1) Unfavorable conditions for crop production (low soil fertility and lack of water); (2) Occurrence of pests and diseases; (3) Young generations are no longer interested in farming; (4) No space for planting food crops; (5) Low price of copra; (6) Illnesses in raised animals; and (7) Food insufficiency.

Organic matter management is crucial when using marginal uplands for the production of annual crops. Many farmers in these problematic lands complain that applying animal manure or other organic fertilizers to the soil does not usually produce the desired crop yield. Experiments conducted by Ratilla et al. (2014) and Lina et al. (2014) in the marginal uplands of Inopacan, Leyte, revealed that the combined application of manure and inorganic fertilizer was better than the sole application of manure. This implies that in soils of very low soil fertility, the application of organic fertilizer alone may not be effective in improving the performance of crops.

Table 8.3 Site qualities of two marginal upland soils from volcanic rocks in Leyte (Jahn and Asio, 1998)

Andosol (Ormoc)			Alisol (Baybay)	
Rooting depth	>100 cm	+++	>100 cm	+++
Rootability		+++		++
Erodibility		±		−−
Air capacity (Ø)	28%	+++	18%	++
Available water capacity	160 $1/m^2$	+++	140 $1/m^2$	+++
Drainage		+++	?	++
Organic matter	42 kg/m^2		18 kg/m^2	
Available N	27 g/m^2	++	34 g/m^2	++
Available P	0.7 g/m^2	−−	0.3 g/m^2	−−
P-retention (Ø)	98%	+++	70%	++
Available K	45 g/m^2	±	36 g/m^2	±
Available Ca	7 g/m^2	−−	198 g/m^2	+
Available Mg	15 kg/m^2	±	194 kg/m^2	+++
Reserves of N	2.7 kg/m^2	+++	1.4 kg/m^2	++
Reserves of P	0.5 kg/m^2	++	0.3 kg/m^2	+
Reserves of K	1.6 kg/m^2	++	0.3 kg/m^2	−
Reserves of Ca	2.9 kg/m^2	+++	1.1 kg/m^2	++
Reserves of Mg	3.4 kg/m^2	+++	2.1 kg/m^2	++

−− = very low, − = low, ± = moderate, + = moderately high, ++ = high, +++ = very high

The most widely accepted strategies to rehabilitate marginal uplands include agroforestry, contour farming, and rainforestation farming.

a. Agroforestry

Agroforestry is now accepted as the sustainable land use for fragile sloping lands including marginal uplands. Countless researches in different parts of the humid tropics have revealed that agroforestry, in its various forms, generally increases land productivity by, among others, improving the fertility status or quality of the marginal soil. In fact, DENR-ERDB (2010) has reported that agroforestry is the national strategy for marginal uplands in the country.

One of the earliest definitions of agroforestry was put forward by Nair (1985) who described it as an approach of integrated land use that involves deliberate retention or admixture of trees and other woody perennials in crop/animal production fields to benefit from the resultant ecological and economic interactions. The formal definition of agroforestry was given by Young (1989) as “a collective name for land-use systems in which woody perennials (trees, shrubs, etc.) are grown in association with herbaceous plants (crops, pastures) and/or livestock in a spatial arrangement, a rotation or both, and in which there are both ecological and economic interactions between the tree and non-tree components of the system." The most appropriate form or type of agroforestry depends on the environmental and socio-economic characteristics of the marginal upland.

b. Contour farming

Contour farming is the practice of cultivating sloping lands along contours in order to conserve water and nutrients. Farming along contours greatly reduces surface runoff and thus also soil erosion. Reduced soil erosion in upland areas means decreased sedimentation rate in low-lying areas as well as in water bodies. PCARRD (2009) considers contour farming as a farming system aimed at stabilizing the desired level of crop productivity in marginal uplands. There are two types of contour farming: vegetative barriers and physical barriers. Vegetative barriers include contour hedgerows especially using leguminous trees, natural vegetative strips, contour composting, and contour fences. Physical barriers include contour furrows, contour bunds, bench terraces, and contour rock walls (PCARRD 2009; Asio and Asio 2021; Rai 2023a and 2023b).

c. Rainforestation Farming

Rainforestation (or *rainforestation farming*) is an innovative approach of rehabilitating degraded lands or restoring the forests using native tree species. It is based on the hypothesis that a farming system is more sustainable when its physical structure and species composition are closer to the rainforest (Margraf and Milan 1996a). The technology was developed by the ViSCA-GTZ Applied Tropical Ecology Project, a bilateral project implemented by the Visayas State College of Agriculture or ViSCA (now VSU) in collaboration with the University of Hohenheim in Germany. The project was funded by the German Agency for Technical Cooperation (GTZ) from 1989 to 1999. Details of the history of rainforestation farming can be found in Asio (2017).

Goeltenboth and Hutter (2004) reported that rainforestation farming combines the necessities of rural development, safe natural resource management, and biodiversity restoration. They noted that more than 100 different local forests and fruit tree species were tested and planted in a near-to-nature planting scheme concerning species composition in a degraded area, a former kaingin farm, covered by *Imperata cylindrica*. The recommended planting scheme includes both sun-requiring native trees (for example, *Dracontomelon dao, Terminalia microcarpa*, *Pterocarpus indicus*, *Melia dubia*, etc.) and shade-loving trees (for example, *Dipterocarpus grandiflorus*, *Dipterocarpus warburgii*, *Shorea negrosensis*, etc.), highly valuable timber trees and fruit trees. One key strategy is to plant the sun-loving trees at a close distance of 2 $\times$ 2 m to quickly reach the condition of a closed canopy in order to shade out the grass. During the second year, shade-loving trees are planted under the established first-year pioneers (Fig. 8.10).

Fig. 8.10 An idealized view of a rainforestation farm after about 10 years (Margraf and Milan, 1996b)

The success of this technology in rehabilitating marginal and degraded lands is clearly reflected by the hundreds of rainforestation farms that now exist in various parts of the country. More importantly, it is now also a national DENR (Department of Environment and Natural Resources) policy to integrate rainforestation in rehabilitating denuded areas. In 2004, then DENR Secretary Elisea G. Gozun issued Memorandum Circular No. 2004-06 on the guidelines for the integration of rainforestation farming strategy in the development of open and denuded areas within protected areas and other appropriate forest lands. In 2011, President Benigno C. Aquino III signed Executive Order No. 26 (Declaring an Interdepartmental Convergence Initiative for a National Greening Program). Section 3.3.3 of the EO is on "appropriate incentives shall be developed by the Convergence Initiative to encourage rainforestation, particularly in the protected area."

8.6 Summary

Uplands are defined as undulating and steep lands that range in elevation from near sea level to about 1000 m elevation. This definition is geomorphological and clearly separates the flat lowlands from the elevated uplands. There is a need to change the antiquated 18% slope rule for at least research and development purposes.

Marginal uplands can be defined as undulating and steep lands ranging in elevation from near sea level to about 1000 m elevation and having low crop productivity due to poor soil fertility.

Marginal uplands occur under the four climate types of the Philippines although the eastern side of the archipelago (Types II and IV) is expected to have faster soil erosion and degradation rates. They also form under various rock formations although soils derived from sedimentary rocks appear to be more prone to degradation ("marginalization") than soils from volcanic rocks.

The common plant species in marginal uplands which can be used as indicator plants are (A). Cogon (*Imperata cyclindrica*), (B). Bugang or Talahib (*Saccharum spontaneum*), (C). Carabao grass (*Paspalum conjugatum*), and (D). Hantutuknaw (*Melastoma malabathricum*). Native guava (*Psidium guajava*) is almost always present in marginal uplands.

The marginal upland soils possess various soil constraints to crop production. Depending on the genesis and land use history of the soil, the constraints may include physical such as shallow depth, compaction, unfavorable slopes, and low infiltration rates as well as chemical constraints such as acidic or alkaline pH, low OM, and nutrient contents. The fertility constraints vary widely among marginal upland soils.

Agroforestry is widely considered as the sustainable form of land use for marginal upland soils. An innovative form of agroforestry which has been formally endorsed by the Philippine government to rehabilitate denuded lands is *rainforestation farming*. Contour farming is also widely practiced by upland farmers.

Acknowledgements This paper is partly based on the senior author's lecture for the SEARCA Regional Professorial Chair Award in 2014. The authors also used data derived from their research projects funded by the Philippine Commission on Higher Education (CHED) and the Australian Center for International Agricultural Research (ACIAR).

References

Alfarisy FK, Andryani I, Bowo C (2020) Evaluation of water quality due to the use of intensive fertilizers on farmer level in the upstream of Bedadung Jember watershed, East Java, Indonesia. J Degrad Min Lands Manag 7(4):2301–2312

Asio VB (2017) Rainforestation farming to rehabilitate marginal uplands: history of its development. Ann Trop Res 39:125–133

Asio VB, Asio LG (2021) Properties, fertility status, and sustainable management of degraded tropical soils. In: Jatav HS, Singh SK, Rajput VP, Minkina T (eds) Sustainable soil fertility management. Nova Science Publishers Inc., New York, pp 1–31

Asio VB, Cabuños CC, Chen ZS (2006) Physiochemical characteristics and fertility status of soils formed from Quaternary limestone in Leyte Philippines. Soil Sci 171(8):648–661

Asio VB, Lina SB, Maranguit DS, Bolledo AB, Doguiles RJT, Sabijon JR, Demain KLB (2014) Characteristics of soils in the marginal uplands of Inopacan Leyte. Ann Trop Res 36(Supplement):1–15

Asio VB, Demain KLB, Olguera DT, Villasica LJD (2015) Characteristics and nutrient status of two degraded upland soils in Samar, Philippines. Ann Trop Res 37:142–166

Asio VB, Jahn R, Perez FO, Navarrete IA, Abit SM Jr (2009) A review of soil degradation in the Philippines. Ann Trop Res 31:9–4

Asio VB, Cagasan EG, Belonias BS (eds) (2014a) Enhancing farm productivity and environmental quality inclimate change vulnerable marginal uplands in Eastern Visayas. Ann Trop Res 36:1–166

Asio VB, Lina SB, Salas R, Añonuevo GC, Harper S (2015a) Progress report, ACIAR Soil Project. In: PCAARRD-ACIAR Annual Meeting, 11–13 August 2015, Davao City (Unpublished)

Asio VB (1996) Characteristics, weathering, formation and degradation of soils from volcanic rocks in Leyte, Philippines, vol 33. Hohenheimer Bodenkundliche Hefte, Stuttgart, Germany

Asio VB (2012) Characteristics and management of degraded upland soils in Leyte, Philippines. Paper presented during the international uplands conference, University of Hohenheim, 16–18 April 2012, Stuttgart, Germany

Bales MC, Cagasan EG, Tabada MATW, Asio MVSG, Baculfo JD, Flandez LG, Ramos AD, Come WD (2014) Participatory assessment of resources and needs of upland communities: a case in Inopacan, Leyte Philippines. Ann Trop Res 36(Supplement):258–276

Belonias BS, Platino CS, Malanguis JT (2014) Agrobiodiversity of home gardens in selected marginal upland villages of Inopacan, Leyte, Philippines. Ann Trop Res 36:48–69

Boado EL (1988) Incentive policies and forest use in the Philippines. In: Public policies and the misuse of forest resources: a world resources report (R. Repetto and M.Gillis, eds.). Cambridge University Press, England, pp 165–198

Briones RM (2009) Land degradation and rehabilitation in the Philippines. The Philippines country environmental analysis. Final Report. Philippine Institute for Development Studies, Manila

Brototi B, Singh A, Rai PK, Kumar J, Walker S (2022) GIS based study of reclamation of degraded semi-arid soil: a case study of Rajasthan from India. Indian J Environ Prot 42(3):302–315

Coronas J (1920) The climate and weather of the Philippines. Bureau of Print, Manila (Philippines)

Cramb RA, Garcia JNM, Gerrits RV, Saguiguit GC (2000) Conservation farming projects in the Philippine uplands: rhetoric and reality. World Dev 28:911–927

Cramb RA (2001) Soil conservation technologies for smallholder farming systems in the Philippine Uplands: a socioeconomic evaluation. ACIAR, Canberra, Australia

Cruz MCJ, Zosa-Feranil, Goce CL (1989) Population pressure and migration: implications for upland development in the Philippines. J Philipp Dev XV:15–46

David WP (1988) Soil and water conservation planning: policy issues and recommendations. J Philipp Dev XV:47–84

Department of Environment and Natural Resources-Ecosystem Research Development Bureau (DERN-ERDB) (2010) A research compendium for marginal uplands. Los Banos, Laguna

FAO (1998) Soil resource depreciation and deforestation: Philippine case study in resource accounting. Rome

Garrity DP, Kummer DM, Guiang ES (1993) The Philippines. In: Committee on sustainable agriculture and the environment in the humid tropics. Board on Agriculture and Board on Science and Technology for International Development and National Research Council (eds) Sustainable agriculture and the environment in the humid tropics. National Academy Press, Washington

Garrity DP (1993) Agroforestry and the sustained productivity of Asia's humid uplands. Upland Agriculture in Asia. In: Proceedings of a Workshop in Bogor, Indonesia, 6–8 April 1993. pp 225–233

Goeltenboth F, Hutter CP (2004) New options for land rehabilitation and landscape ecology in Southeast Asia by "rainforestation farming." J Nat Conserv 12:181–189

Jahn R, Asio VB (1998) Soils of tropical forest of Leyte, Philippines. I. Weathering, soil characteristics, classification and site qualities. In: Schulte A, Ruhiyat D (eds) Soils of tropical forest ecosystems. Springer, Berlin-Heidelberg-New York, pp 29–36

Kang S, Post W, Wang D, Nichols J, Bandaru V, West T (2013) Hierarchical marginal land assessment for land use planning. Land Use Policy 30:106–113

Kummer DM (1992) Upland agriculture, the land frontier and forest decline in the Philippines. Agrofor Syst 18:31–46

Lal R (1998) Soil quality and sustainability. In: Methods for assessment of soil degradation. Advances in soil science. CRC Press, Boca Raton, pp 17–30

Lapar MLA, Pandey S (1999) Adoption of soil conservation: the case of the Philippine uplands. Agric Econ 21:241–256

Lightfoot CO, de Guia Jr, Ocado F (1988) A participatory method for systems problem research: rehabilitating marginal uplands in the Philippines. Expt Agric 24:301–309

Lina SB, Maranguit DS, Asio VB, Sabijon JR, Demain KLB, Bolledo AB (2014) Growth performance of corn as influenced by the combined application of organic and inorganic fertilizers in a marginal upland. Ann Trop Res 36 (Supplement) 36:16–29

Margraf J, Milan PP (1996a) Ecology of Dipterocarp forest and its relevance for island rehabilitation in Leyte, Philippines. In: Schulte A, Schoene D (eds) Dipterocarp forest ecosystems. World Scientific Publishers, Singapore, pp 124–154

Margraf J, Milan PP (1996b) Rainforestation Farming. A farmer's guide to biodiversity management for the Philippines. Philippine-German Applied Tropical Ecology Program, ViSCA, Baybay, Leyte

Mishra VN, Rai PK, Singh P (2021) Geo-information technology in earth resources monitoring and management (edit. Book), Nova Science Publishers, U.S.A., pp 1–395. ISBN: 978-1-53619-669-6

Nair PKR (1985) Classification of agroforestry systems. Agrofor Syst 3:97–128

National Action Plan (NAP) (2004) The Philippine National Action Plan to combat desertification, land degradation, drought, and poverty for 2004–2010. Department of Agriculture, Department of Environment and Natural Resources, Department of Science and Technology, and Department of Agrarian Reform, Manila, Philippines

Nelson RA (1994) Soil erosion and conservation in the Philippine uplands. A review and literature. SEARCA-UQ uplands research project. Working paper No. 3, Los Baños, Laguna

Oraiz K, Saz VO, Cascante MD, Galvez SM, Garrido YYM, Asio VB (2021) Characteristics and nutrient status of limestone soils in Leyte and Samar, Philippines. In: Mishra VN, Rai PK, Singh F (eds) Geo-information technology for earth resources monitoring and management. Nova Science Publishers, New York, pp 211–227

Pailagao CT, Espaldon MVO, Dorado MA, Catacutan DC, Rebancos CM (2010) Drivers of land use change in Lantapan, Bukidnon, Philippines. J Environ Sci Manag 13:1–11

Paningbatan EP, Ciesolka CA, Coughlan KJ, Rose CW (1995) Alley cropping for managing soil erosion of hilly lands in the Philippines. Soil Technol 8:193–204

PCARRD (2009) Philippine S&T agenda on climate change 2010–2016. Agriculture, Forestry & Natural Resources Section. PCARRD, Laguna

Poudel DD, West LT (1999) Soil development and fertility characteristics of a volcanic slope in Mindanao, the Philippines. Soil Sci Soc Am J 63:1258–1273

Poudel DD, Midmore DJ, West LT (1999) Erosion and productivity of vegetable systems on sloping volcanic ash-derived Philippine soils. Soil Sci Soc Am J 63:1366–1376

Poudel DD, Midmore DJ, West LT (2000) Farmer participatory research to minimize soil erosion on steepland vegetable systems in the Philippines. Agric Ecosyst Environ 79:113–127

Prajapati G, Rai PK, Mishra VN, Singh P, Shahi AP (2021) Remote based assessment of waterlogging and soil salinity: a case study of Kerla. Results Geophys Sci 7(100024):1–18

Presbitero AL, Escalante MC, Rose CW, Coughlan KJ, Ciesiolka CA (1995) Erodibility evaluation and the effect of land management practices on soil erosion from steep slopes in Leyte, the Philippines. Soil Technol 8: 205–213

Rai PK Mishra VN, Singh P (2021) Recent technologies for disaster management & risk reduction-sustainable community resilience & responses (edit. Book), Springer Nature, Switzerland, pp 1–471. ISBN: 978-3-030-76116-5. https://doi.org/10.1007/978-3-030-76116-5

Rai PK Mishra VN, Singh P (2022) Geospatial technology for landscape and environment management: sustainable assessment & planning (edit. Book). Springer Nature, Singapore, pp 1–431. ISBN: 978-981-16-7373-3. https://doi.org/10.1007/978-981-16-7373-3

Rai PK (2023a) River conservation and water resource management. Springer Nature, Singapore, pp. 1–352. ISBN No.: 978-981-99-2604-6. eBook ISBN: 978-981-99-2605-3

Rai PK (2023b) Advances in water resource planning & sustainability. Springer Nature, Singapore, pp 1–322. ISBN No. 978-981-99-3659-5

Rambo AT, Sajise PE (1985) Developing a regional network for interdisciplinary research on rural ecology: The Southeast Asian Universities Agroecosystem Network (SUAN) Experiences. Environ Prof 7:289–298

Ratilla BC, Toledo SP, Capuno OB (2014) Evaluation of organic-based fertilizers for corn production in the marginal uplands of Inopacan Leyte. Ann Trop Res 36(Supplement):107–123

Roa JR (2007) Food security in fragile lands. PhD Dissertation, Wageningen University, The Netherlands

Sajise PE, Ganapin DJ (1991) An overview of upland development in the Philippines. In: Technologies for sustainable agriculture in marginal uplands in Southeast Asia (Blair, G. and Lefroy, R, eds). Proceedings of a workshop, 10–14 December 1990. Ternate, Cavite, Philippines, pp 31–44

Shahid SA, Al-Shankiti A (2013) Sustainable food production in marginal lands-case of GDLA member countries. Int Soil Water Conserv Res 1:24–38

Singh A, Rai PK, Deka G, Biswas B, Prasad D, Rai VK (2021) Management of natural resources through integrated watershed management in Nana Kosi Micro Watershed, District Almora, India. Ecol Environ & Conserv 27 (February Suppl. Issue):S260–S268

Snelder DJ (2001) Soil properties of Imperata grasslands and prospects for tree-based farming systems in Northeast Luzon, The Philippines. Agrofor Syst:27–40

Steiner KG (1996) Causes of soil degradation and development approaches to sustainable soil management. Margraf Verlag, Weikersheim

The World Bank (1989) Philippine environment and natural resource management study. A World Bank ountry report. Washington, DC

Young A (1989) Agroforestry for soil conservation. CAB International, Wallingford, UK

Chapter 9
Dynamic Transitions in Wular Lake's Land Use and Land Cover: Analyzing the Role of Urban Development and Agricultural Activities

Gurleen Kour, Rashid Latief Bhatt, and Shashi Prabha

Abstract Wular Lake is one of the largest freshwater lakes in Asia and is essential to the region's ecological and socio-economic structure. However, due to a number of natural and anthropogenic factors, the lake and the lands around it have been experiencing major spatio-temporal changes. This study makes use of geospatial techniques to look into and evaluate the dynamic changes that have occurred in the Wular Lake region during the last twenty-seven years. In order to examine changes in land cover, land use, and water levels, the research integrates a large dataset of remotely sensed satellite photos, field observations, and geographic information system (GIS) technologies. The attempt is to identify the patterns and trends of land use land cover using time-series analysis, geospatial techniques, and interpolation. This research shows that a number of complicated and interrelated factors, including urbanization, agricultural development, climate change, and hydrological changes, are altering the Wular Lake region. However, urbanization and agriculture are the most prominent factors in the land use and land cover changes around the Wular Lake. The study also pinpoints locations that could be particularly sensitive to land-use change and degradation, providing information about the most vulnerable regions. The regions that have experienced drastic changes are west and east regions in terms of increase in built up area, south eastern region in terms of increase in agricultural area. The cumulative impact of increase in built up and agricultural area has led to shrinkage of the Wular Lake which in turn has created a space for swampy land in the central part of it. This study advances knowledge of the changing landscape near Wular Lake which offers crucial data for local communities and policymakers to make well-informed choices regarding conservation and sustainable land management.

Keywords Wular lake · Landuse-landcover · Geo-spatial analysis · Remotely sensed satellite photos

G. Kour (✉) · R. L. Bhatt · S. Prabha
Department of Geography, University of Jammu, Jammu, India
e-mail: gurleen0195@gmail.com

P. K. Rai et al. (eds.), *Natural Resources Management and Sustainable Livelihoods in the Mountainous Region*, Advances in Geographical and Environmental Sciences,
https://doi.org/10.1007/978-981-97-2100-9_9

9.1 Introduction

9.1.1 Conceptual Framework of Land Use and Land Cover (LULC)

The concept of land cover has gone under continuous change for millennia. This change has occurred through the use of fire for game hunting and clearance of patches of land for agriculture and livestock production, with the advent of plant and animal domestication. This is because of the fact that human production demands cannot be fulfilled without modification or conversion of land covers. In the past two centuries, the impact of human activities on land has grown enormously because of population increase, technological development and the requirements thereafter, altering entire landscapes, and ultimately impacting the biodiversity, nutrient and hydrological cycles as well as climate (Sherbinin 2002). The concepts of land use land cover are interrelated in such a way that it is difficult to differentiate one from another. In fact, the use of land depends on its nature of coverage and land coverage depends on the nature of land use.

Land cover is the physical material at the surface of the earth. It is the biophysical state of the earth surface and immediate subsurface (Turner et al. 1995). The term land cover originally referred to the kind and state of vegetation, such as forest or grass cover but it has broadened in subsequent usage to include other things such as human structures, soil type, biodiversity, surface, and groundwater (Meyer 1995). This definition gives the idea that Land cover is the physical material at the surface of the earth.

Every parcel of land on the earth's surface is unique in the cover it possesses. Land use and land cover are distinct yet closely linked characteristics of the earth's surface. The use to which we put land could be grazing, agriculture, urban development, logging, and mining among many others. Land cover categories could be cropland, forest, wetland, pasture, roads, and urban areas among others. Land cover has emphasis on physical material, at the same time a connection between land use and land cover has been made (Meyer 1999).

Land cover refers to the natural vegetative cover types that characterize a particular area and the emphasis is being laid upon natural vegetative cover and other natural phenomena have been ignored (Sherbinin 2002).

Land use and land cover change is commonly divided into two broad categories, (Alun and Clark 1997; Forney 2010):

(a) Conversion of land and
(b) Modification of land
(i) Conversion of land refers to a change from one cover or use category to another (e.g., from agricultural land to settlement).
(ii) Modification of land represents a change within one land use or land cover category (e.g., from rainfed cultivated area to irrigated cultivated area) due to changes in its physical or functional attributes.

According to European Commission (2000) Intensification and Extension are another two vital terms which are related to land use land cover change. These are listed below.

Intensification: A flow representing the transition of land use land cover types associated with low intensity use to a higher intensity (e.g., semi-natural—arable land).

Extension: The process of making something (more) extensive or the geographic spread and distribution of any technology, especially agriculture.

Land use is the human use of land. Even though natural processes may also contribute to changes in land cover, the major driving force is human-induced land uses (Allen and Barnes, 1985). In order to understand the various implications of land cover change, understanding of the concept of land use change is essential. Different human driving forces mediated by the socio-economic setting and influenced by the existing environmental conditions lead to an intended land use of an existing land cover through the manipulation of the biophysical conditions of the land (Turner et al. 1995; Mishra et al. 2021; Rai et al. 2021 and Rai et al. 2022).

According to Sherbinin (2002), land use is the term that is used to describe human uses of land, or immediate actions modifying or converting land cover. It is a description of how people utilize the land and socio-economic activity urban and agricultural land uses are two of the most commonly known land use classes (Fisher 2005).

Fisher in his scholarly and pioneer work has mentioned that land use involves the management and modification of natural environmental or wilderness into built environment such as fields, pastures, and settlements. Mr. Fisher in his definition of land use has excluded the use of land by nature; he only lays emphasis on human use of land. It has also been defined as "the arrangements, activities and inputs people undertake in a certain land cover type to produce, change or maintain it." Land use involves the manner in which biophysical attributes of the land are manipulated, as well as the intent underlying that manipulation. Therefore, land use is a more complicated term (Ellis et al. 2010). Pelorosso et al. (2009) have mentioned that LU and LC are not identical.

According to Ellis and Pontius (2010), natural scientists define land use in terms of syndromes of human activities such as agriculture, forestry, and building construction that alter land surface processes including biogeochemistry, hydrology, and biodiversity. So it is clear that land use and land cover are distinct yet closely linked characteristics of the earth's surface. Land-use change is the proximate cause of land-cover change. The driving forces to this activity could be economic, technological, demographic, scenic, and or other factors. Hence, land use and land cover dynamics is a result of complex interactions between several biophysical and socio-economic conditions which may occur at various temporal and spatial scales (Reid et al. 2000; Arora et al. 2021; Singh et al. 2021; Prajapati et al. 2021).

9.1.2 Aspects of Land Use and Land Cover Change and Factors Responsible

According to Skole and Tucker (1993), land use and land cover (LULC) change on a regional scale is one of the major research endeavors in global change studies. Land use change is probably the single most important factor influencing the conservation of natural environments (Vitousek et al. 1997). In recent times, LULC change is emerging as a key environmental issue. Among the many concerns about global environmental change today, there is an increasing awareness of the importance of issues involved in LULC and its changes over time, since land use is a key descriptor of human influence. The LULC changes will continue to be the most important components of global change over the next several decades. Yet our understanding of the patterns and processes of LULC change is highly inadequate (Abrul et al. 1999). Land use affects land cover and changes in land cover affect land use. A change in either, however, is not necessarily the product of the other. Changes in land cover by land uses do not necessarily imply degradation of the land. However, many shifting land use patterns driven by a variety of social causes, result in land cover changes that affect biodiversity, water and radiation budgets, trace gas emissions, and other processes that come together to affect climate and biosphere. There are few landscapes remaining on the earth's surface that have not been significantly altered or are not being altered by humans in some manner. Mankind's presence on the earth and his modification of the landscape have a profound effect upon the natural environment. These anthropogenic influences on shifting patterns of land use are a primary component of many current environmental concerns as land use and land cover change is gaining recognition as a key driver of environmental change (Riebsame et al. 1994). Sometimes LULC changes in parts of the World indicate a close relation between slope and land cover land use change. Specific forest-related studies (Muller and Zeller 2002) found that dense forests decreased irrespective of slope angle, whereas high steepness favors the return of forests in areas previously occupied by agriculture.

The fact that human beings are the major contributors to LULC change and are the ones experiencing the consequences of these changes will be of paramount importance to understand the interaction between humans and the terrestrial environment. This need becomes more imperative as changes in land use become more rapid affecting the livelihoods of societies. Growing human populations exert increasing pressure on the landscape as demands multiply for resources such as food, water, shelter, and fuel. These socio-economic factors often dictate how land is used regionally. Accelerating human alteration of landscapes around the World has led to increasing concern about the ecological consequences of land use change (Turner et al. 1990; Brototi et al. 2022; Rai et al. 2023a and 2023b).

Urban expansion has increased the exploitation of natural resources and has changed land use land cover patterns. The most obvious LULC changes occurred in the category of urban or built-up land. Others, such as paddy field, other cropland, and orchards also underwent substantial changes (Ding Hang et al. 2007). Studies related with LULC change studies are used for varied reasons such as updating existing or new inventories, proper resource use, restoration of landscapes and ecosystems, analysis of landscape ecology, implementation of better management strategies, assessing the land-transformation processes, and other purposes.

The purpose of this study is to analyze the land cover changes around the Wular Lake, its spatio-temporal analysis and to examine the factors responsible for this change.

9.2 Study Area

Wular Lake is the largest freshwater lake, located in Bandipora district in the Kashmir Valley, 40 km northwest of Srinagar City between 34°20′N 74°36′E and 34°20′N 74°36′E in the Northwest of India. With a size of 189 sq. km, Wular Lake is also one of the largest freshwater lakes in Asia. The lake lies at an altitude of 1,580 m. Its maximum depth is 14 m, and it has a length of 16 km and a breadth of 10 km. Wular Lake plays a significant role in the hydrographic system of the Kashmir Valley by acting as huge absorption basin for annual floodwater. The lake and its surrounding extensive marshes have an important natural wildlife. The rivers Bohnar, Madamati, and Erin from the mountain ranges and the rivers Vetasta (Jhelum) and the Ningal from the south bring hundreds of tons of silt into the lake every year. This rampant siltation and the human encroachments have devastating effects on the lake (Fig. 9.1).

In recognition of its biological, hydrological, and socio-economic values, the lake was included in 1986 as a Wetland of National Importance under the Wetlands

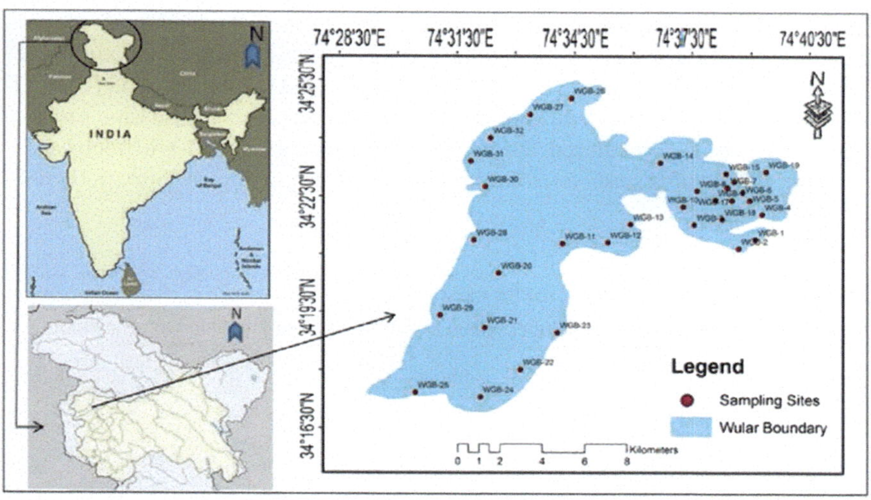

Fig. 9.1 Location of the study area

Programme of the Ministry of Environment and Forests, Government of India for intensive conservation and management purposes. Subsequently in 1990, it was designated as a Wetland of International Importance under the Ramsar Convention.

Wular Lake is a sustainable wintering site for a number of migratory waterfowl species such as Little Egzet (Egretta garzetta), Cattle Egzet (Bubulcus ibis), Shoveler (Anas clypeata), Common Pochard (Aythya farina), and Mallard. Birds like Marbled Teal (Marmaronetta angustirostris) and Pallas's Fish-eagle (Haliaeetus leucoryphus) are species listed in the Red List of IUCN. Many terrestrial bird species observed around the lake are Short-toed Eagle (Circaetus gallicus), Little Cuckoo (Piaya minuta), European Hoopoe (Upupa epops), Monal Pheasant (Lophophorus impejanus), and Himalayan Pied Woodpecker (Dendrocopos himalayensis albescens).

Wular Lake is also an important habitat for fish and contributes about 60% of the fish yield of the Kashmir Valley. The dominant fish species found in the lake are Cyprinus carpio, Barbus conchonius, Gambusia affinis, Nemacheilus sp., Crossocheilus latius, Schizothorax curvifrons, S. esocinus, S. planifrons, S. micropogon, S. longipinus, and S. niger. More than 8,000 fishermen earn their livelihood from Wular Lake. The pollution from fertilizers and animals as well as human wastes, the conversion of vast catchment into agriculture land and the hunting pressure on waterfowl and migratory birds are the biggest problems in the region.

9.3 Data Source and Methodology

9.3.1 Data Collection

First the buffer of 5 km was created around the lake to show the land use land cover changes. The data used in this study include four Landsat images; including three Thematic Mapper (TM) images and a Landsat Operational Land Imager (LANDSAT-OLI) image were downloaded from the USGS Earth Explorer. These time series of Landsat images were acquired during the growing seasons and are freely available from the Landsat archive from the United States Geological Survey (USGS) (Tables 9.1 and 9.2).

Table 9.1 Information of satellite used for the study

Satellite	Sensor	Number of bands	Resolution	Year of acquisition
LANDSAT 5	TM	149,36	30	1991, 2001, 2011
LANDSAT 8	OLI	11	30	2018

Table 9.2 Classification scheme for land use/land cover analysis (NRSC)

Sl. No	Level I	Level II
1	Built-up land or urban land	1.1 Dense urban 1.2 Moderate urban 1.3 Sparse urban
2	Agriculture land	2.1 Crop land 2.2 Fallow land
3	Forest land	3.1 Evergreen forest
4	Wasteland	4.1 Gullied/eroded land 4.2 Land with scrub 4.3 Land without scrub/mine dump yards 4.3 Rock outcrop
5	Water bodies	5.1 River/streams 5.2 Tanks/lakes
6	Others	6.1 Habitation with plantation 6.2 Mining

9.3.1.1 Landsat

The Landsat Program is a series of Earth-observing satellite missions jointly managed by NASA and the U.S. Geological Survey. In a 1966 press release, Secretary of the Interior Stewart L. Udall, announced the start of "Project EROS," a program "aimed at gathering facts about the natural resources of the Earth from earth-observing satellites carrying sophisticated remote sensing observation instruments." Secretary Udall named Dr. William T. Pecora, the Director of the U.S. Geological Survey, to lead the EROS program.

In cooperation with NASA, on July 23, 1972, the Earth Resources Technology Satellite (ERTS-1) was launched. It was later renamed Landsat 1. Additional Landsat satellites followed in the 1970's and 1980's, and in 1999, Landsat 7 was launched. Landsat 8 (initially named Landsat Data Continuity Mission—LDCM) was the most recent satellite to be launched on February 11, 2013. Landsat 9 is in development, with a launch scheduled for late 2020. The Landsat Missions Timeline page contains more details on each of the missions. The instruments aboard the Landsat satellites have acquired millions of images through the course of the missions, and the data are a valuable resource for global change research and applications in agriculture, forestry, geology, regional planning, and education.

Landsat data are received and downlinked to ground stations worldwide, and are archived at the USGS EROS Center in Sioux Falls, South Dakota. Landsat data products are processed and made available for download to all users at no charge via Earth Explorer, GloVis, and the Landsat Look Viewer.

Landsat 5 TM:

Landsat thematic mapper has been used for the study. In general Landsat 4, 5 are carrying both MSS and TM sensors. In my instance, Landsat 5 TM has been used with its resolution of 30 m and has 7 bands. Some characteristics and applications of this sensor, namely, are.

- *Band 1*
  (Blue-green, 0.45–0.52 μm)
  This is useful for analyzing and identifying bathometry, mapping of coastal waters, Chlorophyll absorption, distinguishing between coniferous and deciduous vegetation, and penetration of clear water.
- *Band 2*
  (green, 0.52–0.60 μm)
  This band records green radiation reflected from healthy vegetation, assesses plant vigor, and records reflectance from turbid water.
- *Band 3*
  (red, 0.63–0.69 μm)
  Since vegetation absorbs nearly all red light, it's often called the chlorophyll absorption band. This band can be useful for distinguishing between vegetation and soil and in monitoring vegetation health.
- *Band 4*
  (near infrared, 0.76–0.90 μm)
  This band is an indicator of plant cell structure biomass, plant vigor, and delineation of shoreline.
- *Band 5*
  (Mid-infrared, 1.55–1.75 μm)
  It is indicative of vegetation moister and soil moister mapping, differentiating snow from clouds, and penetration of thin clouds.
- Band 6
  (far infrared, 10.4–12.5 μm)
  This band has 120m resolution; this band is valuable for analyzing vegetation stress, soil moisture, brightness temperature, and some other applications.

Landsat 8 Oli:

The Landsat Operational Land Imager (OLI), built by the Ball Aerospace & Technologies Corporation, measures in the visible, near-infrared, and short-wave infrared portions of the spectrum. Its images have 15-m (49 ft.) panchromatic and 30-m multi-spectral spatial resolutions along a 185 km (115 miles) wide swath, covering wide areas of the Earth's landscape while providing sufficient resolution to distinguish features like urban centers, farms, forests, and other land uses. The entire Earth will fall within view once every 16 days due to Landsat 8's near-polar orbit.

OLI's design is an advancement in Landsat sensor technology and uses an approach demonstrated by the Advanced Land Imager sensor flown on NASA's experimental EO-1 satellite. Instruments on earlier Landsat satellites employed scan mirrors to sweep the instrument fields of view across the surface swath width and transmit light to a few detectors. The OLI instead uses long detector arrays, with over 7,000 detectors per spectral band, aligned across its focal plane to view across the swath. This "push-broom" design results in a more sensitive instrument providing improved land surface information with fewer moving parts. With an improved signal-to-noise ratio compared to past Landsat instruments, OLI is more reliable and provides improved performance.

9.3.2 Data Pre-Processing

Pre-processing of satellite images prior to change detection is essential and has the unique goal of establishing a more direct linkage between the data and biophysical phenomena (Coppin et al. 2004).

Prior to data analysis, initial processing on the raw data is usually carried out to correct for any distortion due to the characteristics of the imaging system and imaging conditions. Depending on the user's requirement, some standard correction procedures may be carried out by the ground station operators before the data is delivered to the end-user. These procedures include radiometric correction to correct for uneven sensor response over the whole image and geometric correction to correct for geometric distortion due to Earth's rotation and other imaging conditions (such as oblique viewing). The image may also be transformed to conform to a specific map projection system. Furthermore, if accurate geographical location of an area on the image needs to be known, ground control points (GCP's) are used to register the image to a precise map (geo-referencing).

9.3.3 Image Classification

The objective of image classification is to assign labels to unknown pixels. The image classification process includes several key points, namely, ground truth and classification schema, extraction training sample (TS), and classification method.

National Remote Sensing Centre (NRSC) in May 2006 devised a Land use land cover classification system for use with remote sensing data in India (NRSC Manual 2006). The primary objective was to familiarize the scientific personnel involved in the mapping process with the definitions, classification scheme and procedural steps for interpretation and mapping so as to maintain standard operational procedures.

Data manipulation and analysis and linking the spatial data with the attribute data for creation of topology was carried out using GIS software. Creation of GIS output in the form of land use/land cover map showing various land use/land cover patterns (Masta mark et al. 2010).

Ground truth data are very important in order to identify the most accurate land class and create a training sample. In this study open source data such as Google Earth and previous studies, my geological and geographical familiarities with the area, National Remote Sensing Centre (NRSC) Land use land cover system have been employed to identify the most accurate land classes within the area of study. Per that scheme the land use and land cover categories of focus were urban, barren, forests, agricultural/crop, water, and swamps.

9.3.4 Classification Techniques

Digital image classification techniques group pixels to represent land cover features. Land cover could be forested, urban agricultural, and other types of features. The purpose of Image classification is to categorize all pixels in a digital image into different land use/land cover classes. Depending on the interaction between computer and interpreter during classification process, there are two types of classification. These two main categories used to achieve classified output are called Supervised and Unsupervised Classification techniques.

9.3.4.1 Unsupervised Classification

In unsupervised classification, it first groups pixels into "clusters" based on their properties. In order to create "clusters," analysts use image clustering algorithms such as K-means and ISODATA.

After picking a clustering algorithm, you identify the number of groups you want to generate. For example, you can create 8, 20, or 42 clusters. To be clear, these are unclassified clusters because in the next step, you manually identify each cluster with land cover classes. For example, if you want to classify vegetation and non-vegetation, you'll have to merge clusters into only 2 clusters.

Overall, unsupervised classification is the most basic technique. Because you don't need samples for unsupervised classification, it's an easy way to segment and understand an image.

In unsupervised classification, the software does most of the processing on its own generally resulting in more categories than the user is interested in. This is the point where the user has to make decisions on which categories can be grouped together into a single land use category. In either case additional image processing may be used to determine which method is better for a given situation. It must be kept in mind that maps are simple attempts to represent what actually exists in the world and are never completely accurate.

9.3.4.2 Supervised Classification

In supervised classification, you select representative samples for each land cover class. The software then uses these "training sites" and applies them to the entire image.

Supervised classification uses the spectral signature defined in the training set. For example, it determines each class on what it resembles most in the training set. The common supervised classification algorithms are maximum likelihood and minimum-distance classification.

Supervised classification is based on the idea that a user can select sample pixels in an image that are representative of specific classes and then direct the image processing software to use these training sites as references for the classification of all other pixels in the image. Training sites are selected based on the knowledge of the user. The user also sets the bounds for how similar other pixels must be to group them together. These bounds are often set based on the spectral characteristics of the training area, plus or minus a certain increment. The user also designates the number of classes that the image is classified into. Many analysts use a combination of supervised and unsupervised classification processes to develop final output analysis and classified maps.

9.3.5 Land Use and Land Cover Change Analysis

Landsat images of 1991, 2001, 2011, and 2018 were collected from USGS Earth Explorer for the generation of LULC in and around the area of study. Many techniques have been employed to analyze and extract information from remotely sensed data. ArcGIS 10.2 software package has been used for the image processing and for classified land use/land cover map generation. Supervised classification uses the spectral signatures obtained from training samples to classify an image, with the assistance of image classification toolbar, training samples were created to represent the classes to be extracted. A signature file was created from the training samples which was then used by the spatial analyst multivariate classification tools to classify the image.

To validate the classification precision in this study, a stratified random sample strategy was used to select 35 samples for each class totaling 280 or more points per image.

9.3.6 Accuracy Assessment

In the context of information extraction by image analysis, accuracy "measures the agreement between a standard assumed to be correct and a classified image of unknown quality" (Campbell 2007). Precision defines the level of detail found within the classification. It is possible to increase the accuracy of a classification by decreasing the amount of detail or by generalizing it to broad classes rather than very specific ones. For example, a scheme which generally categorizes trees vs. crops presents less opportunity for classification error than one that distinguishes many types of trees and many types of crops. In this case, lower precision provides the potential for higher accuracy. However, the user of the map that offers only general classes cannot make precise statements about any given point on the map.

Classification error occurs when a pixel (or feature) belonging to one category is assigned to another category. Errors of omission occur when a feature is left out of the category being evaluated; errors of commission occur when a feature is incorrectly included in the category being evaluated. An error of omission in one category will be counted as an error in commission in another category.

Classification error occurs when a pixel (or feature) belonging to one category is assigned to another category. Errors of omission occur when a feature is left out of the category being evaluated; errors of commission occur when a feature is incorrectly included in the category being evaluated. An error of omission in one category will be counted as an error in commission in another category.

9.4 Results and Discussions

The Land use and Land Cover classes used in this study are based on NRSC Classification Scheme Level I and II, which are designed for remote sensing data. A buffer of 5 km was created around the Wular Lake to show the Land use and land cover changes. The Land use and land cover classification around Wular Lake during 1991–2018 can be generalized in the following five classes. They are Water body, Swamp, Agriculture, Horticulture, Dense forest, Sparse forest, and Built up.

9.4.1 Land Use and Land Cover and Change Analysis in 1991 & 2001

In Table 9.3, the LULC classes are presented which show the changes over a period of 10 years from 1991 to 2001. The total area in 1991 was 43,295.94 hectares.

The area under water body in 1991 was 14,685.7 ha which has come down to 7205.22 hectares in 2001. So, there has been a negative change of 17.23% in this class particularly due to the low rainfall, conversion of swamps into agricultural lands and low water level.

Table 9.3 LULC classes and change analysis 1991–2001 (A period of 10 years)

LULC categories	Area (hectares) 1991	Percentage	Area (hectares) 2001	Percentage	Change (ha) 1991–2001	Change % age
Water body	14,685.7	33.9	7205.22	16.67	−7480.48	−17.23
Swamp	6184.44	14.39	3361.32	7.89	−2823.12	−6.5
Dense forest	6695.28	15.46	3947.58	9.19	−2747.7	−6.27
Horticulture	268.65	0.6	1031.85	2.38	763.2	1.78
Agriculture	10,994.9	25.3	21,881	50.8	10,886.1	25.5
Sparse forest	4344.39	10.06	5350.59	12.57	1006.2	2.51
Built up	122.58	0.29	217.35	0.5	94.77	0.21
Total	43,295.94	100	42,994.91	100		

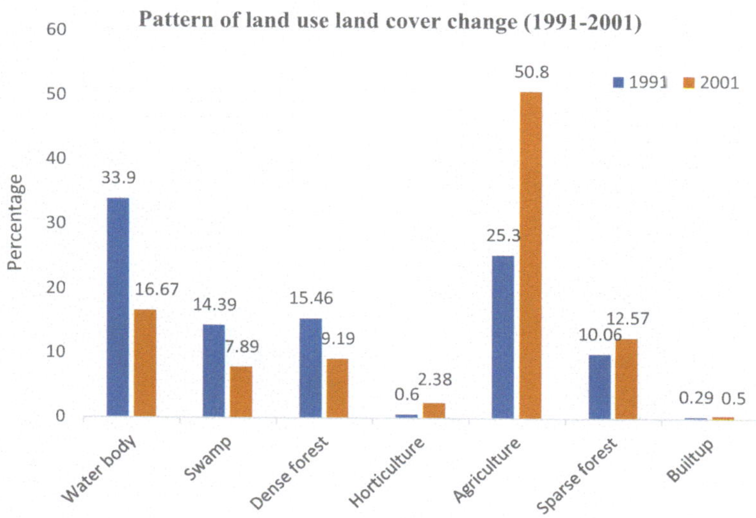

Fig. 9.2 Patterns of land use and land cover change around Wular Lake (1991–2001)

Agricultural land obtained an area of 10,994.9 hectares in 1991 and it has increased till 2001 to 21,881 hectares, showing a positive change of 25.5%.

Swamps and dense forest cover had area of 6184.44 hectares and 6695.28 hectares in 1991 respectively. In 2011, the area of swamps was reduced to 3361.32 hectares and the area under dense forest cover decreased to 3947.58 hectares.

Hence both classes show a negative change of 6.5% and 6.27% respectively. Horticulture is the cultivation, processing, and sale of fruits, nuts, vegetables, ornamental plants, and flowers as well as many additional services. The area under it in 1991 was 286.65 hectares which has increased in 2001 to 1031.85. Horticulture also has shown a positive change of 1.78% from 1991 to 2001.

The area under Sparse forest and built-up classes have also increased from 4344.39 hectares and 122.58 hectares in 1991 respectively. In 2001, the area of sparse forest has increased to 5350.59 hectares, showing a positive change of 2.51% and area of built-up area improved by 217.35 hectares, displaying a positive change of 0.21% (Figs. 9.2 and 9.3).

9.4.2 Land Use and Land Cover and Change Analysis in 2011 & 2018

Land Use/land Cover dynamics involves the complete replacement of one type of land use by another and is the result of interplay between socio-economic, institutional, and environmental factors (Lesschen et al. 2015). Change detection is the process of identifying differences in the state of an object/phenomenon by observing it at

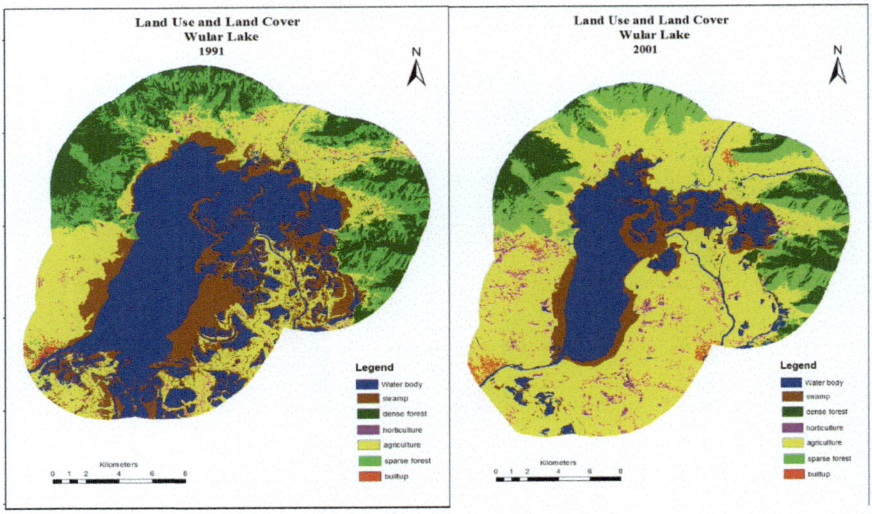

Fig. 9.3 Land use/Land cover map of Wular Lake (1991 & 2001)

different periods of time. It involves the ability to quantify temporal effects using Multi-Temporal datasets from Satellite imageries (Othman et al. 2013).

Over a period of 7 years (2011–2018), drastic changes have undergone in terms of land use and land cover in and around Wular Lake. Table 9.4 displays the area under water body, swamps, agriculture, built up, etc.

The total area in 2011 was 43,295.81 hectares and 43,241.05 hectares in 2018. The area under water body in 2011 was 5249.79 ha which has come down to 3773.63 hectares in 2018. So, there has been a negative change of 3.42% in this class particularly due to the low rainfall in Kashmir valley, conversion of swamps into agricultural lands.

Table 9.4 LULC classes and change analysis 2011–2018 (a period of 7 years)

LULC categories	Area (hectares) 2011	Percentage	Area (hectares) 2018	Percentage	Change (ha) 2011–2018	Change %age
Water body	5249.79	12.14	3773.63	8.72	−1476.16	−3.42
Swamp	8867.52	20.48	8449.97	19.55	−417.55	−0.93
Dense forest	3151.8	7.28	4069.17	9.41	917.37	2.13
Horticulture	1123.92	2.59	1299.92	3.02	176	0.43
Agriculture	16,596.5	38.33	14,221.3	32.88	−2375.2	−5.45
Sparse forest	7475.13	17.26	8352.32	19.31	877.19	2.05
Built up	831.15	1.92	3074.74	7.11	2243.59	5.19
Total	43,295.81	100	43,241.05	100		

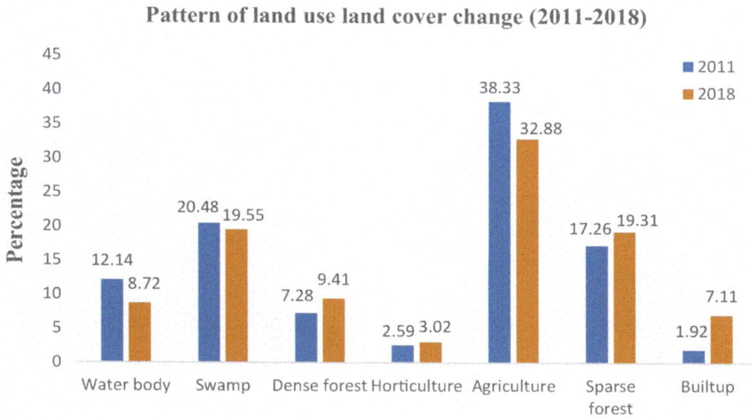

Fig. 9.4 Patterns of land use and land cover change around Wular lake (2011–2018)

Agricultural land had an area of 16,596.5 hectares in 2011 and it has decreased till 2018 to 14,221.3 hectares, showing a negative change of 5.45%. The reason is the conversion of agricultural land into the settlements as a result of growing urbanization.

In Table 9.4, the area under Swamps has also gone down from 8867.52 hectares in 2011 to 8449.97 hectares in 2018, showing a negative transformation of 0.93%. This is due to the trend of converting agricultural land and swamps into orchards.

Dense and sparse forest cover has increased during the 7 years of study. Area under dense forest in 2011 was 3151.8 hectares which has increased to 4069.17 hectares in 2018, showing a positive change of 2.1%. while as sparse forest cover has also increased from 7475.13 hectares in 2011 to 8352.32 hectares in 2.18, showing a positive trend of 2.05%.

Area under horticulture has been increasing since 1991 at a slow pace. In 2011 the area under horticulture was 1123.92 hectares which increased till 2018 to 1299.92 hectares, displaying a positive change of 0.43%.

The built-up has been increasing since the early years of twenty-first century. In 2011, the area under settlements was 831.15 hectares which has shown a positive trend by increasing in 2018 to 3074.74%. The change between 2011 and 2018 has been 5.19% (Table 9.4 and Figs. 9.4 and 9.5).

9.4.3 Accuracy Assessment

The accuracy of LULC maps produced was evaluated using the overall accuracy (OA), producer's accuracy (PA), user accuracy (UA), and Kappa statistics. Producers accuracy quantifies the errors of omission while user accuracy quantifies error of commission. Kappa is another method of expressing classification accuracy as it

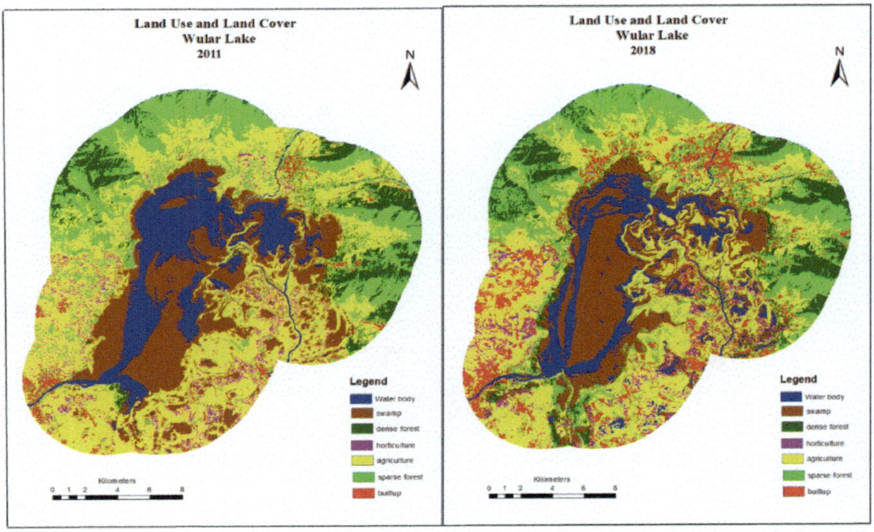

Fig. 9.5 Land use/land cover map of Wular lake (2011 & 2018)

Table 9.5 Results of accuracy assessment for land use and land cover classification

Year	Overall accuracy (%)	Kappa coefficient
1991	88.67	0.85
2001	86.67	0.82
2011	90.39	0.89
2018	94.48	0.94

measures the chance agreement and has been found to be more robust than the overall accuracy of images (Jensen 2005; Lillesand and Kaefer 1999). During the preprocessing step, the imageries were rectified and georeferenced to achieve high overall accuracy.

Table 9.5 shows the accuracy assessment of the study area along with Kappa Coefficient.

9.5 Conclusion

The result showed that the urbanization, deforestation, reforestation, and farming new agricultural land, climate change, etc., have led to the overall change of land use land cover of the area. All of these have an important role in understanding the ecological system of the area and have an impact on its natural balance. The results of this study provided important information in support of developing a deeper understanding of the nature of Land Use Land Cover and Land Use Land Cover changes in the area of study.

Based on the results of the above analysis it is quite clear that the rate of change of Land Use Land Cover classes is high. It was observed that the area of the land cover like water body, swamps, dense forest tend to decrease, but the area of the land use types like agriculture, horticulture, and human settlements tend to increase.

Wular Lake is fed by river Jhelum and other small rivers like Madhumati, Arin Nallah, and many other small streams, which are shrinking day by day. The Wular Lake has shriveled considerably and if the current trend persists, it will die out very soon. For decades the important role of Wular Lake in supporting biodiversity and livelihoods had gone unnoticed. Widespread reclamation and drainage of the lake and its marshes for agriculture, settlements, and plantations took place. This has contributed to shrinkage of the lake area and degradation of the natural resources it provides to wetland dependent communities. During last few decades, the overall water spread and marsh areas shrunk by over half, reducing one-fifth of the area's water storage capacity. Floods and droughts have increased as the water regulation capacity of the wetland has been diminished both due to shrinkage in area as well as loss in water-holding capacity. This has increased the risk of floods and drought for both local and downstream communities.

One-fifth of the water holding capacity has been lost over the last three decades due to siltation from degraded catchments and wetland conversions for agriculture and willow plantations.

Direct discharge of solid and liquid wastes from the settlements all along River Jhelum mainly from Srinagar city and other towns in the upstream area have led to degradation of water quality and health hazards to the communities living around the Wular Lake. Also, lack of well-defined policy and regulatory mechanisms for integrated management has led to cross-sectoral conflicts and overall degradation of wetland and its resource.

Thus, need of the hour is to take these aspects into consideration and come out with some positive and concrete steps in order to save this wetland from further degradation.

References

Abd El-Kawya OR et al (2011) Land use and land cover change detection in the western Nile delta of Egypt using remote sensing data

Arora A, Pandey M, Mishra VN, Kumar R, Rai PK, Costache R, Punia M, Di L (2021) Comparative evaluation of geospatial scenario-based land change simulation models using landscape metrics. Ecol Ind 128:107810

Babamaaji RA, Lee J (2013) Land use/land cover classification of the vicinity of Lake Chad using NigeriaSat-1 and Landsat data

Bhat SA et al (2014) Surface water quality assessment of Wular lake, a Ramsar site in Kashmir Himalaya, using discriminant analysis and WQI

Brototi B, Singh A, Rai PK, Kumar J, Walker S (2022) GIS based study of reclamation of degraded semi-arid soil: a case study of Rajasthan from India. Indian J Environ Prot 42(3):302–315
Campbell BA, Campbell DB, Margot JL, Ghent RR, Nolan M, Chandler J, Carter LM, Stacy NJ (2007) Focused 70-cm wavelength radar mapping of the Moon. IEEE Trans Geosci Remote Sens 45(12):4032–4042
Comprehensive Management Action Plan for Wular Lake by Wetlands International-South Asia (2007)
Coppin P, Jonckheere I, Nackaerts K, Muys B, Lambin E (2004) Review ArticleDigital change detection methods in ecosystem monitoring: a review. Int J Remote Sens 25(9):1565–1596
de Sherbinin A (2002) A CIESIN thematic guide to a CIESIN thematic guide to land-use and land use and land use and land-cover change (LUCC) cover change (LUCC). Center for International Earth Science Information Network (CIESIN)
Ellis E, Pontius R (2010) The encyclopedia of earth. Retrieved 30 March 2011
Ellis EA, Baerenklau KA, Marcos-Martínez R, Chávez E (2010) Land use/land cover change dynamics and drivers in a low-grade marginal coffee growing region of Veracruz, Mexico. Agrofor Syst 80:61–84
Fisher P, Comber AJ, Wadsworth R (2005) Land use and land cover: contradiction or complement. Re-presenting GIS 85:98
Jones A, Clark J (2024) The spatialities of Europeanization: power, governance and territory in Europe. Routledge
Joseph G (2003) Fundamentals of remote sensing. Universities Press, Hyderabad
Kangabam RD et al (2017) Assessment of land use land cover changes in Loktak lake in Indo-Burma biodiversity hotspot using geospatial techniques
Kendall MS, Jensen OP, Alexander C, Field D, McFall G, Bohne R, Monaco ME (2005) Benthic mapping using sonar, video transects, and an innovative approach to accuracy assessment: a characterization of bottom features in the Georgia Bight. J Coas Res 21(6):1154–1165
Li SH et al (2016) Analysis of the Spatio-temporal land-use/land-cover change and its driving forces in Fuxian lake watershed
Lillesand TM, Keifer RW (1999) Remote sensing and image interpretation, 4th edn. John Wiley and Sons Inc., New York
Mendoza ME et al (2010a) Analyzing land cover and land use change processes at watershed level: a multitemporal study in the Lake Cuitzeo watershed, Mexico
Mendoza ME et al (2010b) Analyzing land cover and land use change processes at watershed level: a multitemporal study in the lake Cuitzeo Watershed, Mexico (1975–2003)
Meyer JW (1999) The changing cultural content of the nation-state: A world society perspective. In: State/culture: State-formation after the cultural turn, pp 123–143
Mishra VN, Rai PK, Singh P (2021) Geo-information technology in earth resources monitoring and management (edit. Book). Nova Science Publishers, U.S.A., pp 1–395. ISBN: 978–1–53619–669–6
Muller D, Zeller M (2002) Land use dynamics in the central highlands of Vietnam: a spatial model combining village survey data with satellite imagery interpretation. Agric Econ 27(3):333–354
Neena PT et al (2018). Evaluation of land use-land cover changes in and around Vellayani lake, Thiruvananthapuram, over the last thirty years-an analysis using remote sensing and GIS
NRSC (2006) Natural ResourceCensus-LandUseLandCover Database. Department of Space, ISRO, GOI, Hyderabad
Othman AA, Al-Saady YI, Al-Khafaji AK, Gloaguen R (2014) Environmental change detection in the central part of Iraq using remote sensing data and GIS. Arab J Geosci 7:1017–1028
Oyedotun TDT (2018) Land use change and classification in Chaohu lake catchment from multi-temporal remotely sensed images
Parsa VA, Salehi E (2016) Spatio-temporal analysis and simulation pattern of land use/cover changes, case study: Naghadeh, Iran
Pelorosso R, Leone A, Boccia L (2009) Land cover and land use change in the Italian central Apennines: a comparison of assessment methods. Appl Geogr 29(1):35–48

Prajapati G, Rai PK, Mishra VN, Singh P, Shahi AP (2021) Remote based assessment of waterlogging and soil salinity: a case study of Kerla. Results Geophys Sci 7(100024):1–18

Rai PK, Mishra VN, Singh P (2021) Recent technologies for disaster management & risk reduction-sustainable community resilience & responses (edit. Book). Springer Nature, Switzerland, pp 1–471. ISBN: 978–3–030–76116–5. https://doi.org/10.1007/978-3-030-76116-5

Rai PK, Mishra VN, Singh P (2022) Geospatial technology for landscape and environment management: sustainable assessment & planning (edit. Book). Springer Nature, Singapore, pp 1–431. ISBN: 978–981–16–7373–3. https://doi.org/10.1007/978-981-16-7373-3

Rai PK (2023a) River conservation and water resource management. Springer Nature, Singapore, pp 1–352. ISBN No.: 978–981–99–2604–6. eBook ISBN: 978–981–99–2605–3

Rai PK (2023b) Advances in water resource planning & sustainability. Springer Nature, Singapore, pp 1–322. ISBN No. 978–981–99–3659–5

Reid RS, Kruska RL, Muthui N, Taye A, Wotton S, Wilson CJ, Mulatu W (2000) Land-use and land-cover dynamics in response to changes in climatic, biological and socio-political forces: the case of southwestern Ethiopia. Landscape Ecol 15:339–355

Riebsame WE, Meyer WB, Turner BL (1994) Modeling land use and cover as part of global environmental change. Climatic Change 28(1):45–64

Singh A, Rai PK, Deka G, Biswas B, Prasad D, Rai VK (2021) Management of natural resources through integrated watershed management in Nana Kosi micro watershed, district Almora, India. Ecol Environ & Conserv 27 (February Suppl. Issue):(S260−S268). ISSN 0971–765X

Skole DL, Tucker CJ (1993) Tropical deforestation and habitat fragmentation in the Amazon: satellite data from 1978 to 1988. Science 260:1905–1910. Prog Phys Geogr 32(5):575–580

Turner BL, Skole D, Sanderson S, Fischer G, Fresco L, Leemans R (1995) Land-use and land-cover change: Science/research plan [No source information available]

Verburg PH, Lesschen JP (2006) Agriculture, climate and future land use patterns: potential for a simulation-based exploration. In Agriculture and climate beyond 2015: A new perspective on future land use patterns. Springer Netherlands, Dordrecht, pp 5–32

Vitousek PM, Mooney HA, Lubchenco J, Melillo JM (1997) Human domination of Earth's ecosystems. Science 277(5325):494–499

Wasigea JE et al (2012) Monitoring basin-scale land cover changes in Kagera Basin of lake Victoria using ancillary data and remote sensing

Zhao Y et al (2012) Examining land-use/land-cover change in the lake Dianchi watershed of the Yunnan-Guizhou plateau of southwest China with remote sensing and GIS techniques: 1974–2008

Chapter 10
Petrography and Major Oxide Geochemistry of Dolerite Dykes from Kunnam Region Villupuram, Tamil Nadu, Southern India

Mohamed Sherief Mohammed, Ariputhiran Ramachandran, Viswasam Stephen Pitchaimani, Shaik Mohammad Hussain, and Pappukutty Adarsh

Abstract The mafic dolerite dyke crosscut the charnockite basement in the Kunnam region. The dykes throughout the area were highly fractured and there are various faults and joints are seen. The proper methodology was followed for collecting the unaltered rock samples in the study area. The rock reflects ophitic to sub-ophitic textures and is mainly composed of plagioclase feldspar, clino-pyroxene and opaque minerals (magnetite and titanomagnetite), olivine and quartz are observed as accessory minerals. The geochemical studies indicate that the rock has basic composition based on the percentage of silica. The TAS classification diagram for the mafic dyke shows that the samples fall under basaltic composition in the classification of the volcanic rock, and gabbroic composition in the classification of the plutonic rock. The K_2O versus silica diagram shows that the rock samples fall under the category of calc-alkaline series. The rock weathering was examined with A–CN–K plot indicating that the dolerite rock samples fall under low weathering margin.

Keywords Petrography · Major oxide · Dolerite dyke · Southern Granulitic Terrain · Kunnam · Tamilnadu

---

M. S. Mohammed · S. M. Hussain
Department of Geology, School of Earth and Atmospheric Sciences, University of Madras, Guindy Campus, Chennai 600 025, India

A. Ramachandran (✉) · V. S. Pitchaimani
Department of Geology, V.O. Chidambaram College, Tuticorin, Tamilnadu, India
e-mail: sankarramchand@gmail.com

P. Adarsh
Department of Geology, University of Calicut, Malappuram, Kerala, India

P. K. Rai et al. (eds.), *Natural Resources Management and Sustainable Livelihoods in the Mountainous Region*, Advances in Geographical and Environmental Sciences,
https://doi.org/10.1007/978-981-97-2100-9_10

10.1 Introduction

10.1.1 Background

In the Indian subcontinent, dolerites are the intrusive rocks that have received the least research attention. A variety of details regarding geodynamic and geochemical evolution can be found in these mafic rocks. According to Srivastava (2011a, b), dolerite dyke swarms are a common characteristic of different terranes and serve as a key for geodynamical interpretation (Tarney 1992; Worthing 2005). According to Seymour and Kumarapeli (1995) and Marzoli et al. (2006), fluid circulation, magma mixing and crustal contamination are the main factors influencing fractional crystallisation processes. A lot of study was made possible by the mafic dyke bodies. These intrusive tabular structures of mafic composition, known as mafic dykes, typically have very high length-to-width ratios and penetrate pre-existing crustal cracks. They are good targets for paleo-magnetic studies (Evans 1968, 1999; Li et al. 1996, 2004; Wingate and Evans 2003; Pesonen et al. 2003; Li and Evans 2011; Piispa et al. 2011; Pisarevsky et al. 2015; Liu et al. 2018; Mishra et al. 2021; Rai et al. 2021; Rai et al. 2022; Rai 2023a, 2023b). In addition to being stable land for 100 million years, the Southern Peninsula of India is home to numerous giant dyke swarms that are recognised in the Achaean craton of the Indian shield. These swarms are comprised of the oldest rocks in the world, known as "Archaean rocks" (Murthy 1989, 1995; Halls et al. 2007; French et al. 2008; Ernst and Srivastav 2008; Srivastave et al. 2008; French and Heaman 2010). Mafic magmatism, especially mafic dykes, has occurred in Southern Granulitic terrain multiple times during the Precambrian (Crook Shank 1963; Ramakrishnan 1990; Srivastava 2006). The southern granulitic terrain of Peninsular India is home to the mafic dyke in the Kunnam area. Along with some areas of Kerala, Karnataka, Andhra Pradesh, the host rock is dispersed throughout Tamil Nadu. The mafic dyke can be up to 8 km long and is heavily fractured and jointed. The study area also has several boulders and worn rock. Almost all of the weathered rocks in the study region had experienced spheroidal weathering. Srivastava et al. (2008) have characterised the mafic dykes, including dolerite, in special publications. Dolerite dykes in southern granulite terrain have been described by Devaraju (1995), Srivastava and Ahmed (2008, 2009) and Srivastava (2011a, b). The paleomagnetism and isotopic age of the dykes of the southern granulite terrain have been described by Radhakrishna and Joseph (1993). Radhakrishna (2007) has written about the geochemical characteristics of the dykes in northern Kerala, Coimbatore and Dharmapuri.

10.1.2 Geological Setting

The northern Himalayas, the Indo-Gangetic basins and the Peninsular Shield are the geological divisions of the Indian subcontinent. Many geological events have occurred in Peninsular India over the course of Indian history. It has a varied geological history ranging from the Archaean to the present. The major geological features are the Deccan basalt from the Paleocene to the Eocene, Gondwana formations, Cuddapah, Vindhyan, and Kurnool formations, the Dharwar supergroup, the Aravalli and Delhi groups, the Bijawar group, the Singbhum and Bundhelkhand cratons, the Sausar and Sakoli groups, peninsular gneissic complexes, and charnockite intrusions. Except for the Gondwana sediments and the Deccan traps, all of the groups and formations mentioned above are precambrian in age. The southern Indian shield is usually thought to be a composite continental section created by the accretion of diverse crustal blocks from the Mid-Archean to the Neoproterozoic (Radhakrishna 1989; Harris et al. 1994; Jayananda and Peucaut 1996). In accordance with previous theories, a unitary model of a single continent divided by steep faults has also been put out (Mahadevan 1998; Fermer 1936). Based on geochronology and isotope data, multiple age provinces with diverse precrustal histories are hypothesised within the composite unit of the Southern Indian shield (Harris et al. 1994; Bartlet et al. 1995; Jayananda et al. 1995; Cenki et al., 2004). One of the most notable geographical features of the Indian peninsula is the southern granulite landscape. A large area of high-grade rocks, ranging from ultra-high temperature (UHT) to ultra-high pressure (UHP) granulites, is exposed in the Southern Granulite Terrain (SGT). The current study area is in the southern granulitic terrain's core portion. There are many well-exposed mafic dykes in the vicinity, which are intruded by charnockites and feldspathic gneissic rock formations (Selvaraj et al. 2012). The intrusion of these mafic dyke bodies may clearly be observed in the field. The mafic dykes identified in the research area have a NW–SE trend. The age of the dyke is unknown, however prior research (Selvaraj et al. 2012) suggests that these rock bodies were deposited during the Paleoproterozoic. Spheroidal weathering was seen in many areas of the study area. Throughout its body, the Kunnam region's mafic dyke has seen several deformations, including cracks and joints. It swarms across the Kunnam region and traverses the charnockite and gneisses basement rocks. The dyke in this location is 60 m broad overall, with a 15-m-wide outcrop visible in different areas of the research area at 25-m intervals (Selvaraj et al. 2012). In addition, there are a few areas with charnockite outcrops. The mafic dyke in this region can be seen as a boulder in some spots. There were ten fresh, unmodified samples obtained (Fig. 10.1). Although some samples also exhibit aphanitic texture, the hand specimen primarily exhibits phaneritic texture. The fresh rock samples from this region are formed of plagioclase, augite and opaque minerals, which may be observed in the hand specimens.

10.1.3 Methodology

After a thorough geological field visit, we have mapped the research area. The studied area has extensively fragmented and worn bodies of dolerite dykes (Fig. 10.2). Mafic dolerite boulders can be seen in several areas of the study region (Fig. 10.3). We visited around 16 locations along the dyke extension direction and collected 10

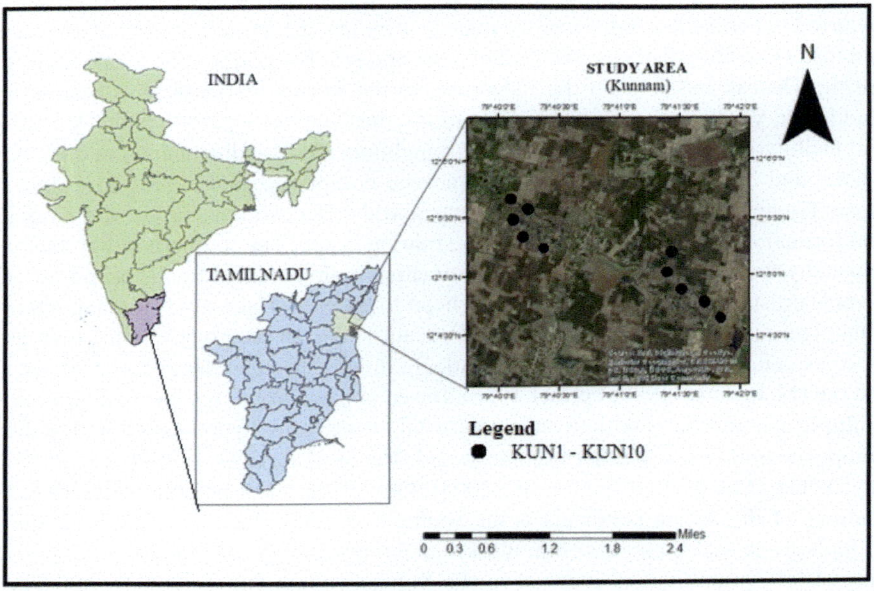

Fig. 10.1 Location map (sampling) of the study area

selected unmodified samples from ten separate locations of the same dolerite body. Each sample was labelled with its geographic coordinate system (latitude and longitude) and was well maintained. After dividing the ten samples into thin sections, the grinding and polishing machine was used to carefully cut and grind the sections. Ten unaffected typical samples were selected for the total rock main oxides investigation utilising X-ray fluorescence techniques using 80–800 (micron) mesh-sized carborundum powder. X-ray fluorescence (XRF) is a well-established technology used for the investigation of geological materials due to its precision, accuracy, adaptability, automation, sensitivity and selectivity. In order to achieve elemental studies that are both qualitative and quantitative, this is one of the fastest and most relevant geochemical analysis procedures available. The rock samples were ground and sieved through a 230 mesh sieve. Collapsible aluminium cups were used to make pressed pellets (Govil 1985). A m-diameter pel pellet was obtained by pressing 1 gramme of the finely powdered sample on top of the boric acid in these cups, which were then filled with boric acid as a binder and compressed at 20 tonnes of pressure using a hydraulic press. Following the preparation of the pellets, the automatic loading system available at the X-RF Laboratory, Department of Geology, University of Madras will be used to measure different peaks and background counts for the major and trace elements using the X-RF Instruments Rigaku model S4 Pioneer sequential wavelength-dispersive X-ray spectrometer equipped with a goniometer (which holds seven analysing crystals) with a 4 kWh X-ray tube and 10 samples. The supporting software used for computer analyses can handle matrix effects, dead time correction,

Fig. 10.2 Highly fractured and jointed mafic dyke

Fig. 10.3 Boulders (mafic dyke) in the study area

background and line overlap corrections. After using calibration curves to convert counts into concentration, the output is given directly as the concentration in weight percentage or ppm. The XRF intensities measured from elements in dry samples vary according to the XRF absorption rates (Tertian and Claisse 1982) (Fig. 10.4).

Fig. 10.4 Photomicrographs showing main characteristic intersertal textures of Kunnam dolerite. The scale bar equals 500 μm. pl = Plagioclase, olv = olivine, qtz = quartz, Aug = Augite, opq = opaques. **a** Showing a sub-ophitic texture of a dolerite rock in which the intergrowth of plagioclase and augite are seen. **b** Plagioclase feldspar surrounded by quartz and olivine crystals. **c** The intergrowth of Plagioclase in the olivine crystal. **d** The larger grains of plagioclase feldspar showing polysynthetic twinning with the opaque mineral

10.2 Petrography

The petrographical investigation was carried out using a Leica (Model DM750) microscope, and all thin-section microphotographs were captured for further examination. All of the samples were inspected under a microscope to determine their mineralogical composition. The research area's mafic dyke is composed of medium- to coarse-crystal dolerite rock. Plagioclase feldspar, pyroxene, and several opaque minerals were discovered in the rock sample. There is a substantial proportion of olivine and quartz in the rock. In modest numbers, the rock also contained accessory minerals such as the amphibole group of minerals and sheets of mica biotite. These have a sub-ophitic to ophitic texture with aguite and plagioclase feldspar intergrowth. The plagioclase in the rock-thin part is labradorite, and some examples have the appearance of bytownite crystals. The optical features of the aforesaid intermediate, calcic plagioclase, confirm its existence. Thin parts of these crystals display lamellar twinning. Plagioclase feldspar appears to be prevalent in many rock samples, accounting for approximately 50% of the rock. Plagioclase crystals are discovered to be phenocrysts in the ophitic matrix in certain other samples. Clinopyroxene, particularly augite, was among the pyroxene minerals discovered in the rock. Aguite minerals are common in all dolerite rocks. The augite was pale brown in plane-polarised light, with pleochroism of pale brown to light brown colour shifts throughout. The relief is rated as low to medium. In terms of cleavage, practically all of the rock samples include an augite crystal with a parallel cleavage. The augite is vivid blue in crossed polar and has anisotropic properties. It has biaxial optical properties and goes extinct at an angle of 35 to 40 degrees, indicating inclined extinction. The rock contains more than 30% augite, which dominates after the plagioclase feldspar. The majority of the augite crystals are sub-hedral in shape, with some exhibiting modest modification at the grain margins, resulting in anhedral crystals. Magnetite and titano magnetite are typical opaque minerals found in rock. Its colour is reddish-brown or dark. They are found as subhedral to anhedral crystals. The presence of microfractures within the grains and the high relief of the olivine in the samples helped to identify it. Olivine is typically pale yellow in plane-polarised light and vivid green in crossed-polarised light. The cleavages are not visible in nearly all of the samples. The crystal turns extinct at zero degrees due to a parallel extinction. After augite, olivine crystals occupy 12.00% to 15.00% of the rock. There is a substantial amount of quartz crystals after the olivine present in the samples. The anhedral shape and undulatory extinction of quartz crystals distinguish them. Conchoidal fractures and low relief are visible in the quartz crystals. The fracture generated by recrystallisation causes a colour play in some samples of quartz crystal. The emplacement of the dykes was probably aided by plume uplift-assisted crustal extension and associated fracturing to develop a dyke swarm along the direction of plate movement over the active mantle plume head, Jayabalan et al. (2012).

10.3 Geochemistry

According to the major oxide geochemical investigation results, the mafic dyke rock exhibits a diversity of compositions. For the main oxide analysis, 10 unmodified fresh samples were chosen (Table 10.1). The dolerite rock from the research area is acidic, with a silica level ranging from 50.36% to 51.63%, with an average of 50.78%. Alumina content ranges from 16.02% to 17.22%, with an average of 16.75% and a range of 16.02% to 17.22%. Fe_2O_3 has a range of 11.94% to 13.08%, with a mean of 12.48%. The percentage of magnesia (MgO) in the rock ranges from 4.04% to 4.78%, with an average of 4.53%. The high ferromagnesian components are consistent with the rock's melenocratic origin and rich mafic minerals. The rock's CaO content ranges from 2.63% to 3.15%, with an average of 2.92%. Following a thorough study of these samples, the potted Harker's variation diagram (Fig. 10.5a–d and e–h) was used to observe the crystallisation behaviour of the current sample. The majority of the oxides in this variation diagram have a negative connection with SiO_2. With a modest enrichment in rock with SiO_2, alumina exhibits a negative tendency against silica, and comparable patterns are observed for FeO and MgO. CaO exhibits significant scatter due to large variation with silica, and similar scatter is also observed in the case of soda, however its trend is negative. The MgO versus silica trend indicates a considerable decrease in MgO content at low silica content and is relatively stable in silica-rich rock. The rock is further classified using a total alkali versus silica (TAS) diagram. According to the foregoing classification, the investigated rock is a basalt (Fig. 10.6a). The chemical classification and nomenclature of plutonic rocks using the total alkalis versus silica (TAS) diagram (Cox et al. 1979) indicate that the rock is Gabbro (Fig. 10.6b). The SiO_2 versus K_2O plot (Fig. 10.7a) depicts the igneous genesis of the Karur-Kangayam belt rocks. The diagram depicts Le Maitre et al. (1989) subdivision (broken lines with nomenclature in italics) and Rickwood (1989) subdivision (nomenclature in brackets). As described by Rickwood (1989), the bands are the fields in which the boundary lines of Peccerillo and Taylore (1976), Ewart (1982), Innocenti et al. (1982), Carr (1985) and Middlemost (1985) fall. The rocks that belong to the tholeiitic and calc-alkaline series. The calc-alkaline series includes dolerite rocks with a medium quantity of potash. The dolerite dyke from the Kunnam region rocks of igneous origin plots in the fields of volcanics of the tholeiitic and calc-alkaline series on the SiO_2–K_2O/Na_2O diagram (Roser and Korsch 1986) (Fig. 3.5) for the determination of the tectonic settings of the Kunnam dolerite. All the figurative composition points of Kunnam dolerite fall in the Oceanic Island arc. Furthermore, the weathering trend (Na_2O + CaO)–Al_2O_3–K_2O diagram of Nesbitt and Young (1984, 1989) for the examined rock (Fig. 10.8) reveals a limited weathering tendency. The dolerites emplacement may have been favoured by a thinner continental crust in extension after a regional uplift, in an active continental margin-type environment, where early rifting led to the formation of the marginal basin, Mey Atour et al. (2021).

10.4 Conclusion

The current study of Kunnam region mafic dykes classed them as subalkalic basalts based on the Calc-alkaline series fractionation pattern seen in the TAS (after Le Martire et al.) and K_2O versus Silica diagram. Petrographic studies show that the rock is made up of major minerals such plagioclase feldspar (Labradorite), clino-pyroxene (Augite) and opaques, as well as minor minerals like olivine and quartz. The rock has an ophitic texture, which is typical of dolerite found in rocks from various locations. The whole-rock geochemical investigation of the mafic dyke reveals that the majority of the oxides in Harker's variation diagram have a negative association with SiO_2, indicating that the dyke body is mafic. The primary oxide analysis reveals a large amount of Fe_2O_3 and MgO, and these high ferromagnesian elements are consistent with the rock's melenocratic character and high mafic minerals. When compared to other oxides, the amount of SiO_2 and Al_2O_3 is larger; these oxides contain a significant number of quartz and feldspar crystals, causing the mineralogical composition of the mafic dyke in the Kunnam area to fluctuate. According to the ACNK diagram, the rocks in the Kunnam region have a low weathering tendency. It

Table 10.1 Whole-rock geochemical analysis of Kunnam mafic Dolerites.

Components	KUN1	KUN2	KUN3	KUN4	KUN5	KUN6	KUN7	KUN8	KUN9	KUN10
Major oxides										
SiO_2	50.77	51.63	50.89	50.55	50.53	50.56	50.92	50.36	50.67	50.88
TiO_2	1.1	1.02	0.94	1.07	0.9	0.88	0.93	1	0.9	0.89
Al_2O_3	16.27	16.03	16.95	17.02	16.91	16.55	16.99	16.64	17.23	17.54
Fe_2O_3	12.86	12.22	12.62	12.53	12.37	11.34	12.27	13.08	11.94	10.95
MnO	0.2	0.17	0.2	0.2	0.19	0.18	0.21	0.17	0.15	0.17
MgO	4.59	4.23	4.05	4.65	4.74	4.47	4.49	4.75	4.79	4.41
CaO	9.87	10.09	9.95	9.23	9.72	9.82	9.66	9.57	9.75	10.05
Na_2O	2.8	2.64	2.74	3.13	3.08	3.14	2.94	2.87	3.15	2.91
K_2O	0.93	1.29	0.95	1.07	0.96	0.9	0.94	0.93	0.92	0.94
LOI	0.95	0.98	0.97	0.94	0.95	0.96	0.95	0.98	0.95	0.94
Total	99.38	99.32	99.29	99.46	99.4	98.8	99.35	99.37	99.5	99.68
Na_2O + CaO	12.67	12.73	12.69	12.36	12.81	12.96	12.61	12.45	12.9	12.96
Na_2O + K_2O	3.73	3.93	3.69	4.2	4.05	4.04	3.88	3.8	4.08	3.85
Trace elements										
NiO	231	0	180	0	140	0	0	210	147	106
CuO	274	227	216	267	184	190	249	224	172	161
ZnO	168	128	139	116	160	153	0	146	137	116
SrO	149	192	170	159	198	152	195	180	173	206
ZrO_2	95	103	118	83	131	127	87	104	62	98
WO_3	0	1032	531	0	462	0	595	0	0	277

is concluded that the Kunnam region's mafic dykes are made up of calc-alkaline dykes that intruded during the Archaean continental crust's passive rifting. The economic value of these dykes from the locations of Kunnan, its first rank variety of granite in the classification of petrography and geochemistry. The mafic minerals from the dolerite bodies mostly from pyroxene group of minerals and accessories of feldspar, quartz and mica group of mineral, the economic grade of this doleritic granite is black in colour the mafic >75% of minerals from this rocks. The discriminate composition diagram (SiO_2–K_2O/Na_2O diagram) for determining tectonic settings for Kunnam region rocks reveals that all of the rocks developed in the oceanic island arc boundary.

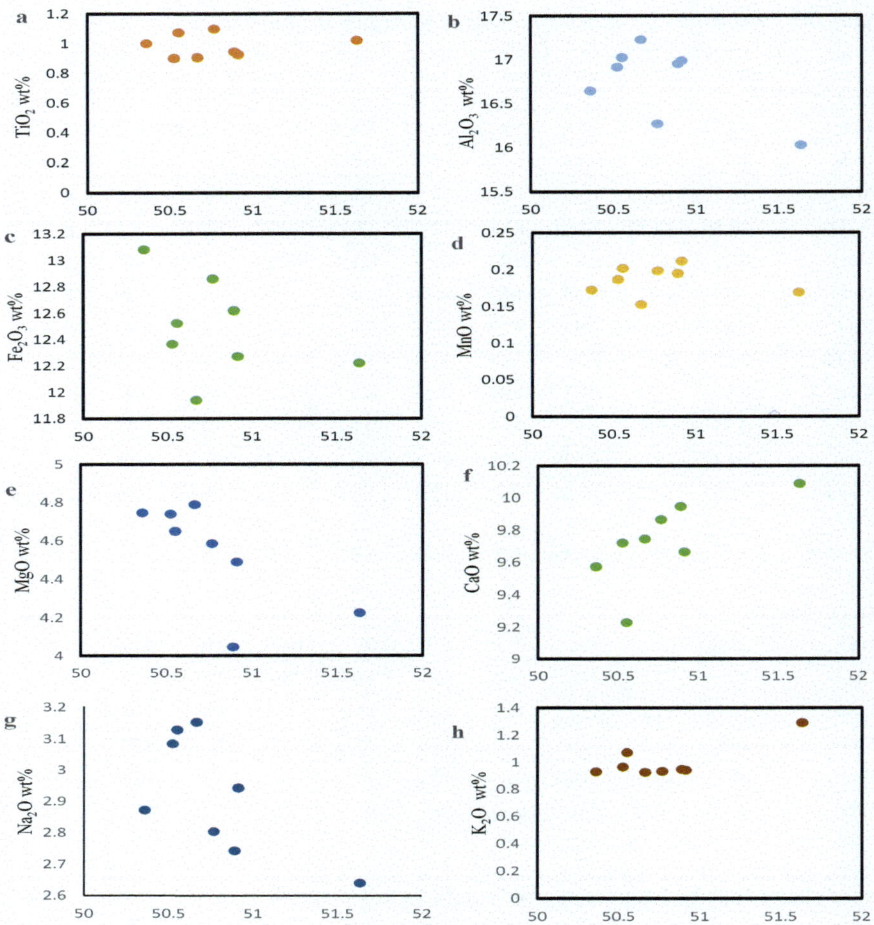

Fig. 10.5 a–h Harker's variation diagram for dolerite dyke of Kunnam area

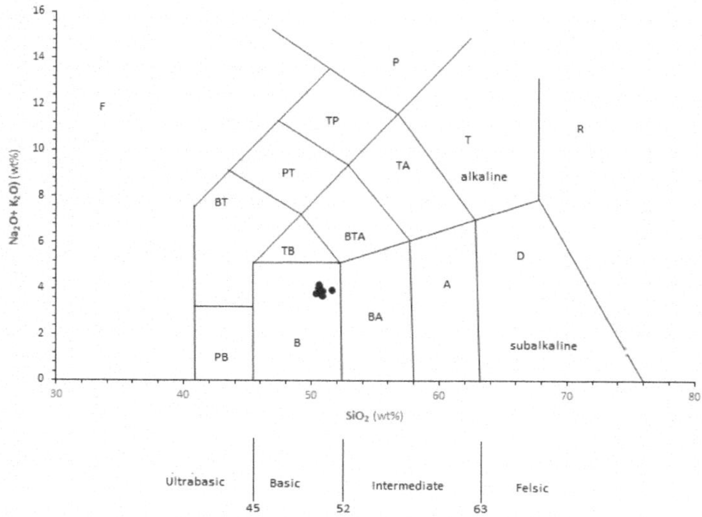

(a)

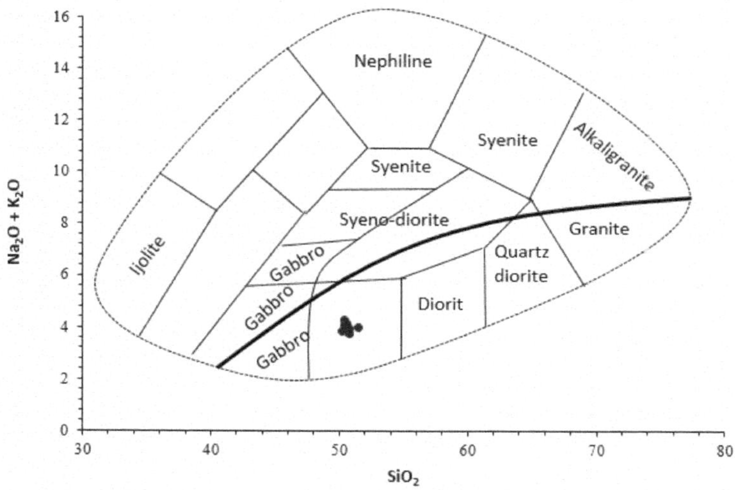

(b)

Fig. 10.6 **a** TAS diagram for the dolerite dyke (after Le Maitre 1989) of Kunnam area. PB: Pico basalt, B: Basalt and Basanite, TB: Trachybasalt, BTA: Basaltic trachyandesite, BA: Basaltic andesite, A: Andesite, D: Dacite, TA: Trachyandesite, T: Trachyte, R: Rhyolite, P: Phonolite, TP: Tephriphonolite, PT: Phonotephrite and F: Foidite. **b** The chemical classification and nomenclature of plutonic rocks using the total alkalis versus silica (TAS) diagram of Cox et al. (1979) adapted by Wilson (1989) for plutonic rocks. The curved solid line subdivides the alkali from subalkalic rocks

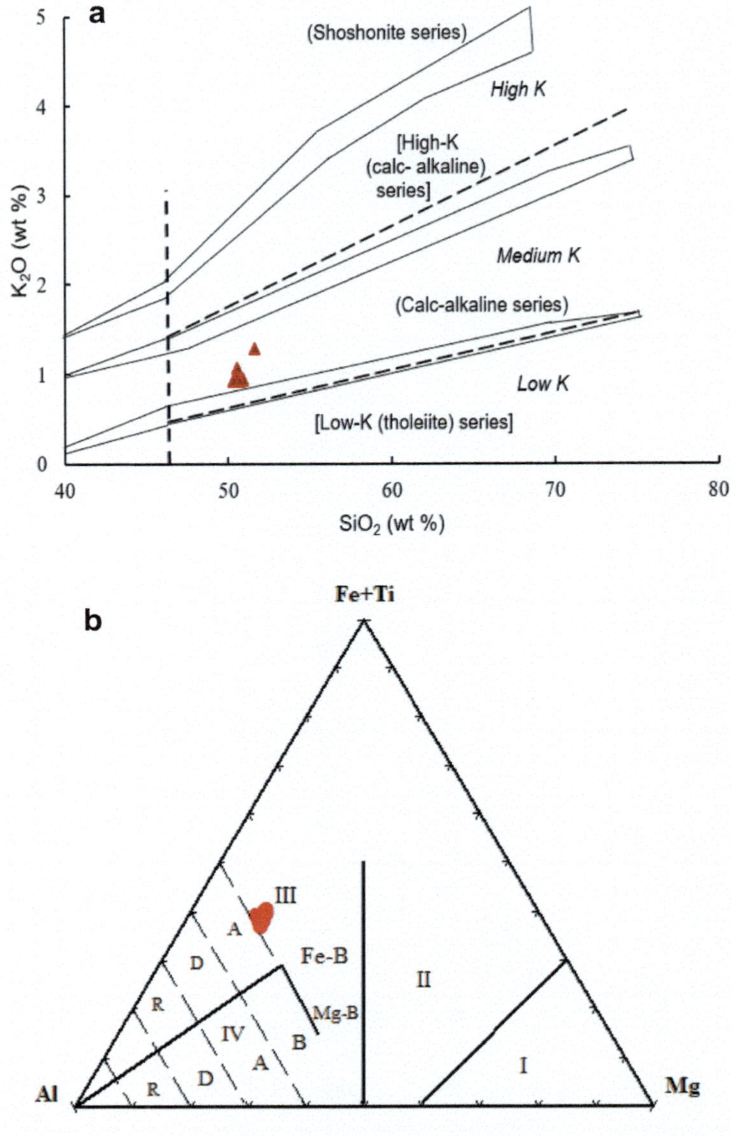

Fig. 10.7 **a** The subdivision of subalkalic rocks using the K_2O versus silica diagram for Kunnam dolerite dyke. The diagram shows the subdivision of Le Maitre et al. (1989) (broken lines with nomenclature in italics) and of Rickwood (1989) (nomenclature in parentheses). The bands are the fields in which fall the boundary lines of Peccerillo and Taylore (1976), Ewart (1982), Innocenti et al. (1982), Carr (1985) and Middlemost (1985) as summarised by Rickwood (1989). **b** Al–(Fe + Ti)–Mg classification plot (L. S. Jensen). The plots denote the composition of Kunnam dolerite dyke. The Roman numerals denote the fields of (I) komatiites; (II) komatiitic and high Mg basalts and (III–IV) volcanics of the tholeiitic and calc-alkaline series, respectively. The letters denote the field of (R) rhyolites, (D) dacites, (A) andesites and (B) basalts (FeB and MgB are high FE and high Mg basalts, respectively)

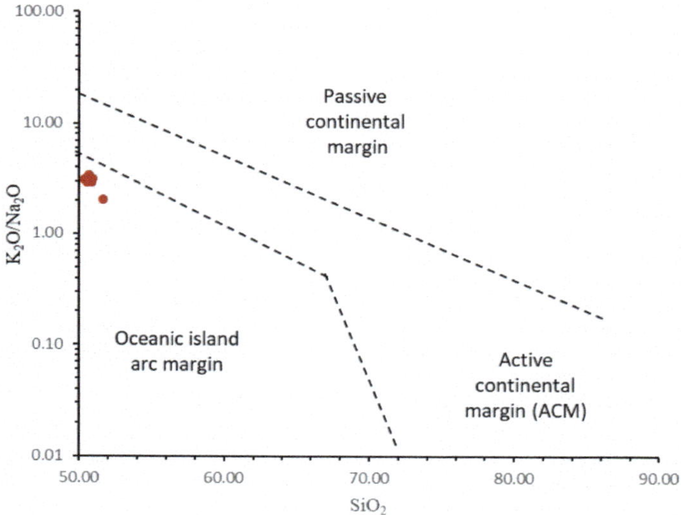

Fig. 10.8 Discrimination composition diagram for rocks of the dolerite dyke of Kunnam region. SiO_2–K_2O/Na_2O diagram (Roser and Korsch 1986) for determination of tectonic settings of the rocks in the Karur-Kangayam belt

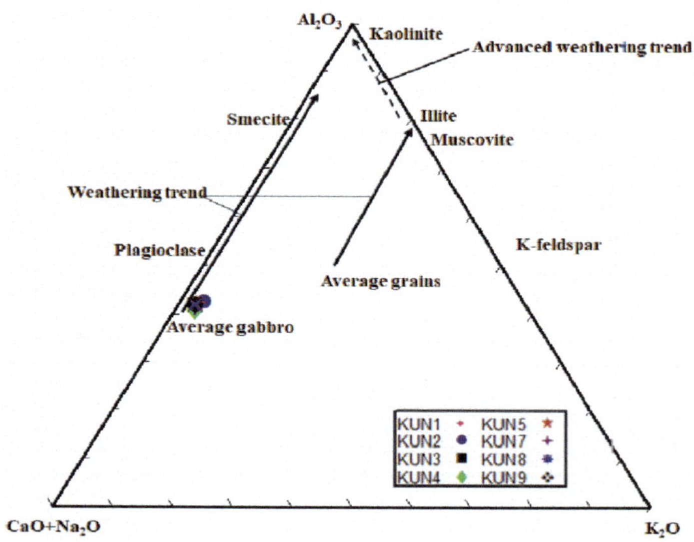

Fig. 10.9 The (Na_2O + CaO)–Al_2O_3–K_2O diagram of Nesbitt and Young (1984, 1989) showing the weathering trend for average granite and average gabbro. The advanced weathering trend for granite is also shown. Compositions are plotted as molar proportions and the composition of plagioclase, K-feldspar, muscovite and kaolinite are shown. CaO represents the CaO* associated with the silicate fraction of the sample

References

Bartlet JM, Harris NBN, Hawkesworth CJ, Santosh M (1995) New isotope constrains on the crustal evolution of South India and Pan-African granulites metamorphism. Mem Geol Soc India 34:391–397

Carr PF (1985) Geochemistry of late Permian shoshonitic lavas from the southern Sydney Basin. In: Sutherland FL, Franklin BJ, Waltho AE (eds) Volcanism in Eastern Australia. Geol. Soc. Aust. N.S.W. Div. Publ. No. 1, pp165–186

Cenki B, Barun I, Broker M (2004) Evolution of the continental crust in the pan African mobile belt of southern most India: evidence from Nd isotope mapping combined with U-P and Rb-Sr geochronology. Precambrian Res 134:1–20

Cox, K.G., J.D. Bell and R.J. Pankhurst, (1979). The interpretation of igneous rocks. George, Allen and Unwin, London.

Crook Shank H (1963) Geology of southern Nastar and Jaypore from the Bailadila range of Eastern Ghats. Mem Geol Surv India 87:150

Devaraju TC (1995) Dyke swarms of Peninsular India. Geol Soc Ind Memoir 33:451

Ernst RE, Srivastava RK (2008) India's place in proterozoic world; constrains from the large igneous province (LIP) record

Evans ME (1968) Magnetization of dikes: a study of the paleomagnetism of the Widgiemooltha dike suite, Western Australia. J Geophys Res 73:3261–3270

Evans T (1999) Extent and nature of the 1.2 Ga Wheatbelt dyke swarm, Yilgarn Craton, Western Australia. B.Sc. thesis, Univ. West. Aust. Perth

Ewart A (1982) The mineralogy and petrology of Tertiary-Recent orogenic vocanic rocks with special reference to andesitic-basaltic composition range. In: Thorpe RS (ed) Andesites. Wiley, Chichester, pp 25–87

Fermer LL (1936) An attempt at the correlation of the ancient schistose formation of peninsular India. Mem Geol Surv India 70:1–52

French JE, Heaman LM, Chacko T, Srivastava RK (2008) 1891–1883 Ma Southern Bastar Cuddapah mafic igneous events, India; A newly ecognized large igneous province. Precambrian Res 160:308–322

French JE, Heman LM (2010) Precise U-Pb dating of paleoproterozoic mafic duke swarms of the dharwar craton, India. Implications for the existence of the neoarchean supercraton Sclavia. Precambraian Res J, 2010.05.003

Govil PK (1985) X-ray fluorescence analysis of major, minor and selected trace elements in new IWG reference rock samples. J Geol Soc India 26:38–42

Halls HC, Kumar A, Srinivasan R, Hamilton MA (2007) Paleomagnetism and U- Pb geochronology of eastern trending dykes in the Dharwar craton, India; Feldspar clouding, radiating dyke swarms and the position of India at 2.3 Ga. Precamb. Res. 155:47–68

Harris NBW, Santosh M, Taylor PN (1994) Crustal evolution in south India; constraints from Nd isotoe. J Geol 102:139–150

Innocenti F, Manetti P, Mazzuuoli R, Pasquare G, Villari. (1982) Anatolia and north-western Iran. In: Thorpe RS (ed) Andesites. Wiley, Chichester, pp 327–349

Jayananda M, Janardhanan AS, Sivasubramanian P, Peucat JJ (1995) Geochronologic and isotopic constraints on granulite formation of the Kodaikanal area, Southern India. Mem Geol Soc India 34:373–390

Jayananda M, Peucaut JJ (1996) Geochronological framework of Southern India. Gondwana Res. Mem. 3:53–75

Jayabalan M, Umamaheswaran G, Suresh A (2012) Petrology and geochemistry of dolerite dykes of Dharmapuri and Salem districts of Tamil Nadu. J Appl Geochem 14(1):52–68

Le Maitre RW, Bateman P, Dudek A, Keller J, Lameyre Le Bas MJ, Sabine PA, Sehmid R, Sorensen H, Streckeisen A, Woolley AR, Zanettin B (1989) A classification of igneous rocks and glossary of terms. Blackwell, Oxford

Li Z-X, Evans DAD (2011) Late Neoproterozoic 40 intraplate rotation within Australia allows for a tighter-fitting and longer-lasting Rodinia. Geology 39:39–42

Li ZX, Evans DAD, Zhang S (2004) A 90° spin on Rodinia: possible causal links between the Neoproterozoic supercontinent, super plume, true polar wander and low-latitude glaciation. Earth Planet Sci Lett 220:409–421. https://doi.org/10.1016/s0012-821x(04)00064-0

Li ZX, Zhang L, Powell CM (1996) Positions of the East Asian cratons in the Neoproterozoic supercontinent Rodinia. Aust J Earth Sci 43:593–604

Liu Y, Li Z-X, Pisarevsky SA, Kirscher U, Mitchell RN, Stark JC (2018) Palaeomagnetism of the 1.89 Ga Boonadgin dykes of the Yilgarn Craton: Possible connection with India. Precambrian Res. https://doi.org/10.1016/j.precamres.2018.05.021

Mahadevan TM (1998) Aunitary model of evolution of the precambrian Indian shield. In: Inter. Sympo on Charnockite and granulite facies rocks. Geol. Assoc. Tamil Nadu, India, pp 153–174

Marzoli A, Chiaradia M, Jourdan F, Bussy F (2006) Parental magmas and crustal contamination of continental tholeiitic basalts from the Central Atlantic magmatic province as revealed by mineral major and trace elements and Sr isotopes. In: Goldschmidt conference abstracts.https://doi.org/10.1016/j.gca.06.801

Mey AM, Fagny BJM, Mefire AN, O, F. Tchameni R. Dinamou A. (2021) Petrology and geochemical framework of dolerites dykes of Temte´. North Cameroon, Central Africa, Acta Geochim 40(3):287–299. https://doi.org/10.1007/s11631-020-00432-4

Middlemost EAK (1985) Magmas and magmatic rocks. Longman, London

Mishra, V.N., Rai, P.K., Singh, P. 2021. Geo-information Technology in Earth Resources Monitoring and Management (edit. Book), Nova Science Publishers, U.S.A., pp. 1–395. ISBN: 978-1-53619-669-6

Murthy NGK (1989). Mafic dyke swarms of the Indian shield. In: Halls Hc, Fahriig WF (eds) Mafic dyke swarms. Geol. Association. Canada. Spl. Pap., vol 34. pp 393–400

Murthy NGK (1995) Proterozoic mafic dykes I n southern peninsular India. A review. Geol Soc India Mem 33:81–98

Nesbitt HW, Young GM (1984) Prediction of some weathering trends of plutonic and volcanic rocks based on thermodynamic and kinetic considerations. Geochemica et Cosmochemica Acta 48:1523–1534

Nesbitt HW, Young GM (1989) Formation and diagenesis of weathering profiles. J Geol 97:129–147

Peccerillo R, Taylor SR (1976) Geochemistry of Eocene calc-alkaline volcanic rocks from the kastamonu area, northern Turkey. Contrib Mineral Petrol 58:63–81

Pesonen LJ, Elming S-Å, Mertanen S, Pisarevsky S, D'Agrella-Filho MS, Meert JG, Schmidt PW, Abrahamsen N, Bylund G (2003) Palaeomagnetic configuration of continents during the Proterozoic. Tectonophysics 375:289–324

Piispa EJ, Smirnov AV, Pesonen LJ, Lingadevaru M, Anantha Murthy KS, Devaraju TC (2011) An integrated study of Proterozoic Dykes Dharwar Craton, Southern India. pp. 33–45,https://doi.org/10.1007/978-3-642-12496-9_3

Pisarevsky S, De Waele B, Jones S, Söderlund U, Ernst RE (2015) Paleomagnetism and U-Pb age of the 2.4 Ga Erayinia mafic dykes in the southwestern Yilgarn, Western Australia: Paleogeographic and geodynamic implications. Precambrian Res 259:222–231

Radhakrishna BP (1989) Suspect tectono stratigraphic terrane elements in the Indian subcontinent. J Geol Soc India 34:1–24

Radhakrishna T, Joseph M (1993) Proterozoic paleomagnetism of the south India shield and tectonic constraints. Geol Soc Ind Memoir 26:321–336

Radhakrishna T (2007) Temporal variations and mantle sources of the mafic dyke swarms in South India Deep Continental Studies. News Lett 17:2–7

Ramakrishnan M (1990) Crustal development in Southern Bastar Central Indian craton. Geol Surv India Spl 28:44–66

Rickwood PC (1989) Boundary lines within petrologic diagrams which use oxides of major and minor elements. Lithos 22:247–263

Roser BP, Korsch RJ (1986) Determination of tectonic setting of sand-stone–mudstone suites using SiO2 content and K2O/Na2O ratio. J Geol 94(5):635–650

Rai PK, Mishra VN, Singh P (2021) recent technologies for disaster management & risk reduction-sustainable community resilience & responses (edit. Book), Springer Nature, Switzerland, pp 1–471. ISBN: 978-3-030-76116-5. https://doi.org/10.1007/978-3-030-76116-5

Rai PK, Mishra VN, Singh P (2022) Geospatial technology for landscape and environment management: sustainable assessment & planning (edit. Book). Springer Nature, Singapore. pp 1–431. ISBN: 978-981-16-7373-3. https://doi.org/10.1007/978-981-16-7373-3

Rai PK (2023a) River conservation and water resource management. Springer Nature, Singapore. pp 1–352. ISBN No.: 978-981-99-2604-6. eBook ISBN: 978-981-99-2605-3

Rai PK (2023b) Advances in water resource planning & sustainability. Springer Nature, Singapore. pp 1–322. ISBN No. 978-981-99-3659-5

Selvaraj B, Prabhakaran R, Vasudevan S and Kumar R S. (2012). Petrology and geochemistry of mafic dykes from Villupuram district Southern granulite terrain of Tamil Nadu, India. INT J CURR SCI 2012, v.4: pp.75–82

Seymour KS, Kumarapeli PS (1995) Geochemistry of the Grenville Dyke Swarm: role of plume—source mantle in magma genesis. Contrib Mineral Petrol 120:29–41

Srivastava RK (2006) Precambrian mafic dyke swarms from the central bastar craton, temporal evolution of the subcontinental mantle. In: Hansi E, Mertanen S, Ramo T, Vulollo J (eds) Dyke swarms-time markers of crustal evolution. Taylor & Francis Group, London, UK, pp 147–159

Srivastava RK (2011a) Dyke Swarms: keys for geodynamic interpretation. Springer, Berlin

Srivastava RK (ed) (2011b) Dyke Swarms: Keys for Geodynamic interpretation. Springer, Berlin Heidelberg. pp 637

Srivastava RK, Sivaji C, Chalapathi Rao NV (2008) Indian dyke through space and time; retrospect and prospect. In: Srivastava RK, Sivaji C, Chalapthy Rao NV (eds) Indian dyke, Geochemistry, geophysics and geochronology. Narosa publishing House Pvt, Ltd, New Delhi, India, pp 1–18

Srivastava RK, Ahmad T (eds) (2008) Precambrian mafic magmatism in the Indian shield, Part 1, Geol. Soc. India. vol 72, no 1. pp 1–140

Srivastava RK, Ahmad T (eds) (2009) Precambrian mafic magmatism in the Indian shield, Part 2, Geol. Soc. India, vol 73, no 1. pp 1–54

Srivastava RK, Sivaji Ch, Chalapathi Rao NV (eds) (2008) Indian dykes: geochemistry, geophysics, and geochronology. Narosa Publishing House, New Delhi

Tarney J (1992) Geochemistry and significance of mafic dyke swarms in the Proterozoic. In: Condie KC (ed) Proterozoic crustal evolution. Elsevier, Amsterdam, pp 151–179

Tertian R, Claisse C (1982) Principles of quantitative X-ray fluorescence analysis. Heyden, London

Wilson M (1989) Igneous petrogenesis. Unwin Hyman, London

Wingate MTD, Evans DAD (2003) Palaeomagnetic constraints on the Proterozoic tectonic evolution of Australia. Geol Soc London Spec Publ 206:77–91

Worthing MA (2005) Petrology and geochronology of a Neoproterozoic dyke swarm from Marbat South Oman. J Afr Earth Sci 41:248–265

Chapter 11
Analyzing the Driving Factors Causing Urban Sprawl Using Logistic Regression Model in Greater Shillong Planning Area, India

Sanu Dolui and Sayani Chakraborty

Abstract In recent decades sprawling of urban areas in the outskirts has played a significant role in moulding the direction of urban growth. The urban sprawl that has been predominantly concentrated in agricultural lands and forest areas in contradiction of city development plan, has posed a serious threat to the Environmental protection. In order to address the obstacles of sustainable urbanization and comprehend the trajectories of future of urban growth it is imperative to determine the driving force behind sprawl. Shillong, a prominent hill town in north-eastern India has grown exponentially in last two decades towards its periphery and has been chosen as a case study to monitor landuse dynamics & uncover factors that contributing to urban sprawl. The recent urban expansion of shillong was mapped using Landsat images for the year 2000, 2014, 2022 and classified through Random Forest algorithms. Kappa statistics yielded a satisfactory result with overall accuracy at approximately 92.67%, 93.67%, and 97.33% respectively. Following that, a logistic regression model was used to evaluate the influence of twelve urban growth determinants identified through a review of relevant literature and consultation with local experts. The outcome of the study concluded that the distance from major roads, the distance from urban areas, and the distance from C.B.D areas are among the most significant determinants for urban expansion with coefficients values of – 0.83, – 0.86, and – 0.91, correspondingly. The findings indicate that there was an uninterrupted growth of urban land from 2000 to 2022, during this time periods there was a stretching of 1857.60 ha built-up land with an average annual rate of 3.29% and average annual intensity rate of 84.44 ha yr^{-1}, respectively. Present study highlight the growing concerns posed by urban sprawl, which has absorbed numerous peripheral villages into the boundaries of the Greater shillong planning areas.

S. Dolui (✉) · S. Chakraborty
Department of Geography, The University of Burdwan, Burdwan, West Bengal, India
e-mail: sanudolui902@gmail.com

S. Chakraborty
e-mail: sayanichakraborty266@gmail.com

P. K. Rai et al. (eds.), *Natural Resources Management and Sustainable Livelihoods in the Mountainous Region*, Advances in Geographical and Environmental Sciences,
https://doi.org/10.1007/978-981-97-2100-9_11

Keywords Urban sprawling · Greater shillong planning area · Landsat · Driving factors · Logistic regression

11.1 Introduction

The emerging nations of world experiencing frequent changes in LULC due to massive industrialization & phenomenal urban growth in recent decades (Belay and Mengistu 2019; Singh, Bhavna et al. 2022). This urban expansion, one of the prominent drivers of landuse change, has emerged as one of the development themes aimed at lifting people out of poverty (Allan et al. 2022; Arifeen et al. 2021; Nuissl et al. 2021). Despite the fact that rapid urbanization has triggered a considerable boost for the nation's socioeconomic development, the expansion of built-up areas has rendered the obvious conflict between development and conservation of nature (Li et al. 2022). The urban areas are expanding outwards as a result of greenfield area development planned by local governments & entrepreneurs, consuming valuable agricultural land & forest cover to meet urbanization demand (Aijaz 2019; Hutchings et al. 2022; Tyagi et al. 2014). Expansion of impervious surfaces particularly in outskirts & semi-urban areas where industrial regulation, building codes and regulatory were poorly drafted exposed higher risks for this unplanned urban development. This scenario is more critical in hill town where intense shortage of horizontal space & high land value in city proper force the people to cater unauthorized housing development in hill slope & highly vulnerable areas. (Kumar 2015). The growing concentration of people and assets in hill areas increased disaster risk. Specifically, the marginalized people tend to dwell in sensitive hill landscapes are susceptible to various natural hazards such as earthquakes, landslides, and floods (Kumar 2018). Over the past few decades landuse in urban areas has been profoundly altered by interweaving of many physical and socio-economic circumstances (Punia and Singh 2012; Mariye et al. 2022). Urban planners and local government officials need to be regularly updated information on urban expansion to ensure smooth access to municipal services and timely installation of infrastructural facilities in urban areas (Maheshwari et al. 2016). This haphazard and uncontrolled growth has causing sprawling of urban settlements, the loss of forest and lucrative agricultural land (Franco et al. 2017; Cheng 2022), and the deterioration and segmentation of remaining natural areas (Munshi et al. 2014; Miller 2012; Degife 2019). The problem of urban sprawl is one of the most frequently debated topics among decision-makers owing to the spatial expansion of this urban fabric, a higher impact has been exerted on the natural landscape (Osman, Taher et al. 2016). Several authors asserted that urban sprawls began to develop on the outskirts of major cities through the extension of commercial and industrial development, which was then followed by a sizable unchecked urban growth with poor services and accessibility (Resnik, David 2010; Zhang, Mingdou et al. 2022; Shi, Yishao et al. 2023). Therefore, to thwart these development impulses, monitoring LULC changes and identification the drivers of urban sprawls has become the central theme of landuse planner in contemporary studies (Jokar Arsanjani et al. 2013).

The process of urban sprawl has become exceedingly complicated in which numerous factors influence its direction & intensity of sprawling (Hamdy, Omar et al. 2016). Therefore, a strong mathematical framework needs to be applied in order to comprehend the future direction of urban sprawl and reroute in the appropriate direction. In recent years, satellite imagery and remote sensing techniques have emerged as indispensable tools in denoting factors of urban sprawl (Dadras et al. 2013; Mostafa, Eman et al. 2021; Behnisch et al. 2022; Olorunfemi et al. 2020). Quick & Real-time monitoring of large scale landuse data make satellite image based remote sensing technology more powerful & popular. Multitemporal remote sensing has been demonstrated to be an effective approach for extracting data pertaining to urban sprawl, offering a spatially consistent repeated perspective of frequent changes in impervious surfaces (Leinenkugel, Patrick et al. 2019; Falanga Bolognesi, Salvatore, et al. 2020; Rai et al. 2023a and 2023b). The LULC change classification assists local authorities in understanding the dynamics of the natural environment in order foster long-term development (Baig et al. 2022; Mwabumba et al. 2022). Replenishment of tradition image classification algorithms with machine learning & deep learning approach empower us with improving the accuracy of image classification (Cravero et al. 2022) in present study random forest algorithms have been used to monitor LULC changes. The ensemble learning technique used by Random Forest is based on the bagging algorithm. The random forest algorithm constructs a number of different decision trees and then combines the results of those trees in order to provide a prediction that is both more accurate and more reliable. Random forest one of stable algorithms, comparatively less impacted by noise, that is usually robust to outliers and can handle missing values automatically.

Urban growth models are an effective instrument for gaining knowledge of urban expansion and its impact on the ambient environment (Rai et al. 2021; Rai et al. 2022; Gharaibeh, Anne, et al. 2023). Numerous researchers and academics have applied a wide variety of approaches in order to forecast and simulate the growth of urban areas. Modelling the probability of urban expansion is another important topic that has received attention in recent decades due to its importance in future landuse decisions. Particularly in developing nations, there has been a growing interest in determining the repercussions of sprawl as well as reason that driving it, with the goal to establish sustainable urban plans and administration policies (Wu, Yizhou et al. 2017). To make precise forecasts regarding future urban growth, it is extremely vital to identify and investigate the relationship between urban growth and the forces that drive it. Some of the important studies on simulating the factors of urban sprawling already been conducted in India (Padmanaban et al. 2017; Vinayak et al. 2021; Hatab et al 2022; Seevarethnam, Mathanraj et al. 2022; Theres, Linda et al. 2023) and other parts of the world as well (Iizuka, Kotaro et al. 2017; Seevarethnam et al. 2022; Asadi, Milad et al. 2022). Determining and analyzing the factors that propel the urban expansion is crucial to understanding the spatiotemporal dynamics of urban growth (Kim, Youjung et al. 2020). Thus, modelling of urban sprawl is essential for the advanced planning of unregulated urban growth and for rerouting this urban growth in the appropriate directions (Lambin and Geist 2006; Lin, Yong et al. 2021). A number of methods have been used to explore the relationship driving variables and urban sprawling including Artificial intelligence algorithms (Maithani 2009) Multiple linear regression (MLR) (Seto, Karen et al. 2011), bivariate regression (Wu and Zhang 2012), analytic hierarchy process (AHP) (Thapa and Murayama 2010), ordinary least square (OLS) model (Wang et al. 2020) regression and Seasonal ARIMA (Malarvizhi et al. 2022). In this study applied Logistic regression model was used to explain correlations between independent (exogenous) and dependent (endogenous) variables due to its certain advantage over others. Logistic regression model one of most popular method for identifying factors and their effects, capable of handling regression problems and require no linear relationship between dependent and independent variables. Logistic regression models provide us a solid comprehension of the urbanization procedure and presents an unambiguous depiction of the relative importance of the factors & their roles that those variables play. The logistic regression model can linearly integrate several physical and socioeconomic driving variables and producing a single probability map those capabilities are missing from many other models. Additional features, such as the advantage of spatial dimension, correlation between variables, and the ability to handle database heterogeneity, make it a robust modelling technique (Jokar Arsanjani et al. 2013; Devkota et al. 2013). A number of studies across the globe already been conducted with logistic regression, Tehran, Iran (Arsanjani 2013), Tripoli Metropolitan City (Libya), (Alsharif et al. 2014), Five Saudi Arabian Cities (Alqurashi, Abdullah et al. 2016), Abouelreesh, Aswan (Hamdy, Omar et al. 2016), Srilanka (Weerakoon 2017), Chennai (Padmanaban, Rajchandar et al. 2017), Greater Cairo Region, (Salem et al. 2019; Osman, Taher et al. 2016), Ghana (Asempah, Mawuli et al. 2021), Estonia (Mozaffaree Pour et al. 2022), Chengdu Metropolitan Area, China (Xia, Xiaojiang et al. 2022), Delhi (Salem, Muhammad, et al. 2021) respectively.

Present study area Shillong city, grown substantially over the last two decades due commercial activity and losing its natural area to the grasp of an unsustainable and ever-expanding urbanization (Parida et al 2022). But this city lies on the Shillong Plateau (SP) susceptible to multiple disasters such as landslides, urban floods and earthquakes, leading to several urban development challenges. So, it's necessary to future Land use-zoning policy should highlight the hazard-susceptible areas and regulating activities while installation of new development activities. Therefore, this study is crucial for understanding factors of urban sprawls and to know what is the probability of urban expansion in near future. As in this study area detection the probability of urban expansion by Logistic regression model has yet not explore, this study envies the trustworthiness of logistic regression model for modelling urban sprawl. This study highlighted the combined use of Remote Sensing and GIS improves quantification, mapping, and our understanding the process of urban sprawling to ascertain the causes of LULCC.

This study basically conducting to explore the factors of urban sprawl in the Shillong city. Specially, the purposes of this study are to: (i) identify the LULC change in the city by using multitemporal Landsat imagery (2000–2022) (ii) (iii) To determine all those factors that could potentially influence urban sprawl during the period of 2000– 2022 (iii) prepared the urban probability map of shillong city with logistic regression. The results of logistic regression may be used instantaneously to gain a better understanding of the possibilities associated with sprawl and it is an effective way to handle binary dependent variables.

11.2 Study Areas

Shillong, the capital of the State of Meghalaya, lies right in the centre of the Shillong Plateau the with an average altitude of 916–1960 m. The Study area Greater Shillong planning areas (SGPA) encompass an area of 187.85 km^2 including Umiam lake areas (Mipun et al. 2015). As per census of 2011, the total population SGPA 4, 16, 118 that is anticipated to increase around 5 lakhs in 2022. This city is bordered by hills, forest areas, whereas Shillong peak(1960 m) being the highest peak in the Meghalaya State. In the last two decades fast expansion of shillong city in its peripheral villages, led to creation of small urban units *(Lawsohtun, Madanrting, Mawlai, Nongkseh, Nongmynsong, Umlyngka, and Umpling*) those urban areas have now become of part of shillong Urban agglomeration. The temperature of shillong is decent and comfortable throughout the year. Summer are moderate warm, average summer temperature remains around 23 °C whereas winter is freezing cools with average temperature remains 4 °C. Due to its location in the rain shadow region of the Shillong plateau, Shillong receives a moderate amount of rainfall, which is less than the surrounding areas. This area received 2,250.4 mm rainfall annually, most of downpour got during months of June to October (south-west moon). The city of Shillong serves as a transportation hub, as well as a center for business and education, for the neighbouring states and nations of Assam, Manipur, as well as the countries of Bangladesh and Myanmar. This has led to an extremely rapid and unsustainable expansion of the city, which has resulted in the construction of a number of protruding buildings, including residential and commercial hubs and markets, but without adequate planning and seismic design. This expansion of urban sprawl driven by rapid population growth & infrastructural developments is responsible for the loss of fertile agricultural land and unspoiled forest in suburban regions in the city. This area belongs to seismically active zones, vulnerable to frequent earthquake & landslide hazard make it sensitive for construction of urban structure. So, it would be quite interesting to have an overview of urban growth in shillong town due to its sensitiveness & Vulnerability to multiple hazards (Fig. 11.1).

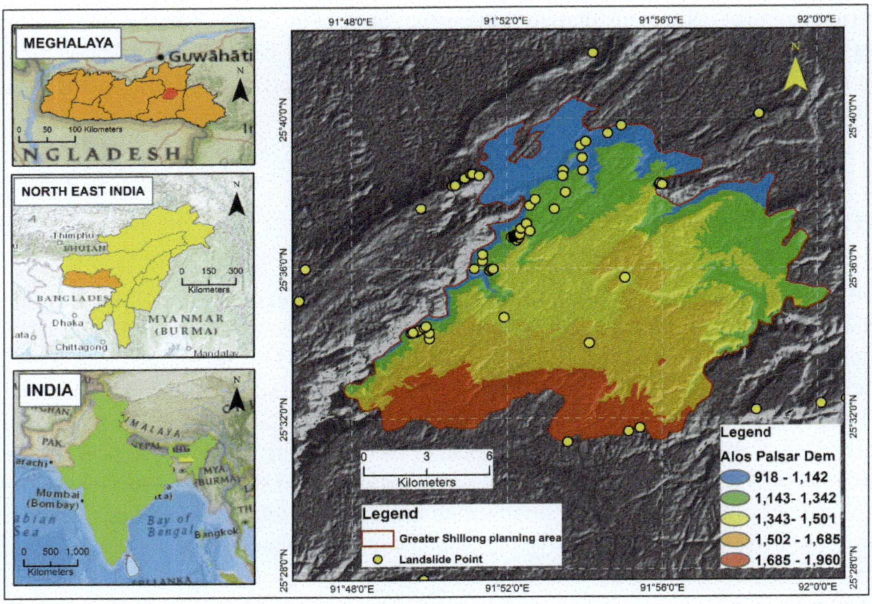

Fig. 11.1 Study area: Greater Shillong planning areas

11.3 Datasets & Methodology

11.3.1 Spatial Dataset

In analyzing land-use changes of shillong city, three sense of Landsat dataset (2000,2014,2022) were downloaded from USGS data portal (USGS, http://glovis.usgs.gov/). Satellite sensors ETM + (Landsat7), OLI-1 (Landsat8), and OLI-2 (Landsat9) were offering 15-m panchromatic images, have been chosen in this research because of their higher spatial, radiometric resolution and narrow bandwidth. To reduce cloud interference, image scenes were collected between November and February with less than 2% cloud cover. All maps in this study were geometrically rectified to a UTM projection systems (Zone 46 N, Datum: WGS-84). Remote sensing data seems susceptible to being influenced by Water vapour, moisture, nanoparticles and sun irradiation while transmitting data resulted in substantial variations that affected the desired outcome during data analysis. Therefore, this study applied FLAASH atmospheric model in ENVI software to correct each image for the purpose of eliminating the part of the effect caused by the atmosphere. Apart from that radiometric corrections such as radiance, reflectance, temperature brightness, haze reduction, noise reduction was performed for all the satellite images in the ENVI(5.4) software which is necessary before classification. Gram-Schmidt Pan Sharpening method have been applied to sharpen multispectral data with 15-m pan

Table 11.1 Information about the used Landsat images in the study

City Name	Satellite	Acquisition date	Sensor	Row/Path	Image Resolution	Cloud cover (%)
Greater shillong planning area, Meghalaya, India	Landsat7	2000/12/21	Enhance Thematic Mapper plus (ETM +)	Path:136 Row:42	15- meter	1.00
	Landsat8	2014/12/20	Operational Land Imager (OLI)	Path: 136 Row: 42	15-m	0.01
	Landsat9	2022/12/18	Operational Land Imager 2 (OLI2)	Path:136 Row:42	15-m	0.82

chromatic dataset. After all the necessary image correction the images were imported to SNAP software for Random Forest Classification (Table 11.1).

11.3.2 *Classification of Satellite Images Through Random Forest (RF)*

After preprocessing all the satellite images, images are imported to SNAP desktop software for classification in Random forest (RF) algorithm. Various studies previously documented the superiority of RF algorithm in classification of heterogenous dataset (Oo, Tin Ko et al. 2022; Talukdar, Swapan et al. 2020). Random Forest (RF), a novel non-parametric classifier-based on ensemble machine learning and classification trees algorithm. Random forest used two types of ensemble techniques, bagging and boosting. A number of decision trees are built throughout the training process as part of this learning technique and the outputs of the modal are anticipated by the individual trees. A number of advantages of RF algorithm including, promising to overcome the overfitting problem of decision tree, can handle large heterogenous dataset with efficacy, compute faster than other algorithms and producing higher accuracy while classifying make it a robust classifier algorithm. Following the class-I, classification system of Anderson (1976) & NRSC 2011 classification guidelines, the images were segmented into following five classes (i) Dense forest (ii) Scattered forest (iii) Agricultural land (iv) River/Water Bodies (V) Urban Area (Table 11.2). To classify the LULC categories, each classes was trained by a decent number of Spectral signatures and verified by the ground truth data. 300 Ground control points were generated purposefully through field Visit (G.P.S survey), Google earth pro (for inaccessible landform) and experience of local residents for unchanged Landscape for accuracy assessments. Standard kappa indices including Producer accuracy (PA),

Table 11.2 LULC classification scheme

Classes	Description
Dense forest (C1)	Dense continuous patches of forest, both natural and planted
Scattered forest (C2)	Degraded, Segmented patches
Agricultural land (C3)	Areas used for crop cultivation
River/Water Bodies (C4)	Permanent open water bodies, Rivers, Lakes, Ponds, Streams, and Canals
Urban areas (C5)	includes areas with all types of artificial surfaces, including residential, commercial, and industrial land uses as well as transportation infrastructure, Road

User accuracy (UA), Overall accuracy (OA) and kappa coefficient(κ) were calculated to validate the classification.

11.3.3 Framework of Model

In last two decades, Shillong one of major cities in north-eastern India has grown exponentially towards its periphery, has been taken as a case study to monitor landuse dynamics & identification factors of urban growth. Therefore, an urban planning strategy should be developed to manage urban growth in a sustainable manner, lessen the negative effects of urban sprawl, and prevent the city from spreading in an unplanned manner. The recent urban expansion of shillong was mapped using Landsat images for the year 2000, 2014, 2022. An application of a logistic regression model was made in this paper to investigate the urbanization process in Shillong, Meghalaya. The objective of the modelling is to uncover the association between urban growth and physical, socioeconomic, and vulnerability variables in order to forecast future urban patterns. After that, three periods of land cover change: 2000–2014, 2014–2022, and 2000- 2022 are extracted using a post classification comparison algorithm. The information gathered from the land cover change detection is then utilized in a subsequent step to determine the trend of urban growth and the corresponding area. Finally, to validate model trustworthiness, Receiver operating characteristic (ROC) statistics was employed. Overall methodology were described below, which outlines the phases of the framework of future urban sprawl prediction, which consists of four sections: (1) Analysis urban sprawling based on Landsat dataset (2000–2022); (2) identifying & explaining the importance of driving factors on urban growth; (3) Modelling urban sprawl using logistic regression (4) *Validate LR model with ROC curve*. In order to provide a concise summary of our research design, we have produced a diagrammatic depiction of the theoretical underpinnings of the methodology that was applied in this investigation (Fig. 11.2). This methodology will be discussed in further detail in the subsections that follow (Fig. 11.2).

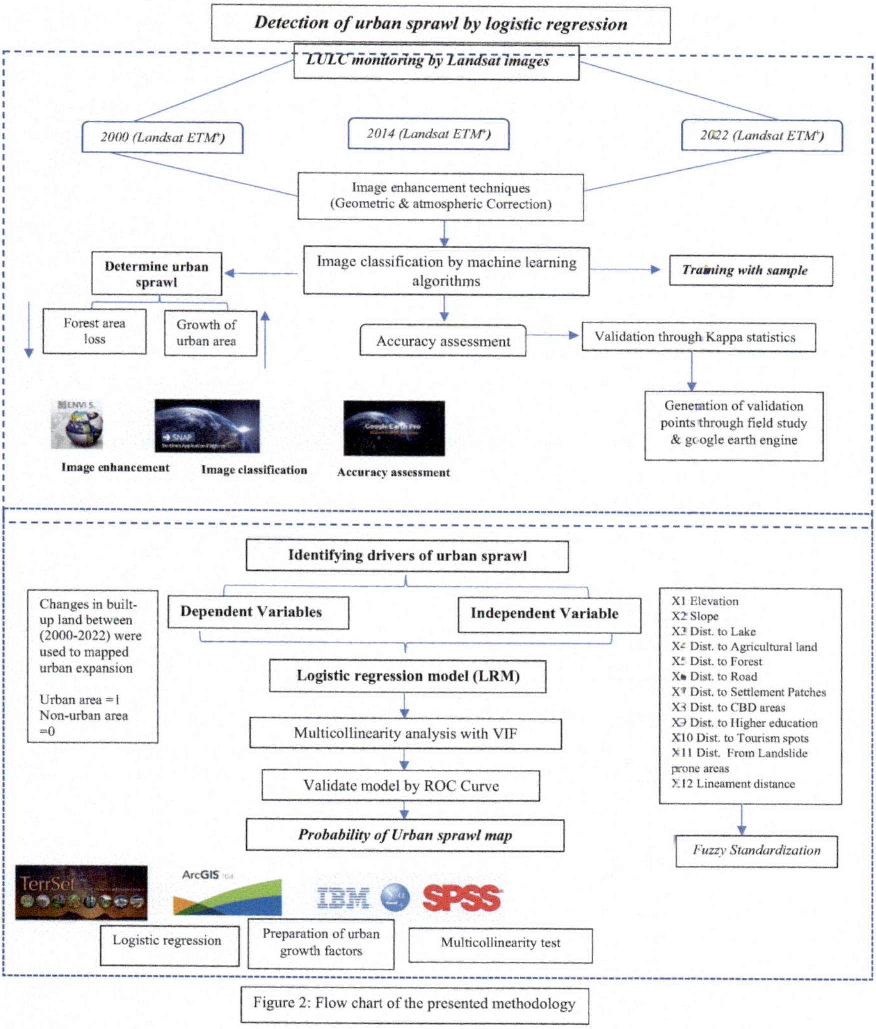

Fig. 11.2 Flow chart of the presented methodology

11.3.4 Logistic Regression

Logistic regression model was applied to develop an urban growth probability map for the Shillong urban agglomeration by analyzing the relationship between urban sprawl & its associated driving variables. The logistic regression method is a popular tool for statistically analyzing contradictory dependent variables and a series of independent variables (Sun, Xiaohui et al. 2018). Over the last two decades, logistic regression analysis has been one of the most extensively used techniques for quantifying the effects of explanatory variables on urban growth as well others vulnerability assessments (Cao, Yu et al. 2020; Nong et al. 2011; Traore et al. 2017; Alqurashi, Abdullah et al. 2016; Huangfu, Wenchao et al. 2021; He, Wancai et al. 2023). Logistic regression, which uses basic linear regression to represent the complicated nonlinear relationship between random events, can be used to estimate the likelihood of a phenomenon occurring. Computation procedure in logistic regression is less time-consuming than other urban prediction methods. And it can experimentally quantify the influence of independent factors on the dependent variable and thus give some assurance about their contribution.

Using the LOGISTICREG module found in the Terrset software and making use of twelve relevant variables (independent) of urban sprawl along with one categorical variable (dependent) the analysis was carried out. The following equation was utilized in the process of computing the binary logistic regression model.

$$\log \text{it}(y) - \log\left(\frac{y}{1-y}\right) = \beta_0 + \beta_1 x_1 + \cdots \beta_n x_n + \varepsilon$$

Here, where y is the likelihood that a cell will become an urban area (1), x_η is a factor or component that operates independently, β_0 is the intercept coefficient, β_η is the coefficient of variable x_η, and ε is the random error term (residuals) (Fig. 11.8).

11.3.5 Identifying the Driving Forces

Modelling the process of urban development and providing guidance for urban planning and administration requires a thorough understanding of the dynamics that influence urban growth (Luo, Ti et al. 2019; Wang et al. 2020). The rate at which urbanization proceeds has been affected by a diverse set of natural and manmade variables. Driving variables were derived after considering previous literature, expert knowledge as well as experience of the researcher on urban growth in shillong urban agglomeration (Jafari, Meisam et al. 2016; Seevarethnam, Mathanraj et al. 2022; Kim, Youjung et al. 2020; Allan, Andrew et al. 2022). Finally, twelve independent variables are found to be more relevant as a factor to determine the urban sprawl in present study. These driving factors were prepared using ArcGIS (10.6.1) and then converted into the 15-m resolution raster format and exported to Terrset software for logistic regression. Factors are classified into two categories dependent variables (where y = 1 signifies urban expansion and y = 0 reflects no urban expansion) and independent variables (Continuous), all those factors were discussed separately below (Tilahun, Deneke et al. 2022; Chaturvedi et al. 2021) (Tables 11.3 and 11.4) (Fig. 11.3).

Dependent variables: (Yi) Within the context of this model, the dependent variables for the logistic regression analysis were maps depicting the dichotomous land use of urban sprawl between the years 2002 and 2022. These maps had two landuse classes, non-urban = 0 and urban = 1 (Fig. 11.4).

Independent variables: Listed independent variables (X_1-X_{12}) maps were prepared in arc-GIS software 10.6.1.

Elevation (X_1): Elevation impacted on temperature, soil erodibility, physiographic stability of any regions. Elevation play a crucial role for human habitation and the overall development of a particular area. Rough terrain always makes it more expensive for development of infrastructure as well as increases the likelihood

Table 11.3 Data used and its source

Serial number	Criteria name	Data source	Techniques used	Spatial resolution
C1	Elevation	Alospalsar Dem (https://search.asf.alaska.edu /)	Elevation Map	30-m
C2	Slope		Slope Map	30-m
C3	Lake	European Space Agency (ESA) Sentinel-2A Imagery, https://scihub.copernicus.eu/	Calculate distance	10-m
C4	Agricultural land			
C5	Forest areas			
C6	Major Road	Open Street Map (OSM) https://www.openstreetmap.org	Proximity analysis	10-m
C7	Settlement Patches	European Space Agency (ESA) Sentinel-2A Imagery, https://scihub.copernicus.eu/	Proximity analysis	10-m
C8	C.B.D areas			
C9	Higher educational institution	Open Street Map (OSM) https://www.openstreetmap.org	Euclidean distance	10-m
C10	Tourism units			
C11	Landslide prone areas	Geological survey of India (GSI) https://bhukosh.gsi.gov.in/	Euclidean distance	30-m
C12	*Faultline Distance*			

* *All those raster images were resampled into 15* 15 m to matches the LULC image resolution*

Table 11.4 List of factors (variables) included in the logistic regression model

Factor types independent (Xi)	Nature of variable	Potential driving factors
Physical	Continuous X_1	Elevation
	Continuous X_2	Slope
	Continuous X_3	Distance from Lake
	Continuous X_4	Distance from Agricultural land
	Continuous X_5	Distance from Forest areas
Socio-economic	Continuous X_6	Distance from major Road
	Continuous X_7	Distance from Settlement Patches
	Continuous X_8	Distance from C.B.D areas
	Continuous X_9	Higher educational institution
	Continuous X_{10}	Distance from Tourism units
Vulnerability	Continuous X_{11}	Distance from Landslide prone areas
	Continuous X_{12}	Faultline Distance
Dependent Y	Dichotomous Y	($Y = 0$) no urban sprawl; ($Y = 1$) urban growth

installations failure. However, relatively flat land is advantageous for farming and development of habitational facilities.

Slope (X_2): Steep slopes are unfavourable and uneconomical for development proposes, should not be used for the expansion of urban areas due to the high costs associated with installation of building in such locations. Development in those areas is often led to rapid erosion and slope instability that promote landslide & earthquake. Therefore, in present study area, sites where steepness of slope < 15 degree were identified as ideal for residential development (Gharaibeh, Anne A., et al. 2023).

Lakes (X3): Excessive urbanization negatively impacted lake ecosystem by contaminating of lake water through urban wastes. Constructing of new housing units, expansion of agriculture, and tourism infrastructure development on wetland or lake surrounding areas causing reduction of overall wetland areas (Alikhani, Somayeh, et al. 2021). Umiam lake, lies in the northern part of shilling planning areas, serves as a major tourist attraction, but recent development of tourist resorts and settlement expansion have harmed the lake's ecosystem.

Agricultural areas (X4): The expansion of impermeable surfaces caused by urbanization consumes agriculturally productive land. Urbanization reduces the economic viability of agriculture, marginalizes farms, and lowers land productivity. It is important to safeguard and preserve areas of fertile agricultural land, in present study area agricultural land were mainly concentrated on southern parts (Gumma, Murali et al. 2017).

Forest areas (X5): Shillong city is surrounded by lush green forest, urban development put astrain on these forest areas; that why a minimum distance from forest areas is a prime concern to plan a sustainable urban development policy. Recent

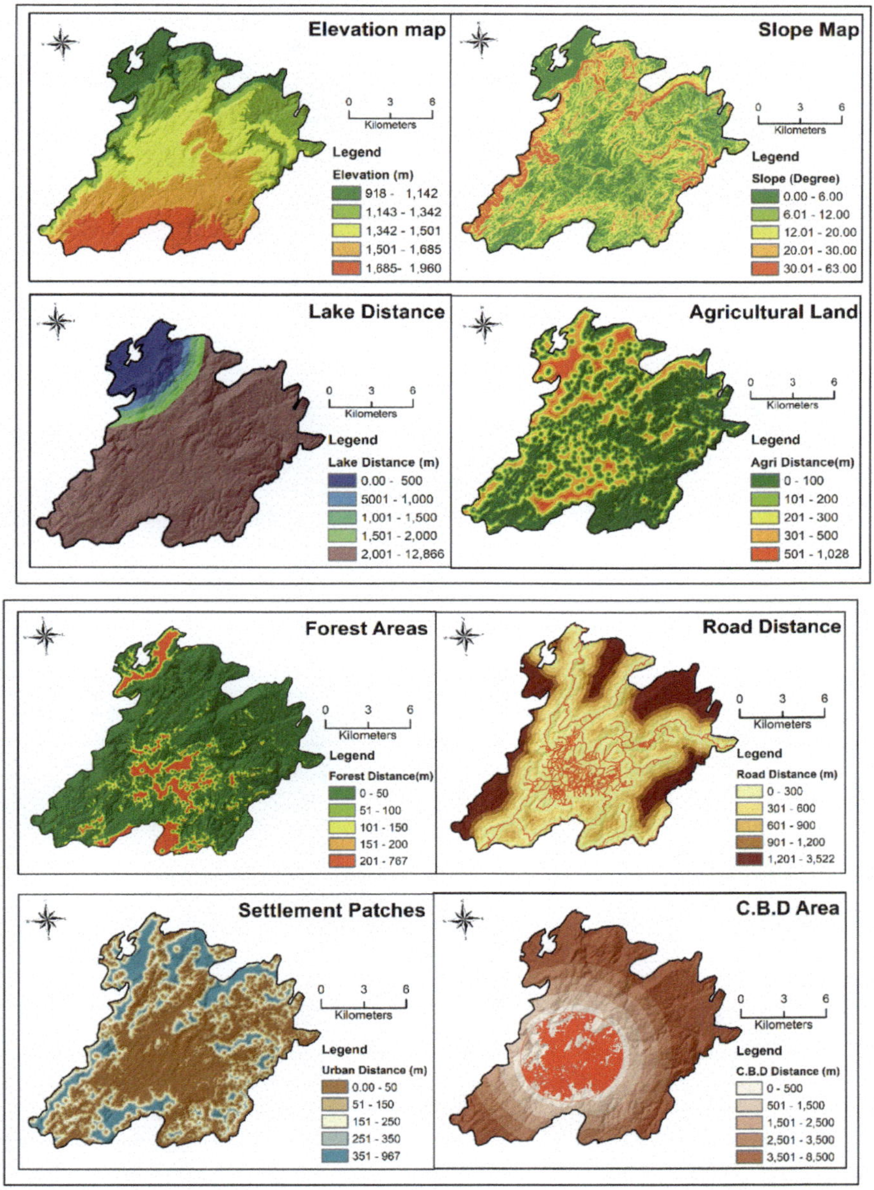

Fig. 11.3 Driving variables for urban growth: C1. Elevation (m), C2. Slope (Degree), C3. Lake Distance (m), C4. Distance from agricultural land (m), C5. Distance from Forest areas (m), C6. Distance from Major road (m), C7. Distance from settlement patches (m) C8. Distance from CBD area (m), C9. Distance from Higher educational institution C10. Tourism units, C11. Distance from Landslide prone areas C12. Distance from Faultline (m)

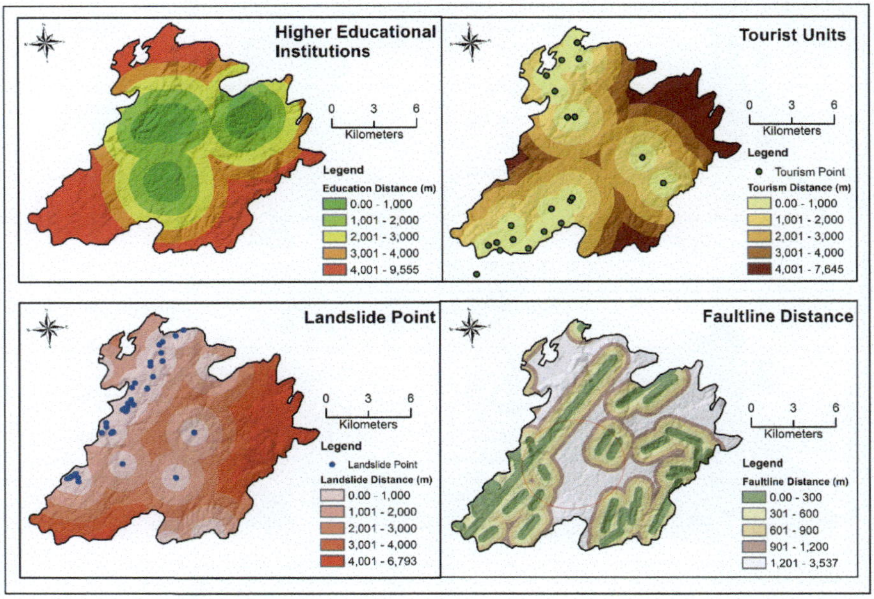

Fig. 11.3 (continued)

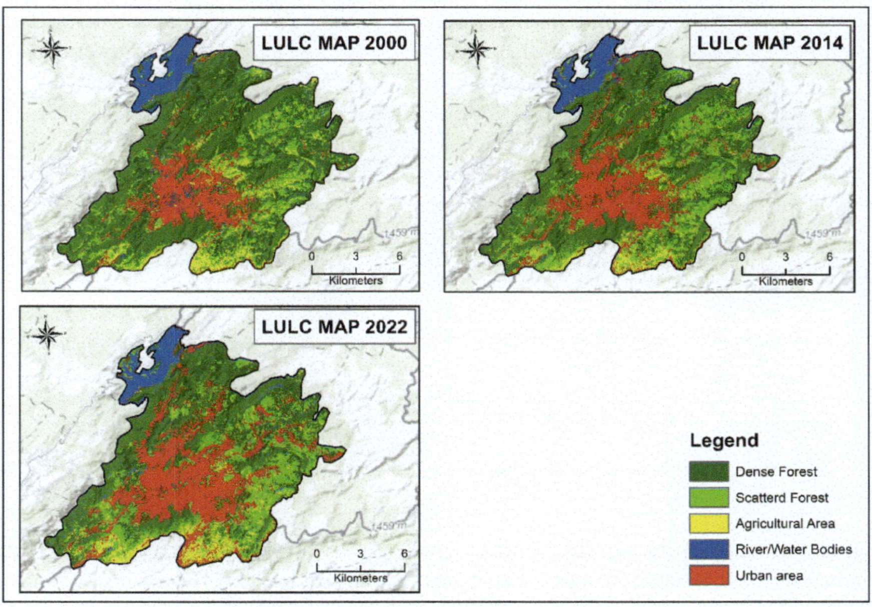

Fig. 11.4 LULC Map for the year 2000, 2014, 2022

research conducted on study area revealed that human activities are responsible for the reduction of forest areas on the urban periphery, as well as the long-term consequences of urbanization results in fragmentation of compact forest patches (Zhou, Weiqi et al. 2017).

Roads (X6): Highway development has speed up urbanization and changed land use directly and indirectly. Increased unplanned settlements near roads bring about a number of issues, such as inaccurate traffic assessments and the encroachment of agricultural land (Song, Jie et al. 2016). Shillong town is well connected to other parts of Meghalaya as well as other major cities of Northeast through National highway (NH 44 & NH40) and others states highway. This staggering highway development has directly ruptured the area environment and indirectly altered the regional landscape by drawing a substantial amount of built-up land transition from agriculture during the last two decades.

Settlement areas (X7): Recent socio-economic & infrastructural development in the outskirts have helped to expand settlement patches, these settlement patches act as a trigger effects for further settlement expansion with the help of existing facilities. Sprawl is noticeable in suburbs of Shillong city, where sparsely spacing village dwellings settlements have gradually supplanted the city's native agricultural and forest mosaic, creating a diverse and ambiguous landscape.

Distance from CBD area (X8): Places that are located in close proximity to the central business district are viewed as more suited and advantageous for development due to the value of their land and the other prospective projects that could take place there. Furthermore, the rate of urban area expansion increases initially and subsequently decreases with distance from the city center where CBD influence lessens (Kim, Youjung et al. 2020). Opportunity of job & others potential developments highly concentrated on CBD areas, in this Shillong town CBD areas specifically located around *Police bazar, Laitumkhrah and Bara Bazaar* areas.

Higher educational institution (X9): Shillong as being a major urban centre, offering high quality educational institute for higher-learning, many students from other parts of Meghalaya state migrated here to take advantage of the educational opportunities. Numerous degree colleges, University (NEHU, 1973), Medical Sciences institute (NEIGRIMS), Management institute (IIM-Shillong), engineering colleges sprouted up outskirts of the city.

Tourism units (X10): The town of Shillong is endowed with an amazing natural landscape, waterfalls, historic buildings, and other picturesque villages. The pleasant weather and convenient transportation connections between Shillong and Guwahati cities help make Shillong a famous tourist destination that draws visitors throughout the year.

Landslide Prone areas (X11) Landslides are a typical geologic hazard that undermines the social-economic fabric of the impacted communities. Landslide vulnerability increase the human anxiety for habitation because of unwanted risk of construction failures and life threat (Sujatha et al. 2021). Landslides frequently cause substantial damage to linear infrastructure, such as roads and railroad, which can impede normal traffic or totally cut off access to the affected areas. Therefore, identifying landslide-prone areas is crucial for ensuring people's safety while also ensuring minimizing economic disturbance in the case of a natural disaster. As shillong is located in highly hazard prone areas severely affected by landslide & earthquake hazard which damaged public and private installations. Village economy is seriously affects due to traffic disruption caused by landslides.

Lineaments (X12): Faultline are detrimental to stability of slope, an increase in the fault density associated with the increasing frequency of earthquakes. Earthquake occurrence is high if faults are active in a complex geological condition. The Shillong town positioned on at the middle of the Shillong plateau is one of the most tectonically active zones (Lee et al. 2003; Saurabh et al. 2019). This region has a number of faults and stress zones along the plate boundaries that might trigger earthquakes alongside destructive ground motion (Carrión-Mero, Paúl et al. 2021).

11.3.6 Model Validation

According to Pontius and Schneider (2001), validating land use change models is a crucial technique for assessing the precision of predictive modelling. The ROC and pseudo-R^2 indices were utilized in this research to validate the accuracy of the regression model (Table 11.5). The range of the ROC value is 0 to 1, with 1 denoting an excellent fit and 0.5 denoting a random fit (Jiang et al. 2015). The accuracy of the model can be split down into three categories according to the ROC values: low precision (ROC between 0.5 and 0.69), satisfactory precision (ROC between 0.70 and 0.9), and excellent precision (ROC more than 0.9). In this investigation, the model was validated by making a comparison between the expected urban expansion shown on the probability map and the actual urban expansion presented on the map for the year 2022. At least a ROC value of > 0.80 ensure that the logistic regression have good fit in explaining the relationship. Another method, For the purpose of determining how precise the model was, the pseudo-R2 index was applied. This index's values ranges between 0 and 1, where 0 denotes a lack of correlation between the dependent and independent variables and 1 denotes a perfect relationship between them. At least the higher than 0.2 values in pseudo-R^2 test is considered a relatively good fit for simulating the model (Seevarethnam et al. 2022; Mahmoudzadeh et al. 2022), irrespective of whether the association between the actual values and the projected values is either positive or negative, the R-squared value will increase according to the amount of the correlation between the actual values and the projected values.

Pseudo R-Squared = (log (Likelihood)/log(L0).

where L0 is the probability function's value if all coefficients—aside from the intercept—are equal to 0.

Table 11.5 Area statistics of LULC of years 2000, 2014 & 2022

Area under different landuse in km^2							Annual Rate of Change LULC changes (2000–2014,2014–2022,2000–2022)					
Class Name	LULC 2000 (Ha)	(%)	LULC 2014 (Ha)	(%)	LULC 2022 (Ha)	(%)	2000- 2014 (Ha)	Rates of change per years	2014–2022 (Ha)	Rates of change per years	2000–2022 (Ha)	Rates of change per years
Dense Forest	10941.30	58.23	9194.71	49.46	7692.85	40.94	−1746.59	−124.76	−1501.86	−187.73	−3248.45	−147.66
Scattered forest	2714.88	14.45	3184.08	17.13	3832.26	20.39	469.20	33.51	648.17	81.02	1117.38	50.79
Agricultural land	1351.00	7.19	1554.52	8.36	1848.89	9.84	203.52	14.54	294.37	36.80	497.89	22.63
River/Water Bodies	1216.88	6.48	1075.44	5.78	992.46	5.72	−141.44	−10.10	−82.98	−10.37	−224.42	−10.20
Urban Areas	2566.79	13.66	3582.09	19.27	4424.39	23.55	1015.31	72.52	842.30	105.29	1857.60	84.44
Total Areas	18790.84	100.00	18590.84	100.00	18790.84	100.00						

11.4 Results

11.4.1 LULC Change Analysis

Over the past few decades, urban sprawl has had a formative effect on the suburbs areas of Shillong. The LULCC maps of the study area for the years 2000, 2014 and 2022 are classified into five classes and presented in Fig. 11.3. In this study, trajectories of LULC change that occurred over the course of twenty years were analyzed using Random Forest algorithms. Cross-tabulated matrix was calculated to identify the variation in the variations in different land use classes over time. The magnitude, percentage change, and annual rate of change of LULCC for the three periods 2000–2014, 2014–2022 and 2000–2022 are presented below in the following (Tables 11.5, 11.6 and 11.7).

11.4.1.1 Forest Cover

During the time period under study (2000–2022), there was a discernible thinning of the forest cover in this study area. In 2000, the extension of forest class was 10,941.30 ha (58.23%) of the entire research area and by 2014, that number had dropped to 9194.71 ha (49.46%) and by the year 2022 it was further declined to 7692.85 ha (40.94%). (Table 11.5) The annual pace of forest loss was estimated to be 124.76 ha $year^{-1}$ during the first phase (2000–2014), but the rate of forest destruction continued to rapidly accelerate about 187.73 ha $year^{-1}$ in the second phase (2014–2022) and the overall rates of deforestation 147.66 ha $year^{-1}$ during overall study periods (2000–2022). During course of the entire study period (2000–2022), the forest cover declined by 29.69% (1.35% annually) of the total area In comparison to the total area covered by forests in the year 2000 (Figs. 11.5, 11.6 and 11.7). The shillong town was once covered by 90% by evergreen pine trees, due its detachable location from mainland India. But this study observed within this 22 year of timespan built-up land are expanded to cater for growing demand in housing and service sectors. In agreement with this, numerous academics noted that the rapid population expansion, growing demand for arable land and need for infrastructure development are all contributing factors for the destruction of forests. The human desire to be lived closer to nature can cause the destruction and fragmentation of forest by expansion of agriculture & other economic activities causing forests to deteriorate, and loss of biodiversity (Miller, 2012; Ortiz, Diana et al. 2021; Gao 2014; Kumari et al. 2020) (Fig. 11.6a).

11.4.1.2 Scattered Forest

Another important category in terms of covering area was scattered forest land which occupied a total of 2714.88 hectares (14.45%), 3184.08 hectares (17.13%), and

Table 11.6 Dynamic LULC changes between 2000 and 2022 in terms of hectares and percentage of changes for each category

Area under different Landuse				Intensity of Landuse change (2000–2014, 2014–2022, 2000–2022)					
Class Name	LULC 2000 *(Ha)*	LULC 2014 *(Ha)*	LULC 2022 *(Ha)*	% (2000–2014)	% Annually	% (2014–2022)	% Annually	% 2000–2022	% Annually
Dense Forest	10941.30	9194.71	7692.85	-15.96%	-1.14%	-16.33%	-2.04%	-29.69%	-1.35%
Scattered forest	2714.88	3184.08	3832.26	17.28%	1.23%	20.36%	2.54%	41.16%	1.87%
Agricultural land	1351.00	1554.52	1848.89	15.06%	1.08%	18.94%	2.37%	36.85%	1.68%
River/Water Bodies	1216.88	1075.44	992.46	−11.62%	−0.83%	−7.72%	−0.96%	−18.44%	−0.84%
Urban Areas	2566.79	3582.09	4424.39	39.56%	2.83%	23.51%	2.94%	72.37%	3.29%
Total Areas	18790.84	18590.84	18790.84						

Table 11.7 Horizontal urban expansion of Shillong Town (2002–2022)

	Greater shillong planning area			
	Built up area (ha)	Expansion of urban areas (Changes in Ha)		
		2000–2014	2014–2022	2000–2022
2000	2566.79 (13.66%)	1015.31		
2014	3582.09 (19.27%)		842.30	
2022	4424.39 (23.55%)			1857.60
Annual rate change LULC changes		72.52	105.29	84.44
Intensity of Landuse change		2.83%	2.94%	3.29%

3832.26 hectares (20.39%) of the entire research area in the year 2000, 2014, and 2022, correspondingly. There is a consistence increase in scattered forest cover due to forest fragmentation. The scattered forest class grew by 17.28% during the first phase (2000–2014), the same trend follows, continued to grow by 20.36% during the second phase (2014–2022), and overall increases of 41.16% during the course of the entire study period (2000–2022) (Tables 11.5 and 11.6). The annual rate of expansion in scattered forest class was 33.51 ha year − 1 in the first period (2000–2014); however, the rate increased to roughly 81.02 ha/year in the second period (2014–2022), and throughout the course of the entire study period (2000–2022) the rates of increase estimated 50.79 ha year − 1. The combined effects of climatic change and anthropogenic activities causing fragmentation and conversion of dense forest to scattered forest. Some of the studies reported that conversion of scattered forest to agricultural & settlement expansion between 2000 and 2022 periods. In recent decades there is shifting in practices from agriculture to plantation of different cash crops pine apple, chestnut, tea estate, pine apples, strawberry, palm and rubber plantation practices in the gentle and moderately steep slopes of hill areas (Zambrano et al. 2019) (Fig. 11.6b).

11.4.1.3 Agricultural Land

Though modern farming methods have been adopted in the areas around Shillong, Meghalaya has historically followed Shifting cultivation methods used by the tribal inhabitants of north-eastern India's mountainous regions. Main contributing factor for deforestation in these areas is agricultural expansion, that is further triggered by haphazard urban expansion. An decent increase in proportion of agricultural land, in the year 2000 approximately 1351.00 ha (7.19%), that is increased to 1554.52 ha (8.36%) in 2014 and further in increase in 2022, about 1848.89 ha (9.84%). (Table 11.5). Between 2000 and 2014, the amount of land under cultivation increased by 15.06% at a pace of 14.54 ha per year, (2000–2014) 18.94% at a rate of 36.80 ha per year, (2014–2022), and 36.85% at a rate of 22.63 ha per year during the full study period (2002–2022) (Tables 11.5 and 11.6 and Fig. 11.3). The data also showed that cultivated land was expanding significantly in the studied area.

Table 11.8 Accuracy assessment of LULC maps of different years

Land use categories	2000				2014				2022			
	User accuracy	Producer accuracy	Overall accuracy	Kappa coefficient (T)	User accuracy	Producer accuracy	Overall accuracy	Kappa coefficient (T)	User accuracy	Producer accuracy	Overall accuracy	Kappa coefficient (T)
Dense forest	91.43%	92.75%	92.67%	90.81%	94.20%	92.86%	93.67%	92.06%	97.14%	97.14%	97.33%	96.66%
Scattered forest	89.09%	96.08%			92.45%	96.08%			96.15%	98.04%		
Agricultural land	92.98%	89.83%			93.22%	91.67%			96.72%	98.33%		
River/Water bodies	91.07%	92.73%			91.23%	92.86%			98.15%	96.36%		
Urban areas	98.39%	92.42%			96.77%	95.24%			98.41%	96.88%		

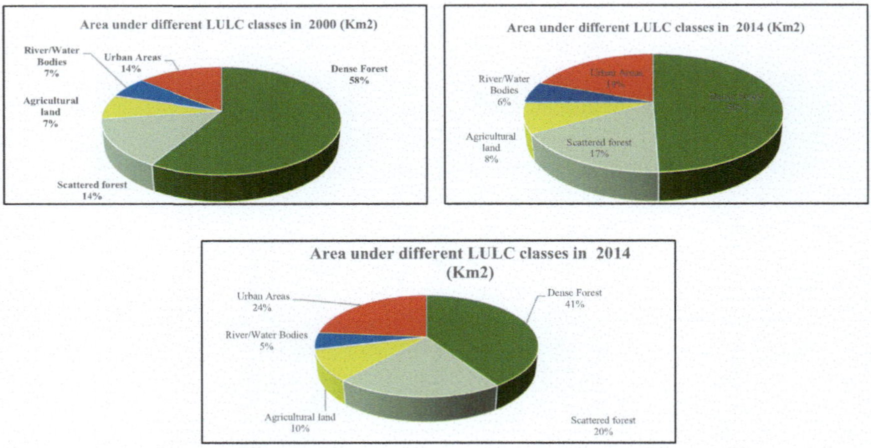

Fig. 11.5 Proportion of different LULC (2000, 2014, 2022)

Fig. 11.6 **a** and **b** Images of forest coverage in the study area Dense forest (upper image) and Scattered forest (lower image). (*Source*: Google earth image of 2022)

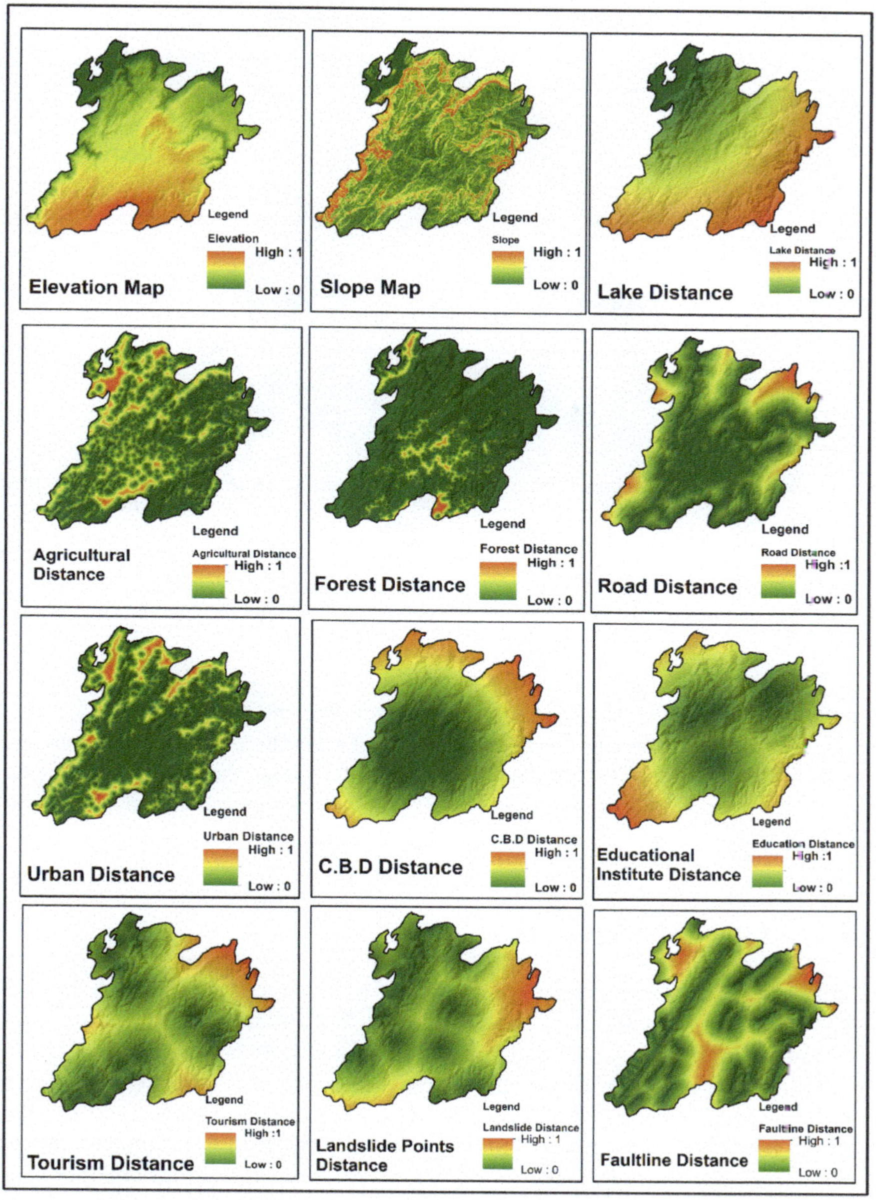

Fig. 11.7 Fuzzified raster map for logistic regression

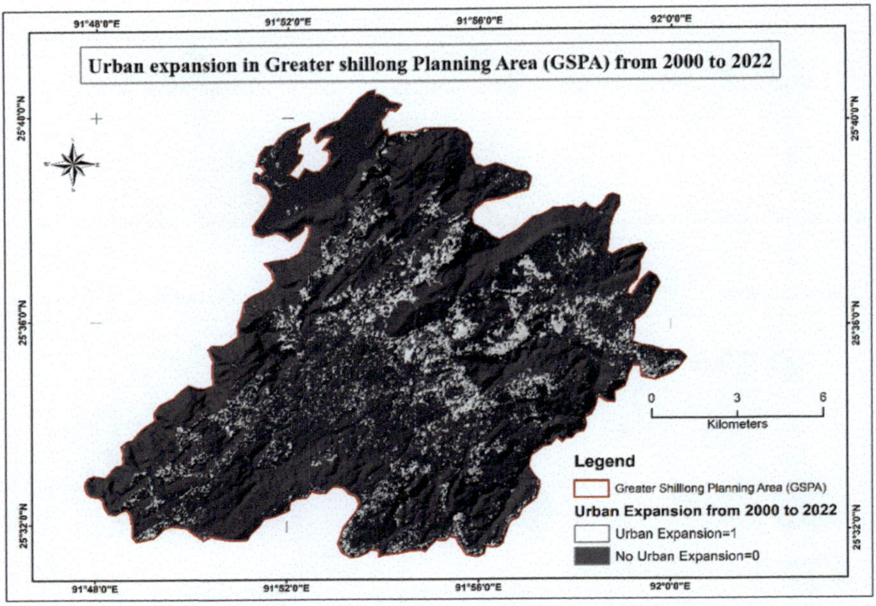

Fig. 11.8 (Y): urban expansion in shillong planning areas between 1990–2018

Population growth in rural and urban area boosted demand for agricultural led to a steady increase in the area under cultivation (Fig. 11.3). The increase of agricultural activities over the past 20 years at the expense of forests, shrub land, and grassland has a significant negative influence on sustainable environmental management (Wang and Qiu 2017).

11.4.1.4 River and Lakes

River & lakes area covered a small proportion of its total area 1216.88 ha (6.48%) in 2000, where as in 2014 1075.44 ha (5.78%) and in 2022 992.46 ha (5.72%), the total water bodies are literally remaining unchanged during study period. Ward's Lake & Umiam Lake are amongst the popular tourist attractions of Shillong characterized by rich flora and fauna. *Umkrah and Umshyrpi* rivers which merges and form the Umiam River, which pass through Shillong enroot to Umiam Lake, are significant sources of drinking water for the inhabitants of the city. Apart from this there are so many small streams, and rills are found in the area following through narrow, steep-sided valleys.

11.4.1.5 Urban Area/Built-Up Expansion

The image classification results revealed that intense LULC changes occurred in built-up classes. Area of built-up lands increased from 2566.79 ha (13.66%) in 2000 to 3582.09 ha (19.27%) in 2014; finally increased to 4424.39 ha (23.55%) in the year 2022, respectively (Table 11.3). During this whole timeframe 2000–2022, total expansion of built-up was 1857.60 he with overall increase of 72.37%. The data presented in Table 11.5 and 11.6 show that the annual rates of built-up area in the period 2002–2014, 2014–2022, and 2002–2022 were 72.52, 105.29, and 84.44 per year per hectares, respectively. The employment opportunities that resulted in the setting up of more residential areas on the town's outskirts were a major factor in the growth of the urban built-up region. Along with a surge in agricultural activity, the area also saw a significant change in the growth of urban built-up areas. (Fig. 11.7). The built-up area largely concentrated in the cantonment and municipal area of Shillong to the north and east part of city. Particularly in Shillong's Mawprem, Laitumkhrah, and Mawkhar regions, experience a quick growth in built-up areas was noticeable. These causes eventually resulted in overcrowding of population in Shillong, which changed the city's land use and land cover. Over time, the region under study became an attractive destination for people to migrate to from far away and nearby locations (including other north eastern states of India and Bangladesh). The spatial trends of LULC changes from 2000 to 2022, the Forest & open land was mostly consumed by urban sprawl, particularly in the northeast part of the study area. Recent satellite images revealed that city's residential growth has expanded well outside boundaries the planning areas as a result of the lack of buildable land within the planning areas. At present, the residentialdevelopment has spread in all the three directions along the major arteries namely, towards Umiam (educational & industrial expansion) & Umsawriang (airport area) in northern portion in the north, Umsawli & Mawpdang in the eastern shillong Mawñianglah, Ritmawniew villages in south Shillong to the Nongkrem & Smit in the southeast.

11.4.2 Accuracy Assessment

Users of LULC maps need to be informed of the degree to which the maps are accurate to make LULC classification more effective. According to Anderson et al. (1976), the minimal degree of accuracy while calculating kappa statistics should at least 85% to ensure the trustworthiness of applied algorithms (Manandhar, Ramita et al. 2009). Overall accuracy, producer's accuracy, user's accuracy and Kappa statistics are generally reported measures of accuracy have been applied in many studies using reference data obtained from other sources (Baig et al. 2022; Foody 2022; Rwanga et al. 2017). The accuracy was assessed using the cross-diagonal confusion matrix, tables calculate the accuracy of classified maps of 2000, 2014 & 2022. The overall accuracy of classified images (2000, 2014, 2022) was 92.67%, 93.67% and 97.33% whereas kappa coefficient values obtained 90.81%, 92.06% and 96.66% respectively.

Using respective indices, the kappa index results indicated that all of the images met the minimum of 85% accuracy.

11.4.3 Urban Expansion Probability Map

Twelve independent variables derived from literature reviews and expert judgment make up the logistic regression model. The relative importance of each and individual variables in urban sprawling throughout the study area were analyzed separately for each of these variables between 2000 and 2022. In order to generate an urbanization probability map, the coefficients from the logistic regression model were utilized and then applied in Eq. 1. In an effort to improve the performance of the model, temporal dynamics were taken into consideration. Among the driving variables Elevation (X1) and Slope (X2) taken as constant don't change over times. Apart from that all other pertaining variables data Distance from Lake (X3), Distance from Agricultural land (X4), Distance from Forest areas (X5), Distance from major Road (X6), Distance from Settlement Patches (X7), Distance from C.B.D areas (X8), Higher educational institution (X9), Distance from Tourism units (X10), Distance from Landslide prone areas (X11), Faultline Distance (X12) were updated by using the database of 2022. Higher probability values indicated those areas has a higher probability of urban conversion whereas lower LR values reflectance low probability of urban conversion. (Figs. 11.9 and 11.10).

The results of the regression model reveal whether a certain variable has a positive or negative impact on the likelihood of any change regarding land use to be converted to urban. The coefficients give an indication of the effects that each of the independent variables has on the dependent variable. It makes sense that pixels close to independent variables with larger coefficient values have greater potential for change. The coefficient's sign (±) denotes whether the dependent variable's response is positively or negatively correlated. For instance, a negative (-) coefficient index indicates that the intensity of urban expansion reduced as the distance increased.

Factors including elevation (X1), slope (X2), and distance from agriculture (X4) have very modest coefficient values which signifies that these criteria has little impact on urban expansion and it does not have much superior effects with increasing or decreasing distances. Urban growth was positively correlated with proximity to lakes (X3), forests (X5), and educational institutions (X9) that indicated the probability of urban expansion is low close to those location than the areas away from it. Most of educational institution recently spurted in outskirt of city in low density population areas that why in present scenario the probability of urban expansion is low. While the correlation was negative with the other factors: the distance to major road (X6), distance from settlement patches (X7), Distance from C.B.D (X8) and Distance from Tourism units (X10), Distance from Landslide prone areas (X11) & Distances from fault line (X12). The majority of the independent factors, on the other hand, have negative coefficients, indicating an inverse relationship between the variables. This implies that the likelihood of urban growth will rise as one gets closer

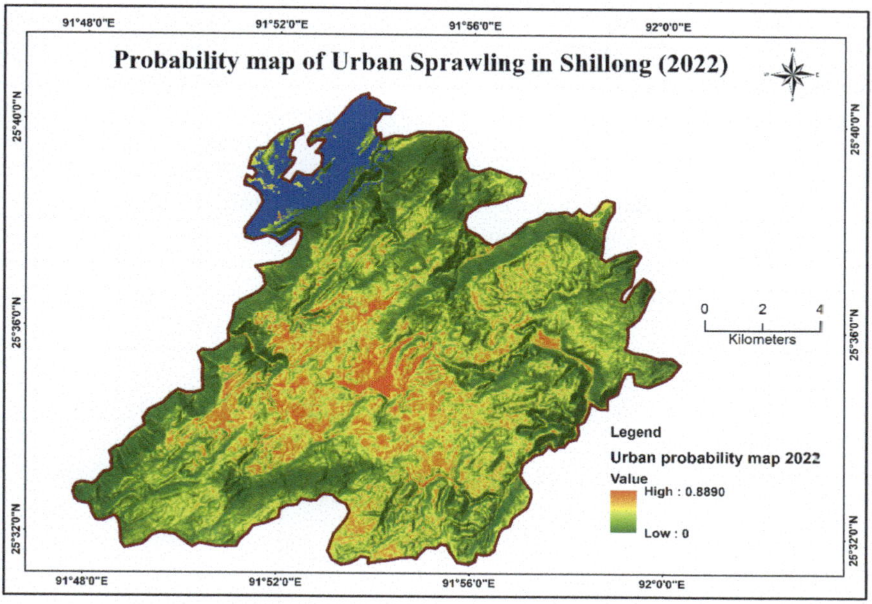

Fig. 11.9 Urbanization probability map of Shillong planning areas in future

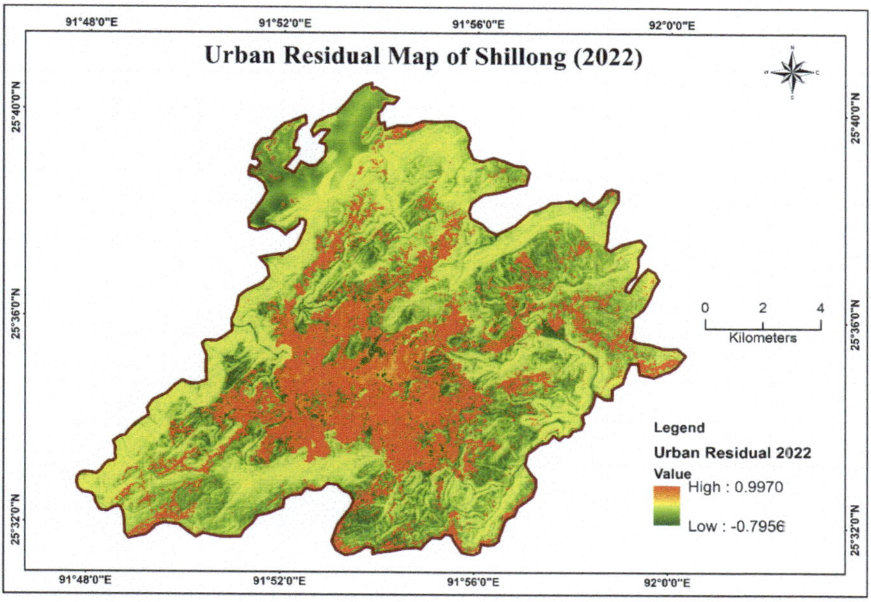

Fig. 11.10 Urbanization Residual map of Shillong planning areas in future

to these variables, and they have an enticing influence on urban sprawl. The results show that elevation (X1) and Slope (X2) have a positive coefficient values 0.0505 & 0.0369, which indicated that high elevation & higher slope unfavourable for urban expansion though, there is very minute positive correlation found between urban growth & those factors. Distance from major road (X6), Distance from settlement (X7) and distance from CBD (X8) have the high negative coefficient values −0.86, −0.83, −0.91 indicated that distance from those places intensity of urban expansion significantly decreases. Distance from Tourism units (X10), Distance from Landslide prone areas (X11), Faultline Distance (X12) also have negative coefficient values of -0.1762, -0.0324, -0.0930, distance from those places intensity of urban expansion also decreases (Table 11.9).

Table 11.9 Effective variables in urban growth with their Coefficients values and Adjusted odds ratios (2000–2022)

Potential driving factors	Coefficient of variables	Collinearity statistics		ROC value	Odd ratio
		tolerance	VIF value		
Elevation (X1)	0.0505	0.9271	1.079	83.45	0.8003
Slope (X2)	0.0369	0.9516	1.051	88.89	1.0901
Lake (X3)	0.4069	0.9471	1.056	89.12	1.0159
Distance from Agricultural land (X4)	0.0503	0.9775	1.023	90.11	1.1175
Distance from Forest areas (X5)	0.5979	0.9324	1.073	91.89	1.8652
Distance from major Road (X6)	−0.8608	0.6751	1.481	90.23	0.2577
Distance from Settlement Patches (X7)	−0.8367	0.6279	1.593	91.11	0.3121
Distance from C.B.D areas (X8)	−0.9115	0.5511	1.815	87.09	0.1991
Higher educational institution (X9)	0.1808	0.8844	1.131	80.04	3.6993
Distance from Tourism units (X10)	−0.1762	0.7599	1.316	78.03	2.7109
Distance from Landslide prone areas (X11)	−0.0324	0.9775	1.023	76.09	1.5331
Faultline Distance (X12)	−0.0930	0.9559	1.046	85.09	1.0285

* Odds ratio meaning the ratio of the probability of success to the probability of failure for the variable, which represents the effect of each variable in the urban expansion process

Table 11.10 Goodness-of-fit statistics of the logistic regression

Step	– 2 Log (Likelihood)	Cox and Snell R^2	Nagelkerke R^2
1	1696.027	0.71	0.83

11.4.4 Goodness of Fit of the Model

Similar to multivariate regression, the coefficient of determination (R2) is used to determine the degree to which the independent variables and dependent variables are related to one another. In order to evaluate how well the logistic regression model reflects association among the variables, the 2log(likelihood) statistic one of the basic statistical measures to determine goodness of fit was calculated and applied. 2log(likelihood) statistic deviance is a measurement of the amount of variation in our logistic regression model that cannot be explained by the data. The – 2log (likelihood) (1696.027) is the value of the likelihood function for the full model as fitted. To further assess the goodness of fit of our model, two additional computations, ***Cox and Snell R^2*** *(*1989) and ***Nagelkerke R^2****(*1991), which are comparable indicators to the $\mathbf{R^2}$ statistics of the linear regression, were performed. R^2 values nearer to 1 show that the model is closer to being accurate. The model's Cox and Snell R2 coefficient value was 0.71, indicating an exceptionally well fit in the study area. By using the Nagelkerke R^2 0.83, the R^2 values indicates an excellent fit of the model (Table 11.10). From the outputs of goodness of fit in both model the significant of driving variables is enough in explaining urban expansion.

11.4.5 Multicollinearity Analysis

Multicollinearity tests of those chosen variables are carried out to strengthen this parameter selection and make methodology more robust. Multicollinearity in the respective dataset may appear when there is strong correlation or association to each other's in such circumstance multicollinearity appear as a challenge in linear regression analysis. The higher degree of multicollinearity, the greater likelihood of inaccurate outcomes. Multicollinearity test helps us to eliminate those insignificant variables to reduced error from results as they decreased precision from the resulted outcome. An analysis of correlation between possible determinants on dependent variables was carried out to check out multicollinearity. Multicollinearity can be detected using various techniques, but most common way to detect multicollinearity is by using the tolerance (Tol) and its reciprocal, known as the variance inflation factor (VIF). The Variance Inflation Factor (VIF) is 1/Tolerance, it is always greater than or equal to 1. Previous literature suggested that if any variables score is greater than 10, they are discarded and excluded from the models because of high degree of collinearity. The influencing variables, which are continuous data, are standardized into the range of 0–1 so that the LR model may obtain the distribution of the

Table 11.11 Variance inflation factor (VIF) scale

VIF	Value conclusion
VIF = > 1	Variables are not correlated, Multicollinearity doesn't exist
1 > VIF ≤ 5	Variables are moderately correlated, Low multicollinearity exists
VIF > 5	Highly corelated among the variables, High Multicollinearity exists

spatial value on the same scale. Than selected driving was converted to ASCII files for multicollinearity analysis under the IBM SPSS 25 statistics software. In order to conduct the multicollinearity analysis, one independent variable was regressed against the twelve residual variables in a repeating mechanism (Tables 11.11 and 11.9).

A multicollinearity test was performed was conducted on each and every set of selected variables to eliminate any superfluous driving factors if present in study in order to maintain the integrity of our coefficients. Tolerance and the variance inflation factor (VIF) statics were used to analyzes the multicollinearity present in variables. The tolerance value ranges from 0.5511 to 0.9775, and the VIF (1/tolerance) varies from 1.023 to 1.815. According to the multicollinearity assessment, none of the model driving factors have a VIF greater than 10, indicating that the model performs exceedingly well (Bui et al. 2011; Salem et al., 2019).

11.4.6 Model Validation Using the ROC Technique

As previously stated, the pseudo-R^2 indices and relative operating characteristic (ROC) curve were used to assess the model's effectiveness in predicting the likelihood of urban expansion in the shillong town. The ROC method is excellent statistical measure to test a model's effectiveness by simulating a probability image that depicts the likelihood that a phenomenon will occur against a binary image that shows where instances of that class can be found (Pontius and Schneider 2001). In this study, 1500 randomly selected cells from each map were used to compare the modelling of future urban development with the actual urban expansion map from 2022. The result of the ROC index was acquired during the period of 2000–2022 with a value of was 0.8968, with a standard error of 0.001, which is an extremely strong value and suggests that which specifies that the probability map of urban expansion is valid in explaining relationship between variables (Fig. 11.11). The ROC (0.89) and pseudo-R^2 (0.76) indices value shown in Table 11.12 show how the referenced and actual urban expansion fits well in terms of probability map.

The maps of the probability of urban development were obtained with values of greater than 0.8 for ROC and 0.7 for pseudo-R^2, which demonstrate that the independent variables used to simulate urban development were carefully chosen. For sensitivity analysis in this study, the logistic regression model was rerun after each independent variable separately with dependent variables from the full set of

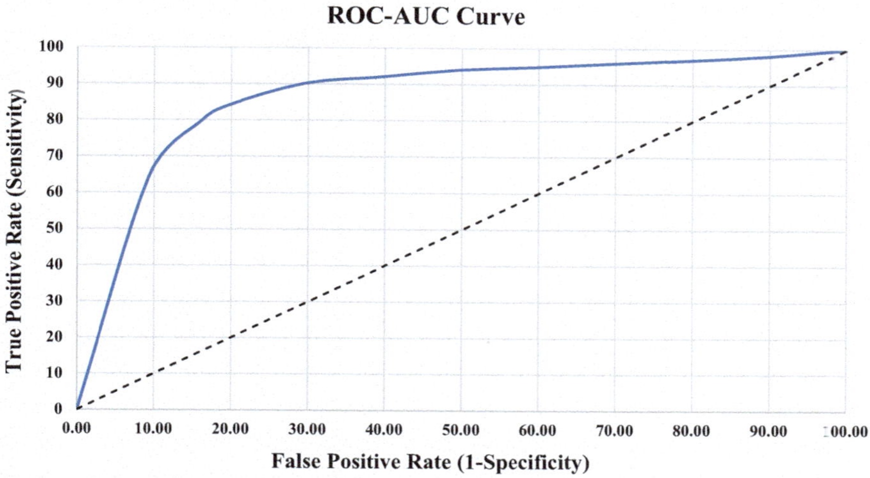

Fig. 11.11 ROC-AUC Curve

Table 11.12 Table for ROC statistics

2000–2022	
ROC	Pseudo-R2
0.8968	0.7636

variables and recording acquired ROCs for the 2000–2022 period. The advantage of using this method is that it allows for the sensitivity of the variables to be determined, as well as the effect that each variable has on the final model (Fig. 11.12).

11.5 Discussion

Understanding the causes and effects of urban sprawl is essential for effective urban planning and management that reduces its negative effects. Vulnerability of hill landscape is another major concern in urban expansion, because urban expansion exerts a strong impact on the hill ecosystem and increase the chances of forest loss. Moreover, due to complicated and interrelated proximate and underlying drivers, the changes in LULC are unavoidable and have a non-linear form. The urban landscape of Shillong city is quite diverse geologically and geomorphologically. In terms of the dispersion of building patterns, most of the places, even freshly built structures, have dense, haphazard construction. In addition, the outcomes of the logistic regression model were used to construct a probability map of urban expansion in the Shillong planning region. The growth tend to take place in shillong planning areas can be segregated into

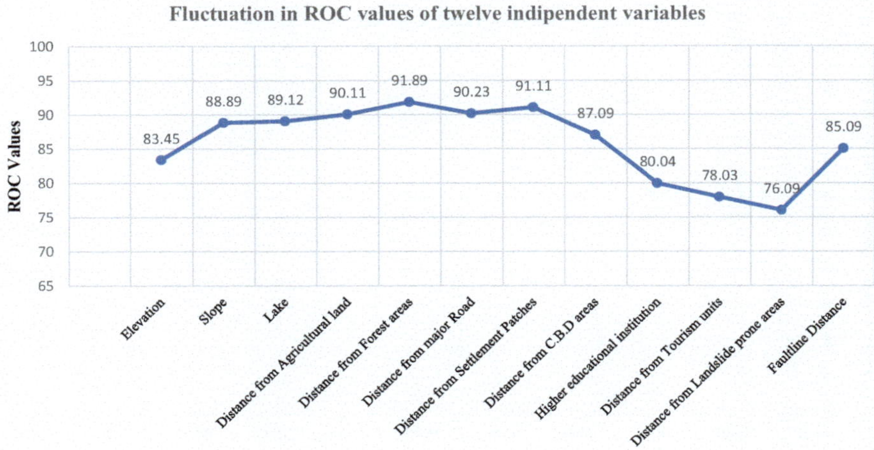

Fig. 11.12 Sensitivity analysis of the logistic regression model by the elimination of independent variables for the 2000–2022 period with ROC fluctuation

two parts one is natural urban expansion following the road & using existing infrastructure. The city has extended towards the north on the Guwahati-Shillong National Highway 6 and south-west on portion on NH 206 along Mawsynram- Cheerapunji road. Another, planned industrial & educational development after 2000, in Umiam region, health & medical science institute at Mawdiangdiang, Indian Institute of Management (IIM) at Umsawli these are some of planned development in the fringe areas of shillong. High rates of change that have resulted in unsustainable development in recent decades have caught the attention of policymakers and planners, which has increased the need to comprehend the underlying causes. The unchecked growth of cities, which is a regular occurrence in developing countries, not only puts strain on the outskirts of a city but also on the core. The increase in built up areas is 72.37% with 22 years signifies an ala situation for urban planner. On the other hand, due to the other considerations, the expansion will most likely continue to take place in sites that are close to established areas, which will result in more informal areas (Figs. 11.9 and 11.10).

Odd ratio values another good measurement to know the impact of urban expansion. Areas nearer to road (X6) had greater potential for developing to urban uses which have an Odds ratio values of 0.2577 that indicate that the probability of urban expansion in the area nearer to road is estimated as 3.880 (1/0.2577) more than the areas away from it. The impact of the road was a contributing factor in the creation of ribbon and strip developments. New town developments tourism assets, educational institute and industrial areas are located by the side of roads. Various others study also encounter that distance from main roads had a very high impact on urban expansion and urban expansion is more likely to occur close to the main roads (Mozaffaree Pour et al. 2022; Sarkar and Chouhan, 2020).

In the same manner settlement (X7) have an odds ratio value 0.3121 that means, cells near to Settlement patches have 3.204 (1/0.3121) have higher probability of urban expansion. Urban expansion was negatively impacted by distance from urban regions; this suggests that if one moves further away from nearby cities, the likelihood of urban expansion decreases. Apartment areas, Medical institution, small companies, and spontaneously created industries are clustered closer to county urban centers, where unplanned urban sprawl develops. Land developers tended to mimic land-use characteristics in existing urban regions where urban sprawl tended to emerge near existing urban uses (Mozaffaree Pour et al. 2022).

The odds ratio for Distance from CBD area (X8) is 0.1991, which mean that the probabilities of urban expansion in the areas with proximity to CBD area are five times (1/0.1991 = 5.022) more than the probabilities of urban expansion in the areas further away from CBD areas. Accessibility is reduced and transportation expenses rise when distance from the central business area (CBD) increases (Brown, Scott et al. 2014).

Slope (X2), has a positive impact on urban growth, high slope was not suitable for urban growth. The slope is an unpredictable environmental feature that influences urban expansion; building construction and city development on steep-slope terrain have been problematic., low to moderate slope ideal for urban developments (Zhou, Liang et al. 2021).

Distance from Lake (X3), & Faultline distance (X12) That indicates that none of these factors have an effect on the urbanization process taking place within the scope of the study area.

As hill areas are lacking in public transportation systems peoples are generally concentrated on either existing settlement areas or side of the road to have easily access to city centre. As the distance from the CBD increases, the likelihood of urban expansion decreases rapidly, particularly in hill stations. Because the CBD serves as a commercial, entertainment, and educational transportation hub for the surrounding area.

The estimated odds ratio of urban sprawl probability in agricultural lands around existing urban cores is 1.1175, that means the probability of urban lower (1/1.1175 = 0.8948) which is less than the sprawl probability in lands farther away from existing cores. This is because in a hilly area we can hardly found any agricultural tract that no conversion from agricultural to built up land observed during this course of study.

Distance from Dense forest areas have an Odds ration values is 1.8652 that mean dense forest cover have very low probability of conversion (1/1.8652). The odds ratio for the distance from tourism unit is 2.7109, which indicates that the likelihood of urban expansion will rise by with each additional urban cell within the neighbourhood 0.3688.

Some of the previous study that used logistic regression model to determine the effects of urban expansion are discussed below to make a comparable success of this model. Sarkar and Chouhan (2020) studied spatial determinants of urban expansion of Siliguri a metropolitan city by logistic regression, and the model has achieved the accuracy 0.881. Osman et al. (2016) examines how driving factors affect urban development in the Greater Cairo Metropolitan's governorate of Giza. In their research, which considered the factors that cause LULC to shift as well as remotely sensed data encompassing the years 2004–2013, found that the overall correct model percentage was 0.95 (north), 0.88 (middle), and 0.94 (south).

Salem et al. (2019) in their study on peri-urban expansion of Greater Cairo Region applied logistic regression model with eight driving variables, during validation process this model achieved an accuracy of 0.93. Salem, Muhammad, et al. (2021) simulate urban expansion Delhi using Nine driving factors and Landsat images of 1989, 2000, 2010, 2020. The correctness of the simulation was evaluated using the relative operating characteristic approach, and the value that was obtained as a result demonstrated that the simulation was indeed accurate. All of these studies successfully utilized logistic regression model for detecting urban expansion, those studies show the trustworthiness of logistic regression model in different scenarios.

11.6 Conclusions

Observing how urban areas are growing by determining impact of driving variables on urban sprawl which providing assistance to policymakers and urban planners to detangle the nature of urbanization that can be anticipated in the future. Among several selected variables, some of which strongly related to urban sprawl and play a detrimental role on urban growth in Shillong urban areas and others variables are unattractive for growth of urban area. Factor such as s proximity to the existing settlement area (–0.83), distance from the CBD area (–0.91), and proximity to the road (–0.86) are highly influential for urban growth. It was observed that urban sprawl was spread out across the entire region without a particular concentration in certain locations. Though this study includes several physical & socio-economic variables, apart from that geomorphic structure, hillshde, temperature, visibility, rainfall and government future city development some of the factors that's might affects the urban growth in hill area recommended to include for future scope of research. Although the logistic regression model struggles with temporal analysis and quantification of change, it can be combined with a CA–Markov model to produce the best simulated results. Our analysis demonstrated that urban modelling gives a superior explanation for understanding the likely trends of sprawl activities in the near future and delivers vital information to assist policy-makers in developing appropriate laws for attaining urban sustainability in Shillong. The model's predictive power for urban growth was evaluated using the ROC curve. The ROC curve's excellent score (0.8968) demonstrated the model's viability. Finally, the integration of the LR model, remote sensing, and GIS can be effectively employed as a tool to comprehend the motivating elements and to forecast the future urban growth pattern of the Shillong planning regions. Development regulations should be strengthened by inclusion of these findings to safeguard the urban environment from deteriorating further.

Author Contributions: S.D: Conceptualization, Methodology, Formal Analysis, writing original draft, **S.C.** Review and editing, software, validation Writing. The final manuscript has been approved for publication by all authors after reading it.

Funding (optional): This research received no external funding from any organization or institution.

Conflicts of interest/Competing Interests: The authors declare consciously that they have no known competing financial interests or personal relationships that could have appeared to influence the work reported in this paper.

Data Availability Statements All data generated or analyzed during this study are included in this published article (and there is no supplementary information files).

Compliance with Ethical Standards The manuscript has not been published previously partly or in full in any journal, were in accordance with the ethical standards of this journals.

Ethical Approval: This article does not contain any studies with human participants or animals performed by any of the authors.

It is Ensure that all the authors mentioned in the manuscript have agreed for authorship, read and approved the manuscript, and given consent for submission and subsequent publication of the manuscript. The order of authorship agreed by all named authors prior to submission.

References

Aburas MM, Abdullah SH, Ramli MF, As' shari ZH.(2016 June) Landscape analysis of urban growth patterns in Seremban, Malaysia, using spatio-temporal data. In: IOP conference series: earth and environmental science vol 37, no 1. IOP Publishing, p 012055

Allan A, Soltani A, Abdi MH, Zarei M (2022) Driving Forces behind land use and land cover change: a systematic and bibliometric review. Land 11(8):1222. https://doi.org/10.3390/land11081222

Anderson, J.R.; Hardy, E.E.; Roach, J.T.; Witmer, R.E. (1976) A Land Use and Land Cover Classification System for Use with Remote Sensor Data. Vol. 964, US Government Printing Office, Washington DC, USA.

Alqurashi A, Kumar L, Al-Ghamdi K (2016) Spatiotemporal Modeling of Urban Growth Predictions Based on Driving Force Factors in Five Saudi Arabian Cities. ISPRS Int J Geo Inf 5(8):139. https://doi.org/10.3390/ijgi5080139

Asadi M, Oshnooei-Nooshabadi A, Saleh S-S, Habibnezhad F, Sarafraz-Asbagh S, Van Genderen JL (2022) Urban Sprawl Simulation Mapping of Urmia (Iran) by Comparison of Cellular Automata–Markov Chain and Artificial Neural Network (ANN) Modeling Approach. Sustainability 14(23):15625. https://doi.org/10.3390/su142315625

Alikhani S, Nummi P, Ojala A (2021) Urban Wetlands: A Review on Ecological and Cultural Values. Water 13(22):3301. https://doi.org/10.3390/w13223301

Alsharif AA, Pradhan B (2014) Urban sprawl analysis of Tripoli Metropolitan city (Libya) using remote sensing data and multivariate logistic regression model. Journal of the Indian Society of Remote Sensing 42:149–163. https://doi.org/10.1007/s12524-013-0299-7

Asempah M, Sahwan W, Schütt B (2021) Assessment of land cover dynamics and drivers of urban expansion using geospatial and logistic regression approach in Wa municipality. Ghana. Land 10(11):1251. https://doi.org/10.3390/land10111251

Arsanjani JJ, Helbich M, Kainz W, Boloorani AD (2013) Integration of Logistic Regression, Markov Chain and Cellular Automata Models to Simulate Urban Expansion. Int J Appl Earth Obs Geoinf 21:265–275. https://doi.org/10.1016/j.jag.2011.12.014

Arifeen HM, Phoungthong K, Mostafaeipour A, Yuangyai N, Yuangyai C, Techato K, Jutidamrongphan W (2021) Determine the Land-Use Land-Cover Changes, Urban Expansion and Their Driving Factors for Sustainable Development in Gazipur Bangladesh. Atmosphere 12(10):1353. https://doi.org/10.3390/atmos12101353

Baruah et al., (2019): Seismic vulnerability assessment of earthquake-prone mega-city Shillong, India using geophysical mapping and remote sensing, Georisk: Assessment and Management of Risk for Engineered Systems and Geohazards, https://doi.org/10.1080/17499518.2019.1598560

Bui DT, Lofman O, Revhaug I et al (2011) Landslide susceptibility analysis in the Hoa Binh province of Vietnam using statistical index and logistic regression. Nat Hazards 59:1413–1444. https://doi.org/10.1007/s11069-011-9844-2

Brown SC, Lombard J, Toro M, Huang S, Perrino T, Perez-Gomez G, Plater-Zyberk E, Pantin H, Affuso O, Kumar N, Wang K, Szapocznik J (2014) Walking and proximity to the urban growth boundary and central business district. Am J Prev Med 47(4):481–486. https://doi.org/10.1016/j.amepre.2014.05.008

Baig MF, Mustafa MRU, Baig I, Takaijudin HB, Zeshan MT (2022) Assessment of Land Use Land Cover Changes and Future Predictions Using CA-ANN Simulation for Selangor. Malaysia. Water 14(3):402. https://doi.org/10.3390/w14030402

Behnisch M, Krüger T, Jaeger JAG (2022) Rapid rise in urban sprawl: Global hotspots and trends since 1990. PLOS Sustain Transform 1(11):e0000034. https://doi.org/10.1371/journal.pstr.0000034

Belay T, Mengistu DA (2019) Land use and land cover dynamics and drivers in the Muga watershed, Upper Blue Nile basin, Ethiopia. Remote Sensing Applications: Society and Environment 15:100249. https://doi.org/10.1016/j.rsase.2019.100249

Cox DR, Snell EJ (1989) The analysis of binary data, 2nd edn. Chapman and Hall, London

Congalton RG, Green K (2019) Assessing the Accuracy of Remotely Sensed Data: Principles and Practices; CRC Press: Boca Raton. FL, USA

Chen J, Chang KT, Karacsonyi D, Zhang X (2014) Comparing urban land expansion and its driving factors in Shenzhen and Dongguan, China. Habitat Int 43:61–71

Cravero A, Pardo S, Sepúlveda S, Muñoz L (2022) Challenges to Use Machine Learning in Agricultural Big Data: A Systematic Literature Review. Agronomy 12(3):748. https://doi.org/10.3390/agronomy12030748

Carrión-Mero P, Briones-Bitar J, Morante-Carballo F, Stay-Coello D, Blanco-Torrens R, Berrezueta E (2021) Evaluation of Slope Stability in an Urban Area as a Basis for Territorial Planning: A Case Study. Appl Sci 11(11):5013. https://doi.org/10.3390/app11115013

Chaturvedi V, de Vries WT (2021) Machine Learning Algorithms for Urban Land Use Planning: A Review. Urban Science 5(3):68. https://doi.org/10.3390/urbansci5030068

Cao, Y., Zhang, X., Fu, Y., Lu, Z., & Shen, X. (2020). Urban spatial growth modeling using logistic regression and cellular automata: A case study of Hangzhou. Ecological Indicators., 113. https://doi.org/10.1016/j.ecolind.2020.106200

Cheng LL, Tian C, Yin TT (2022) Identifying driving factors of urban land expansion using Google Earth Engine and machine-learning approaches in Mentougou District. China. Sci Rep 12:16248. https://doi.org/10.1038/s41598-022-20478-z

Chakraborty, A., Omrani, H., & Teller, J. (2022). A Comparative Analysis of Drivers Impacting Urban Densification for Cross Regional Scenarios in Brussels Metropolitan Area. Land, 11(12), 2291. https://doi.org/10.3390/land11122291

Dadras, M., Shafri, H. Z., Ahmad, N., & Pradhan, B. (2013). Six Decades of Urban Growth Using Remote Sensing and GIS in the City of Bandar Abbas, Iran. IOP Conference Series: Earth and Environmental Science, 20, Article ID: 012007.https://doi.org/10.1088/1755-1315/20/1/012007

Degife A, Worku H, Gizaw S, Legesse A (2019) Land use land cover dynamics, its drivers and environmental implications in Lake Hawassa Watershed of Ethiopia. Remote Sensing Applications: Society and Environment 14:178–190. https://doi.org/10.1016/j.rsase.2019.03.005

Changjia Li, C. Li, Tong Lu, T. Lu, Bojie Fu, B. Fu, Shuai Wang, S. Wang, & Joseph Holden, J. Holden. (2022). Sustainable city development challenged by extreme weather in a warming world. Geography and sustainability, 3, 114-118. https://doi.org/10.1016/j.geosus.2022.04.001

Chakraborty A, Omrani H, Teller J (2022) A Comparative Analysis of Drivers Impacting Urban Densification for Cross Regional Scenarios in Brussels Metropolitan Area. Land 11(12) 2291. https://doi.org/10.3390/land11122291

Devkota KC, Regmi AD, Pourghasemi HR, Yoshida K, Pradhan B, Ryu IC et al (2013) Landslide Susceptibility Mapping using Certainty Factor, Index of Entropy and Logistic Regression Models in GIS and Their Comparison at Mugling-Narayanghat Road Section in Nepal Himalaya. Nat Hazards 65:135–165

https://doi.org/10.1007/s11069-012-0347-6

Foody GM (2022) Global and Local Assessment of Image Classification Quality on an Overall and Per-Class Basis without Ground Reference Data. Remote Sensing 14(21):5380. https://doi.org/10.3390/rs14215380

Falanga Bolognesi S, Pasolli E, Belfiore O, De Michele C, D'Urso G (2020) Harmonized Landsat 8 and Sentinel-2 Time Series Data to Detect Irrigated Areas: An Application in Southern Italy. Remote Sensing 12(8):1275. https://doi.org/10.3390/rs12081275

Franco S, Mandla VR, Rao KR (2017) Trajectory of Urban Growth and Its Socioeconomic Impact on a Rapidly Emerging Megacity. Journal of Urban Planning and Development-Asce 143:04017002. https://doi.org/10.1061/(ASCE)UP.1943-5444.0000378

Grekousis G, Manetos P, Photis YN (2013) Modeling urban evolution using neural networks, fuzzy logic and GIS: The case of the Athens metropolitan area. Cities 30:193–203

Gharaibeh AA, Jaradat MA, Kanaan LM (2023) A Machine Learning Framework for Assessing Urban Growth of Cities and Suitability Analysis. Land 12(1):214. https://doi.org/10.3390/land12010214

Gumma M, Mohammad I, Nedumaran S, Whitbread A, Lagerkvist C (2017) Urban Sprawl and Adverse Impacts on Agricultural Land: A Case Study on Hyderabad. India. Remote Sensing 9(11):1136. https://doi.org/10.3390/rs9111136

Gao Q, Yu M (2014) Discerning Fragmentation Dynamics of Tropical Forest and Wetland during Reforestation, Urban Sprawl, and Policy Shifts. PLoS ONE 9(11):e113140. https://doi.org/10.1371/journal.pone.0113140

Huangfu W, Wu W, Zhou X, Lin Z, Zhang G, Chen R, Song Y, Lang T, Qin Y, Ou P, Zhang Y, Xie L, Huang X, Fu X, Li J, Jiang J, Zhang M, Liu Y, Peng S, Liu W (2021) Landslide Geo-Hazard Risk Mapping Using Logistic Regression Modeling in Guixi, Jiangxi. China. Sustainability 13(9):4830. https://doi.org/10.3390/su13094830

He W, Chen G, Zhao J, Lin Y, Qin B, Yao W, Cao Q (2023) Landslide Susceptibility Evaluation of Machine Learning Based on Information Volume and Frequency Ratio: A Case Study of Weixin County. China. Sensors 23(5):2549. https://doi.org/10.3390/s23052549

Hu Z, Lo CP (2007) Modeling urban growth in Atlanta using logistic regression. Comput Environ Urban Syst 31(6):667–688

Hamdy O, Zhao S, Osman T, Salheen M, Eid Y (2016) Applying a Hybrid Model of Markov Chain and Logistic Regression to Identify Future Urban Sprawl in Abouelreesh, Aswan: A Case Study. Geosciences 6(4):43. https://doi.org/10.3390/geosciences6040043

Hatab AA, Ravula P, Nedumaran S et al (2022) Perceptions of the impacts of urban sprawl among urban and peri-urban dwellers of Hyderabad, India: a Latent class clustering analysis. Environ Dev Sustain 24:12787–12812. https://doi.org/10.1007/s10668-021-01964-2

Hutchings P, Willcock S, Lynch K et al (2022) Understanding rural–urban transitions in the Global South through peri-urban turbulence. Nat Sustain 5:924–930. https://doi.org/10.1038/s41893-022-00920-w

Iizuka K, Johnson BA, Onishi A, Magcale-Macandog DB, Endo I, Bragais M (2017) Modeling Future Urban Sprawl and Landscape Change in the Laguna de Bay Area. Philippines. Land 6(2):26. https://doi.org/10.3390/land6020026

Ibrahim Mahmoud M, Duker A, Conrad C, Thiel M, Shaba Ahmad H (2016) Analysis of Settlement Expansion and Urban Growth Modelling Using Geoinformation for Assessing Potential Impacts of Urbanization on Climate in Abuja City. Nigeria. Remote Sensing 8(3):220. https://doi.org/10.3390/rs8030220

Jena R, Pradhan B, Alamri AM (2020) Susceptibility to Seismic Amplification and Earthquake Probability Estimation Using Recurrent Neural Network (RNN) Model in Odisha. India. Applied Sciences 10(15):5355. https://doi.org/10.3390/app10155355

Jat MK, Garg PK, Khare D (2008) Monitoring and modelling of urban sprawl using remote sensing and GIS techniques. Int. J. Appl. Earth Obs. Geoinformation 10:26–43. https://doi.org/10.1016/j.jag.2007.04.002

Jafari M, Majedi H, Monavari S, Alesheikh A, Kheirkhah Zarkesh M (2016) Dynamic Simulation of Urban Expansion Based on Cellular Automata and Logistic Regression Model: Case Study of the Hyrcanian Region of Iran. Sustainability 8(8):810. https://doi.org/10.3390/su8080810

Jiang W, Chen Z, Lei X, Jia K, Wu Y (2015) Simulating urban land use change by incorporating an autologistic regression model into a CLUE-S model. J Geogr Sci 25:836–850

K, D., & Angadi, D.P. (2021) Urban expansion quantification from remote sensing data for sustainable land-use planning in Mangaluru, India. Remote Sensing Applications: Society and Environment. https://doi.org/10.1016/j.rsase.2021.100602

Kim Y, Newman G, Güneralp B (2020) A Review of Driving Factors, Scenarios, and Topics in Urban Land Change Models. Land 9(8):246. https://doi.org/10.3390/land9080246

Kumar A (2018) Review of building regulations for safety against hazards in Indian hill towns. Journal of Urban Management 7(2):97–110. https://doi.org/10.1016/j.jum.2018.06.002

Kumar, Ashwani (2015). Impact of building regulations on Indian hill towns. HBRC Journal, (), S1687404815000061–. https://doi.org/10.1016/j.hbrcj.2015.02.002

Kumari R, Banerjee A, Kumar R, Kumar A, Saikia P, Latif Khan M (2020) Deforestation in India: Consequences and Sustainable Solutions. IntechOpen. https://doi.org/10.5772/intechopen.85804

Land-use change models (Verburg et al., 2004) based on geospatial technologies, multi-temporal remote sensing and spatial analysis, have proven to be valuable, efficient and technologically sound ways to analyze land conversion activities across space and over time.

Li G, Fang C, Li Y, Wang Z, Sun S, He S, Qi W, Bao C, Ma H, Fan Y, Feng Y, Liu X (2022a) Global impacts of future urban expansion on terrestrial vertebrate diversity. Nat Commun 13(1):1628. https://doi.org/10.1038/s41467-022-29324-2

Lee HY, Chung SL, Wang JR, Wen DJ, Lo CH, Yang TF, Zhang Y et al (2003) Miocene Jiali Faulting and Its Implications for Tibetan Tectonic Evolution. Earth Planet Sci Lett 205(3–4):185–194

Luo T, Tan R, Kong X, Zhou J (2019) Analysis of the Driving Forces of Urban Expansion Based on a Modified Logistic Regression Model: A Case Study of Wuhan City. Central China. Sustainability 11(8):2207. https://doi.org/10.3390/su11082207

Lin Y, Chen A, Yin Y, Li Q, Zhu Q, Luo J (2021) A framework for sustainable management of the platform service supply chain: An empirical study of the logistics sector in China. Int J Prod Econ 235:108112. https://doi.org/10.1016/J.IJPE.2021.108112

Leinenkugel P, Deck R, Huth J, Ottinger M, Mack B (2019) The Potential of Open Geodata for Automated Large-Scale Land Use and Land Cover Classification. Remote Sensing 11(19):2249. https://doi.org/10.3390/rs11192249

Li P, Hao S, Cui Y, Xu Y, Liao C, Sheng L (2022b) Analysis of the Evolution of the Relationship between the Urban Pattern and Economic Development in Guangdong Province Based on Coupled Multisource Data. ISPRS Int J Geo Inf 11(10):512. https://doi.org/10.3390/ijgi11100512

Lambin, E.F. and Geist, H.J. (2006) Land Use and Land Cover Change-Local Processes and Global Impacts. Springer Publication, 222.

Mariye M, Jianhua L, Maryo M (2022) Land use and land cover change, and analysis of its drivers in Ojoje watershed. Southern Ethiopia. Heliyon 8(4):e09267. https://doi.org/10.1016/j.heliyon.2022.e09267

Maheshwari, B., Singh, V.P., Thoradeniya, B. (2016). Balanced Urban Development: Is It a Myth or Reality?. In: Maheshwari, B., Thoradeniya, B., Singh, V.P. (eds) Balanced Urban Development: Options and Strategies for Liveable Cities. Water Science and Technology Library, vol 72. Springer, Cham. https://doi.org/10.1007/978-3-319-28112-4_1

Mozaffaree Pour, N., Karasov, O., Burdun, I., & Oja, T. (2022). Simulation of land use/land cover changes and urban expansion in Estonia by a hybrid ANN-CA-MCA model and utilizing spectral-textural indices. Environmental Monitoring Assessment 194(8):584

Munshi, Talat; Zuidgeest, Mark; Brussel, Mark; van Maarseveen, Martin (2014). Logistic regression and cellular automata-based modelling of retail, commercial and residential development in the city of Ahmedabad, India. Cities, 39(), 68–86. https://doi.org/10.1016/j.cities.2014.02.007

Mostafa E, Li X, Sadek M, Dossou JF (2021) Monitoring and Forecasting of Urban Expansion Using Machine Learning-Based Techniques and Remotely Sensed Data: A Case Study of Gharbia Governorate. Egypt. Remote Sensing 13(22):4498. https://doi.org/10.3390/rs13224498

Mwabumba M, Yadav BK, Rwiza MJ, Larbi I, Twisa S (2022) Analysis of land use and land-cover pattern to monitor dynamics of Ngorongoro world heritage site (Tanzania) using hybrid cellular automata-Markov model. Current Research in Environmental Sustainability. https://doi.org/10.1016/j.crsust.2022.100126

Mozaffaree Pour N, Oja T (2021) Prediction Power of Logistic Regression (LR) and Multi-Layer Perceptron (MLP) Models in Exploring Driving Forces of Urban Expansion to Be Sustainable in Estonia. Sustainability 14(1):160. https://doi.org/10.3390/su14010160

Mipun, B.S., Hazarika, R., Mondal, M., & Mukhopadhyay, S. (2015). Solid Waste Management in Greater Shillong Planning Area (GSPA) Using Spatial Multi-Criteria Decision Analysis for Site Suitability Assessment. ISPRS - International Archives of the Photogrammetry, Remote Sensing and Spatial Information Sciences, 657-662. https://doi.org/10.5194/ISPRSARCHIVES-XL-7-W3-657-2015

Mahmoudzadeh H, Abedini A, Aram F (2022) Urban Growth Modeling and Land-Use/Land-Cover Change Analysis in a Metropolitan Area (Case Study: Tabriz). Land 11(12):2162. https://doi.org/10.3390/land11122162

Manandhar R, Odeh I, Ancev T (2009) Improving the Accuracy of Land Use and Land Cover Classification of Landsat Data Using Post-Classification Enhancement. Remote Sensing 1(3):330–344. https://doi.org/10.3390/rs1030330

Maxwell, Aaron E., and Timothy A. Warner. 2020. "Thematic Classification Accuracy Assessment with Inherently Uncertain Boundaries: An Argument for Center-Weighted Accuracy Assessment Metrics" Remote Sensing 12, no. 12: 1905. https://doi.org/10.3390/rs12121905

Malarvizhi K, Kumar SV, Porchelvan P (2022) Urban sprawl modelling and prediction using regression and Seasonal ARIMA: a case study for Vellore. India. Model. Earth Syst. Environ. 8:1597–1615. https://doi.org/10.1007/s40808-021-01170-z

Miller MD (2012) The impacts of Atlanta's urban sprawl on forest cover and fragmentation. Appl Geogr 34:171–179. https://doi.org/10.1016/j.apgeog.2011.11.010

Maithani S (2009) A neural network based urban growth model of an Indian city. J Indian Soc Remote Sens 37:363–376. https://doi.org/10.1007/s12524-009-0041-7

Nong Y, Du Q (2011) Urban growth pattern modeling using logistic regression. Geo-Spatial Information Science 14(1):62–67. https://doi.org/10.1007/s11806-011-0427-x

Nandy DR (2001) Geodynamics of Northeastern India and the Adjoining Region. ABC Publications, Calcutta

Nuissl, H., Siedentop, S. (2021). Urbanisation and Land Use Change. In: Weith, T., Barkmann, T., Gaasch, N., Rogga, S., Strauß, C., Zscheischler, J. (eds) Sustainable Land Management in a European Context. Human-Environment Interactions, vol 8. Springer, Cham. https://doi.org/10.1007/978-3-030-50841-8_5

Nagelkerke NJD (1991) A note on a general definition of the coefficient of determination. Biometrika 78:691–692

Osman T, Divigalpitiya P, Arima T (2016) Driving factors of urban sprawl in Giza governorate of the Greater Cairo Metropolitan Region using a logistic regression model. Int J Urban Sci 20:206–225. https://doi.org/10.1080/12265934.2016.1162728

Oo TK, Arunrat N, Sereenonchai S, Ussawarujikulchai A, Chareonwong U, Nutmagul W (2022) Comparing Four Machine Learning Algorithms for Land Cover Classification in Gold Mining: A Case Study of Kyaukpahto Gold Mine. Northern Myanmar. Sustainability 14(17):10754. https://doi.org/10.3390/su141710754

Osman, Taher et al. "Driving factors of urban sprawl in Giza governorate of the Greater Cairo Metropolitan Region using a logistic regression model." International Journal of Urban Sciences 20 (2016): 206 - 225.

Ortiz DI, Piche-Ovares M, Romero-Vega LM, Wagman J, Troyo A (2021) The Impact of Deforestation, Urbanization, and Changing Land Use Patterns on the Ecology of Mosquito and Tick-Borne Diseases in Central America. Insects 13(1):20. https://doi.org/10.3390/insects13010020

Olorunfemi IE, Fasinmirin JT, Olufayo AA, Komolafe AA (2020) GIS and remote sensing-based analysis of the impacts of land use/land cover change (LULCC) on the environmental sustainability of Ekiti State, southwestern Nigeria. Environ Dev Sustain 22:661–692

Padmanaban R, Bhowmik AK, Cabral P, Zamyatin A, Almegdadi O, Wang S (2017) Modelling Urban Sprawl Using Remotely Sensed Data: A Case Study of Chennai City. Tamilnadu. Entropy 19(4):163. https://doi.org/10.3390/e19040163

Parida, B.R., Pandey, A.C., Behera, M.D., & Kumar, N. (Eds.). (2022). Handbook of Himalayan Ecosystems and Sustainability, Volume 2: Spatio-Temporal Monitoring of Water Resources and Climate (1st ed.). CRC Press. https://doi.org/10.1201/9781003265160

Population data was used from the 1 km resolution gridded Landscan population dataset (2008 and 2013) which has been widely used in various studies (McKee et al. 2015; Lu and Guldmann 2012; Roy and Blaschke 2014; Bajat et al. 2013; Sharma et al. 2017; Sharma and Saikia 2018).

Pontius Jr RG, Schneider LC (2001) Land-cover change model validation by an ROC method for the Ipswich watershed, Massachusetts, USA. Agricult Ecosyst Environ 85(1–3):239–248

Punia M, Singh L (2012) Entropy Approach for Assessment of Urban Growth: A Case Study of Jaipur, INDIA. Journal of the Indian Society of Remote Sensing 40:231–244. https://doi.org/10.1007/s12524-011-0141-z

Rumi Aijaz, "India's Peri-Urban Regions: The Need for Policy and the Challenges of Governance", ORF Issue Brief No. 285, March 2019, Observer Research Foundation.

Resnik DB (2010) Urban sprawl, smart growth, and deliberative democracy. Am J Public Health 100(10):1852–1856. https://doi.org/10.2105/AJPH.2009.182501

Rwanga S, Ndambuki J (2017) Accuracy Assessment of Land Use/Land Cover Classification Using Remote Sensing and GIS. Int J Geosci 8:611–622. https://doi.org/10.4236/ijg.2017.84033

Rai, P.K. Mishra, V.N., Singh, P. 2021. Recent Technologies for Disaster Management & Risk Reduction-Sustainable Community Resilience & Responses (edit. Book), Springer Nature, Switzerland, pp. 1–471. ISBN: 978–3–030–76116–5. https://doi.org/10.1007/978-3-030-76116-5.

Mishra, V.N., Rai, P.K., Singh, P. 2021. Geo-information Technology in Earth Resources Monitoring and Management (edit. Book), Nova Science Publishers, U.S.A., pp. 1–395. ISBN: 978–1–53619–669–6.

Rai, P.K. Mishra, V.N., Singh, P. 2022. Geospatial Technology for Landscape and Environment Management: Sustainable Assessment & Planning (edit. Book), Springer Nature, Singapore. Pp. 1–431. ISBN: 978–981–16–7373–3. https://doi.org/10.1007/978-981-16-7373-3.

Rai, P.K. 2023a. River Conservation and Water Resource Management, Springer Nature, Singapore. Pp. 1–352. ISBN No.: 978–981–99–2604–6. eBook ISBN: 978–981–99–2605–3.

Rai, P.K. 2023b. Advances in Water Resource Planning & Sustainability, Springer Nature, Singapore. Pp. 1–322. ISBN No. 978–981–99–3659–5.

Song J, Ye J, Zhu E, Deng J, Wang K (2016) Analyzing the Impact of Highways Associated with Farmland Loss under Rapid Urbanization. ISPRS Int J Geo Inf 5(6):94. https://doi.org/10.3390/ijgi5060094

Sujatha ER, Sridhar V (2021) Landslide Susceptibility Analysis: A Logistic Regression Model Case Study in Coonoor. India. Hydrology 8(1):41. https://doi.org/10.3390/hydrology8010041

Saurabh et al., (2019). Seismic vulnerability assessment of earthquake-prone mega-city Shillong, India using geophysical mapping and remote sensing. Georisk: Assessment and Management of Risk for Engineered Systems and Geohazards, 1–16. https://doi.org/10.1080/17499518.2019.1598560

Seevarethnam M, Rusli N, Ling GHT (2022) Prediction of Urban Sprawl by Integrating Socioeconomic Factors in the Batticaloa Municipal Council, Sri Lanka. ISPRS Int J Geo Inf 11(8):442. https://doi.org/10.3390/ijgi11080442

Saikia, Anup; Thapa, Pankaj (2019). Environmental Change in the Himalayan Region (Twelve Case Studies) || Swelter in the City: Population Distribution, Land Use Change and the Urban Heat Island Effect in Shillong, India. , https://doi.org/10.1007/978-3-030-03362-0(Chapter 9), 173–187. https://doi.org/10.1007/978-3-030-03362-0_9

Sarkar A, Chouhan P (2020) Modeling spatial determinants of urban expansion of Siliguri a metropolitan city of India using logistic regression. Model. Earth Syst. Environ. 6:2317–2331. https://doi.org/10.1007/s40808-020-00815-9

Shamsudeen M, Padmanaban R, Cabral P, Morgado P (2022) Spatio-Temporal Analysis of the Impact of Landscape Changes on Vegetation and Land Surface Temperature over Tamil Nadu. Earth 3(2):614–638. https://doi.org/10.3390/earth3020036

Sun X, Chen J, Bao Y, Han X, Zhan J, Peng W (2018) Landslide Susceptibility Mapping Using Logistic Regression Analysis along the Jinsha River and Its Tributaries Close to Derong and Deqin County, Southwestern China. ISPRS Int J Geo Inf 7(11):438. https://doi.org/10.3390/ijgi7110438

Salem M, Tsurusaki N, Divigalpitiya P (2019) Analyzing the Driving Factors Causing Urban Expansion in the Peri-Urban Areas Using Logistic Regression: A Case Study of the Greater Cairo Region. Infrastructures 4(1):4. https://doi.org/10.3390/infrastructures4010004

Salem M, Bose A, Bashir B, Basak D, Roy S, Chowdhury IR, Alsalman A, Tsurusaki N (2021) Urban Expansion Simulation Based on Various Driving Factors Using a Logistic Regression Model: Delhi as a Case Study. Sustainability 13(19):10805. https://doi.org/10.3390/su131910805

Seto KC, Fragkias M, Güneralp B, Reilly MK (2011) A meta-analysis of global urban land expansion. PLoS ONE 6(8):e23777

Singh B, Venkatramanan V, Deshmukh B (2022) Monitoring of land use land cover dynamics and prediction of urban growth using Land Change Modeler in Delhi and its environs, India. Environ Sci Pollut Res Int 29(47):71534–71554. https://doi.org/10.1007/s11356-022-20900-z

Shi Y, Zhou L, Guo X, Li J (2023) The Multidimensional Measurement Method of Urban Sprawl and Its Empirical Analysis in Shanghai Metropolitan Area. Sustainability 15(2):1020. https://doi.org/10.3390/su15021020

Tyagi S, Garg N, Paudel R (2014) Environmental degradation: Causes and consequences. European Researcher 81(8–2):1491

Theres L, Radhakrishnan S, Rahman A (2023) Simulating Urban Growth Using the Cellular Automata Markov Chain Model in the Context of Spatiotemporal Influences for Salem and Its Peripherals. India. Earth 4(2):296–314. https://doi.org/10.3390/earth4020016

Thapa RB, Murayama Y (2010) Drivers of urban growth in the Kathmandu valley, Nepal: Examining the efficacy of the analytic hierarchy process. Appl Geogr 30(1):70–83

Talukdar S, Singha P, Mahato S, Shahfahad P, S., Liou, Y.-A., & Rahman, A. (2020) Land-Use Land-Cover Classification by Machine Learning Classifiers for Satellite Observations—A Review. Remote Sensing 12(7):1135. https://doi.org/10.3390/rs12071135

Traore A, Watanabe T (2017) Modeling Determinants of Urban Growth in Conakry, Guinea: A Spatial Logistic Approach. Urban Science 1(2):12. https://doi.org/10.3390/urbansci1020012

Tilahun D, Gashu K, Shiferaw GT (2022) Effects of Agricultural Land and Urban Expansion on Peri-Urban Forest Degradation and Implications on Sustainable Environmental Management in Southern Ethiopia. Sustainability 14(24):16527. https://doi.org/10.3390/su142416527

Vinayak B, Lee HS, Gedem S (2021) Prediction of Land Use and Land Cover Changes in Mumbai City, India, Using Remote Sensing Data and a Multilayer Perceptron Neural Network-Based Markov Chain Model. Sustainability 13(2):471. https://doi.org/10.3390/su13020471

Wu W, Zhao S, Henebry GM (2018) Drivers of urban expansion over the past three decades: a comparative study of Beijing, Tianjin, and Shijiazhuang. Environ Monit Assess 191:1–15. https://doi.org/10.1007/s10661-018-7151-z

Weerakoon P (2017) GIS Integrated Spatio-Temporal Urban Growth Modelling: Colombo Urban Fringe, Sri Lanka. J Geogr Inf Syst 9:372–389. https://doi.org/10.4236/jgis.2017.93023

Wang H, Qiu F (2017) Investigating the impact of agricultural land losses on deforestation: Evidence from a peri-urban area in Canada. Ecol Econ 139:9–18. https://doi.org/10.1016/j.ecolecon.2017.04.002

Wang L et al (Nov 2020) Analysing the driving forces and environmental effects of urban expansion by mapping the speed and acceleration of built-up areas in China between 1978 and 2017. Remote Sens 12(23):3929. https://doi.org/10.3390/rs12233929.

Wu, Yizhou, et al. "The Effectiveness of Planning Control on Urban Growth: Evidence from Hangzhou, China." Sustainability, vol. 9, no. 5, May 2017, p. 855. Crossref, https://doi.org/10.3390/su9050855.

Wu Y, Fan P, Li B, Ouyang Z, Liu Y, You H (2017) The Effectiveness of Planning Control on Urban Growth: Evidence from Hangzhou. China. Sustainability 9(5):855. https://doi.org/10.3390/su9050855

Xia X, Zhang Y, Shi X, Chen J, Rao T (2022) Simulation of Dynamic Urban Growth Boundary Combining Urban Vitality and Ecological Networks: A Case Study in Chengdu Metropolitan Area. Land 11(10):1793. https://doi.org/10.3390/land11101793

Wang L, Jia Y, Li X, Gong P (2020) Analysing the driving forces and environmental effects of urban expansion by mapping the speed and acceleration of built-up areas in China between 1978 and 2017. Remote Sens 12(23):3929

Wu Y, Zhang Y (2012) Partially monotone tensor spline estimation of the joint distribution function with bivariate current status data

Zhang M, Li Y, Guo R, Yan Y (2022) Heterogeneous Effects of Urban Sprawl on Economic Development: Empirical Evidence from China. Sustainability 14(3):1582. https://doi.org/10.3390/su14031582

Zhou W, Zhang S, Yu W, Wang J, Wang W (2017) Effects of urban expansion on forest loss and fragmentation in six megaregions. China Remote Sens 9(10):991. https://doi.org/10.3390/rs9100991

Zhou L, Dang X, Mu H, Wang B, Wang S (2021) Cities are going uphill: Slope gradient analysis of urban expansion and its driving factors in China. Sci Total Environ 775:145836. https://doi.org/10.1016/j.scitotenv.2021.145836

Zambrano L, Aronson MFJ, Fernandez T (2019) The consequences of landscape fragmentation on Socio-Ecological patterns in a rapidly developing urban area: a case study of the national autonomous university of Mexico. Front Environ Sci 7:152. https://doi.org/10.3389/fenvs.2019.00152

Chapter 12
Slope Stability Analysis of Road Cut Slopes Using Limit Equilibrium Method in Kumaun Lesser Himalayan Belt

Rajesh Singh, Rahul Kumar Verma, and Prateek Sharma

Abstract The tectonic and anthropogenic activity in lesser Himalayas made it prone to Landslides. Living beings, all development projects such as the construction of highways, tunnels, dams, bridges, etc., and tourism in hilly areas are affected by landslides. The landslides occur in different forms which include rockfall, road subsidence, and complete slope failure. The majority of landslides occur during monsoon time due to the saturation of rock mass; the resistance force decreases than the driving force and the slope is being unstable. Therefore, identification, recognition, characterization, and zonation of blocks or debris, plays vital role in the study of slope stability. The slope stability analysis using a conventional method like the Limit Equilibrium Method (LEM) directly depends on the geological conditions of the slope mass. The FoS calculated by this method should be more precise. The present study focuses on the Limit Equilibrium Method in Lesser Himalayan road cut slopes. The Factor of Safety was estimated for two critical slopes based on the Mohr–Coulomb criteria. Finally, the mitigation and remedial measures are suggested for identified vulnerable slopes.

Keywords Limit equilibrium method · Factor of safety · Slope stability · Road cut slopes

12.1 Introduction

The mountainous area faces Landslide hazards due to various triggering factors like Earthquakes, rainfall, and anthropogenic factors such as the widening of roads, and the construction of dams, bridges, and tunnels (Verma et al. 2021a, b; Onagh et al. 2012a, b; Rai et al. 2014; Mishra et al. 2021; Rai et al. 2021; Singh et al. 2021; Sur et al. 2021; Rai et al. 2022). The construction and widening of roads are very important

R. Singh · R. K. Verma (✉) · P. Sharma
Rock Science and Rock Engineering Laboratory, Department of Geology, University of Lucknow, Lucknow-226007, India
e-mail: getrahul9@gmail.com

P. K. Rai et al. (eds.), *Natural Resources Management and Sustainable Livelihoods in the Mountainous Region*, Advances in Geographical and Environmental Sciences,
https://doi.org/10.1007/978-981-97-2100-9_12

for connecting livelihoods in the hilly terrain. The geotechnical investigation is vital in completing these projects with sustainability prospects (Verma et al. 2019). For this purpose, the slope stability analysis performed using the conventional method gives a reliable outcome and it is directly depending on the geological condition of the slope mass (Huang 2014). The common techniques are methods of slices like Bishop Simplified (1955), Spencer (1967), the ordinary method of slices (Fellenius 1936), and Morgenstern-price method (1965). The results obtained from particular methods can vary due to different assumptions used. These assumptions: (i) The analysis is carried out in a 2-D framework ignoring the effect of 3-D. (ii) A predetermined failure mechanism is assumed. (iii) Shear stress mobilization occurs locally. (iv) Movement of soil mass assumed as a rigid block and along the failure surface itself. The LEM is based on statics principles like vertical force, horizontal force, and summation of moments. The FoS for each method should be greater than 1.0 for the stability of slopes, however in this study, FoS greater than 1 and less than or equal to 1.5 is considered as critical/partially stable slopes and FoS greater than 1.5 is considered as a stable slope.

This study is based on the conventional method like LEM, which estimates the Factor of Safety (FoS) using Mohr–Coulomb criteria. The study focused on the vulnerable zones composed of meta-basic and quartzite rocks with steep slopes lying on the Bhowali-Almora highway (Fig. 12.1) and samples collected from these vulnerable slopes (Fig. 12.2). The laboratory test was performed for the determination of the geotechnical parameters like unit weight, Uniaxial Compressive Strength (UCS) and Geological Strength Index (GSI) value estimated in the field. Rock mass properties of material was estimated using the intact rock properties and field joint conditions i.e., GSI (Hoek et al. 2002).

12.2 Methodology

12.2.1 Limit Equilibrium Method (LEM)

The LEM is a conventional approach for analyzing slope stability and calculating the Factor of Safety (FoS). LEM is based on the slicing method. The material that lies in the rock mass above the slip surface is divided into several vertical slices (Janbu 1954; Bishop 1955; Spencer 1967). The stability of the material above the slip surface is investigated by taking into account the static equilibrium of the individual slices as well as the overall equilibrium of the failing slope. The static equilibrium of slices can be achieved by either removing or considering the interslice forces and the moment equilibrium of the slices. These approaches calculate the FoS along the critical slip surface as the ratio of shear strength to shear stress acting on the surface. In the current work, the simplified Bishop's approach based on the method of slices was utilized to calculate the FoS at mentioned locations (Bishop 1955).

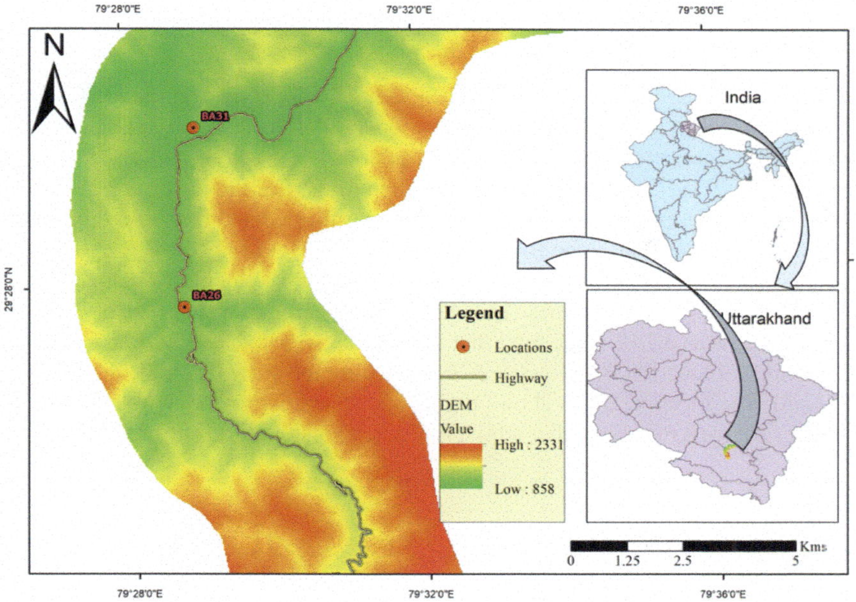

Fig. 12.1 Map of the study area illustrating the locations of samples collected along the highway

Fig. 12.2 Illustrating a wide view of the road-cut slope sections

These methods are based on the assumption that failure will occur with the rotational mass of rock/soil on a circular slip surface centred on a common point. The assumption on the slice side is that the horizontal force has acted, implying that there are no shear tensions between slides (Bishop 1955). The total normal force is considered to be applied at the middle of each slice base and is calculated by summing forces in the vertical direction. Thus, in this work, LEM of the slopes was performed using Bishop's simplified approach with Mohr–Coulomb failure criteria as implemented by Slide v.6 software (Slide 2010).

12.2.2 Results and Discussions

The current study looks into the stability of road-cut slopes in Uttarakhand, India, along the Bhowali-Almora Highway. An examination of the geometry and geomaterials of slopes was carried out in the field such as slope height, slope angle, Geological Strength Index (GSI), and condition of the slope mass. For the purpose of identification of rock type, petrographic study has been performed and based on this study, it was found that road cut-slope BA26 is composed of Meta-basic rock and BA31 is Quartzite rock. The FoS along the critical slip surface were calculated using Bishop simplified method for two slopes (Fig. 12.3). The Factor of Safety (FoS) value at location BA26 was 3.741, indicating that the slope is in a stable condition however, the FoS value at location BA31 was 1.476, indicating the critical condition of slopes (Fig. 12.3). The location BA31 shows higher and steeper slopes, due to its highness the load of the rock mass is higher which is also responsible for its critical condition.

12.3 Conclusions

This study focused on examining the stability of the road cut-slopes along the Bhowali-Almora highway using the Limit Equilibrium Method (LEM). The rock mass properties were determined using Hoek and Brown criteria. FoS has been estimated using Mohr–Coulomb criteria for these road-cut slopes BA26 and BA31. The Slope's FoS depicts that BA26 lies in stable condition and BA31 lies in a critical condition or a partially stable condition. Rock bolting, wire meshing with shotcrete type support system is essential for partially stable slope BA31.

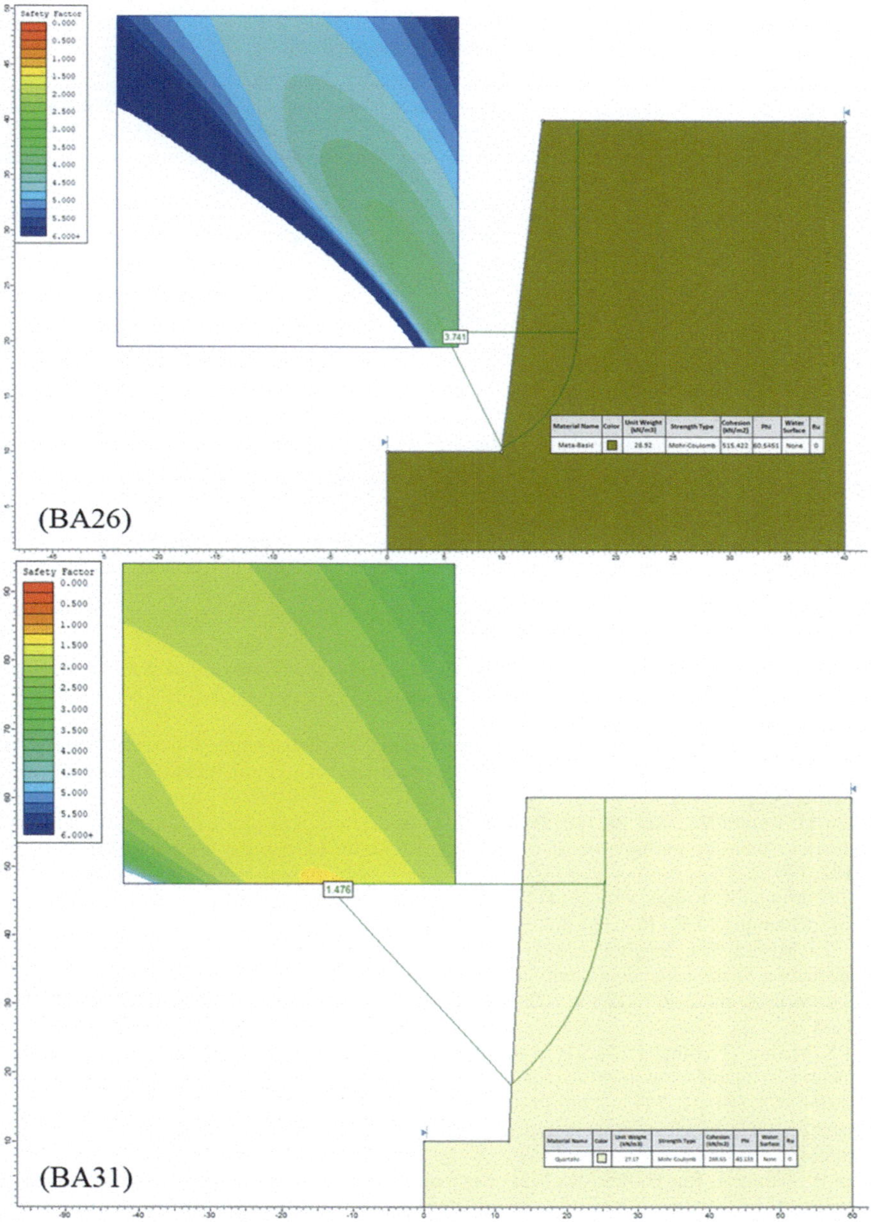

Fig. 12.3 Slopes indicating minimal safety factor for cut slope models, safety factor contours, and slip surfaces

Acknowledgements The authors express gratitude to the Head, Department of Geology, University of Lucknow for providing helpful suggestions and encouragement.

Funding This study was carried out under the research project sponsored by the Natural Resources Data Management System, Department of Science and Technology (Grant No. NRDMS/02/64/017), Government of India, and Center of Excellence project sponsored by Higher Education Government of Uttar Pradesh.

References

Bishop AW (1955) The use of slip circles in stability analysis of slopes. Géotechnique 5(1):7–17

Fellunius W (1936) Calculations of the stability of earth dams. In: Proceedings of the Second Congress of Large Dams. 4:445-63, Washington D. C.

Hoek E, Carranza-Torres CT, Corkum B (2002) Hoek-Brown failure criterion-2002 edition. Proc North Amer Rock Mech Soc

Huang YPH (2014) Slope stability analysis by the limit equilibrium method: fundamentals and methods, ASCE Press

Janbu N (1954) Stability analysis of slopes with dimensionless parameters. Harv Soil Mech Ser, 46

Morgenstern NR, Price VE (1965) The analysis of the stability of general slip surfaces. Geotechnique 15(1):77–93

Rocscience (2010) Slide: stability analysis for soil and rock slopes. Rocscience, Toronto. http://www.rocscience.com

Mishra VN, Rai PK, Singh P (2021) Geo-information technology in earth resources monitoring and management (edit. Book), Nova Science Publishers, U.S.A., pp 1–395. ISBN: 978–1–53619–669–6

Onagh M, Kumra VK, Rai PK (2012a) Application of multiple linear regression model in landslide susceptibility zonation mapping: the case study of narmab basin (Iran). Int J Geol, Earth Environ Sci, 2(2), pp 87–101

Onagh M, Kumra VK, Rai PK (2012b) Landslide susceptibility mapping in a part of uttarkashi district (India) by multiple linear regression method. Int J Geol, Earth Environ Sci, 2(2), pp 102–120

Rai PK, Mohan K, Kumra VK (2014) Landslide hazard and its mapping using remote sensing & GIS techniques. J Sci Res, Fac Sci, Banaras Hindu Univ 58:1–13

Rai PK, Mishra VN, Singh P (2021) Recent technologies for disaster management & Risk Reduction-Sustainable community resilience & responses (edit. Book), Springer Nature, Switzerland, pp 1–471. ISBN: 978–3–030–76116–5. https://doi.org/10.1007/978-3-030-76116-5

Rai PK, Mishra VN, Singh P (2022) Geospatial technology for landscape and environment management: sustainable assessment & planning (edit. Book), Springer Nature, Singapore. pp 1–431. ISBN: 978–981–16–7373–3. https://doi.org/10.1007/978-981-16-7373-3

Spencer E (1967) A method of analysis of the stability of embankments assuming parallel inter-slice forces. Geotechnique No. 1(17):11–26

Singh P, Sharma A, Sur U, Rai PK (2021) Comparative landslide susceptibility assessment using statistical information value and index of entropy model in Bhanupali Beri region. Himachal Pradesh, India, Environ Dev Sustain 23:5233–5250. https://doi.org/10.1007/s10668-020-00811-0

Sur U, Singh P, Rai PK (2021) Landslide Probability Mapping by considering Fuzzy Numerical Risk Factor (FNRF) and landscape change for road corridor of Uttarakhand, India, Environment. Dev Sustain (springer) 23:13526–13554. https://doi.org/10.1007/s10668-021-01226-1

Verma RK, Harshuman Singh R, Singh TN (2019) Highway design without considering landslide problem; In: Dev H and Singh R. (Eds). In: INDOROCK-2019, 8th Indian Rock Conference pp 303–310

Verma RK, Sharma P, Singh R, Singh TN (2021a) Evaluation of slope stability analysis using kinematic and chart method: a case study. J Geol Soc India 97(11):1387–1395

Verma RK, Singh R, Sharma P, Mishra P (2021b) Evaluation of slope stability analysis using conventional methods: A case study. In: Kapil SL (ed). In: Proceedings of International Conference EGCON-2021 pp 452–457

Chapter 13
Assessment of Land Use/land Cover Change and Forest Fragmentation in Kohima District, Nagaland: A Mountainous Region of Northeast India

Varun Narayan Mishra, Seyiekhriezo Vizo, Maya Kumari, Akshay Kumar, Rahul Kumar Misra, and Monika Mishra

Abstract Land use/land cover (LULC) change is the result of numerous interactions between anthropogenic actions and natural processes. It has an impact on ecological processes, biological cycles and biodiversity, making it a significant contributor to global climate change. The loss of ecosystem is linked to forest fragmentation, which also has an adverse effect on biodiversity, forest degradation and the spatial organisation of the landscape. The current work aims to examine the changes in LULC followed by forest fragmentation of Kohima district, Nagaland, India over the time span of 22 years using geospatial approach. A supervised classification method was utilised to prepare LULC maps using Landsat satellite images for the years 2000 and 2022, respectively. The overall classification accuracies were found 80.50% and 81.13% for classified maps of years 2000 and 2022 respectively. The results reveal that during 2000–2022, agriculture land and forest area are decreased by 1.73% and 3.39% while built up, vegetation/grassland and water bodies are increased by 4.33%,

V. N. Mishra (✉) · S. Vizo
Amity Institute of Geoinformatics and Remote Sensing (AIGIRS), Amity University, Sector 125, Noida 201313, India
e-mail: varun9686@gmail.com

M. Kumari
Amity School of Natural Resources and Sustainable Development (ASNRSD), Amity University, Sector 125, Noida 201313, India

A. Kumar
Environmental Science and Engineering Department, Indian Institute of Technology, Bombay, Powai 400076, India

R. K. Misra
Department of Economics, Khwaja Moinuddin Chishti Language University, Lucknow, India

M. Mishra
Pt. Rameshwar Tripathi Degree College, Raibareli, India

P. K. Rai et al. (eds.), *Natural Resources Management and Sustainable Livelihoods in the Mountainous Region*, Advances in Geographical and Environmental Sciences,
https://doi.org/10.1007/978-981-97-2100-9_13

0.70% and 0.09% respectively. By utilising multi-temporal Landsat data, the current work also focused to evaluate and measure the spatial pattern of forest fragmentation during 2000–2022. The forest area is divided into four categories using the Landscape Fragmentation tool: patch, edge, perforated and core forest with 100-m edge width. The study reveals reduction in rich forest cover and is now in a poor condition, while built-up land use has continued to rise. The examination of forest land fragmentation showed a rise in deforestation and the deterioration of the thick forest into scattered, small-scale areas.

Keywords Land use/land cover · Forest fragmentation · Remote sensing · Landsat

13.1 Introduction

Rapid urbanisation, industrialisation and other human activities have significantly changed the LULC over the last few years (Mmbaga et al., 2017; Mishra et al. 2018; Arora et al. 2021; Sajan et al. 2022). Alteration in LULC has emerged as one of the most intense impacts on the Earth's environment and natural resources (Mishra and Rai 2016; Gupta and Sharma, 2020). Analysing the processes and consequences of the changes in LULC is considered as the fundamental research in the field of landscape ecology and sustainable management of earth's resources (Wu and Hobbs, 2002; Arora et al., 2021; Mishra 2023). Large tracts of the forest are currently being reduced because of the rising demand of land for population growth, road construction for greater connectivity, agricultural land and other purposes. The land fragmentation is the decrease and split of any land area into two or more fragments because of LULC changes that leads to decline in forest biodiversity (Reddy et al. 2013). There are several biological and environmental effects of forest loss and fragmentation.

Geospatial technologies have been emerged as a potential tool for the management and conservation of natural resources. Timely and accurate information derived from remotely sensed images are the keys for quantifying and monitoring several aspects of the changes in LULC at different scales (Mishra et al. 2019; Guidigan et al. 2019; Arora et al. 2021; Li et al. 2023). Remote sensing method facilitates the cost affective temporal data for effectively investigating and quantifying the LULC changes over the large areas (Zhang et al. 2019). Landsat series satellites are the major source of freely available multi-date images for detecting the changes in LULC and forest, reported in many studies (Negassa et al. 2020; Arora et al. 2021; Shastri et al., 2020; Singh et al., 2021; Sajan et al. 2022; Mishra 2023).

The mountainous regions in north-eastern India have recently seen population growth as well, which has led to changes in the terrain, with forest regions suffering the most. LULC change detection focuses on identifying and quantifying the changes that occurred in landscape over time, while forest fragmentation analysis specifically examines the spatial patterns and its fragmentation (Batar et al. 2017; Devi and Simrah 2022). Ramchandra et al. (2016) conducted Geospatial analysis-based forest fragmentation for analysing the landscape dynamics. The changing pattern of Burdwan Forest Division (BFD) landscape is examined using satellite images to analyse structural pattern of the forest (Dutta et al. 2020). An integrated analysis of LULC change and forest fragmentation using RS and GIS allows to comprehensively understand the dynamic relationship between human activities and forest ecosystems in such mountainous regions. The present study aims to (1) assess the changing pattern of LULC and (2) examine the forest fragmentation and its pattern during the period of 2000–2022 in Kohima district of Nagaland, India. This kind of analytical study are invaluable and useful for land managers, environmental planners and policymakers in making informed decisions regarding land use planning, forest conservation and sustainable development.

13.2 Study Area

The present work was conducted in Kohima, the capital city of Nagaland, located in Northeast India. Kohima is situated in the eastern Himalayan region, and is known for its rich biodiversity, with lush green forests, rolling hills and valleys, home to a wide range of flora and fauna. The region's forest area can be broadly classified into three types: tropical evergreen forest, subtropical pine forest and temperate broadleaf forest. It is located at an altitude of 1,444 m above sea level. Kohima district's climate is characterised by mild to cool winters and warm to hot summers. The monsoon season begins in May and lasts until September, bringing heavy rainfall to the region. The average annual rainfall in Kohima is around 2,300 mm (90 inches). During the winter months (December to February), the temperature in Kohima can drop to 5 °C, with occasional snowfall in the higher elevations. In the summer months (June to August), the temperature can reach up to 30 °C. The geographical location of the study area is displayed in Fig. 13.1.

13.3 Materials and Methodology

13.3.1 Data Used

In the current study, multi-date remote sensing images (<10% cloud cover) acquired by Landsat 4–5 Thematic Mapper (TM) and Landsat 8 Operational Land Imager/Thermal Infrared Sensor (OLI/TIRS) are used. The specifications of the datasets are given in Table 13.1.

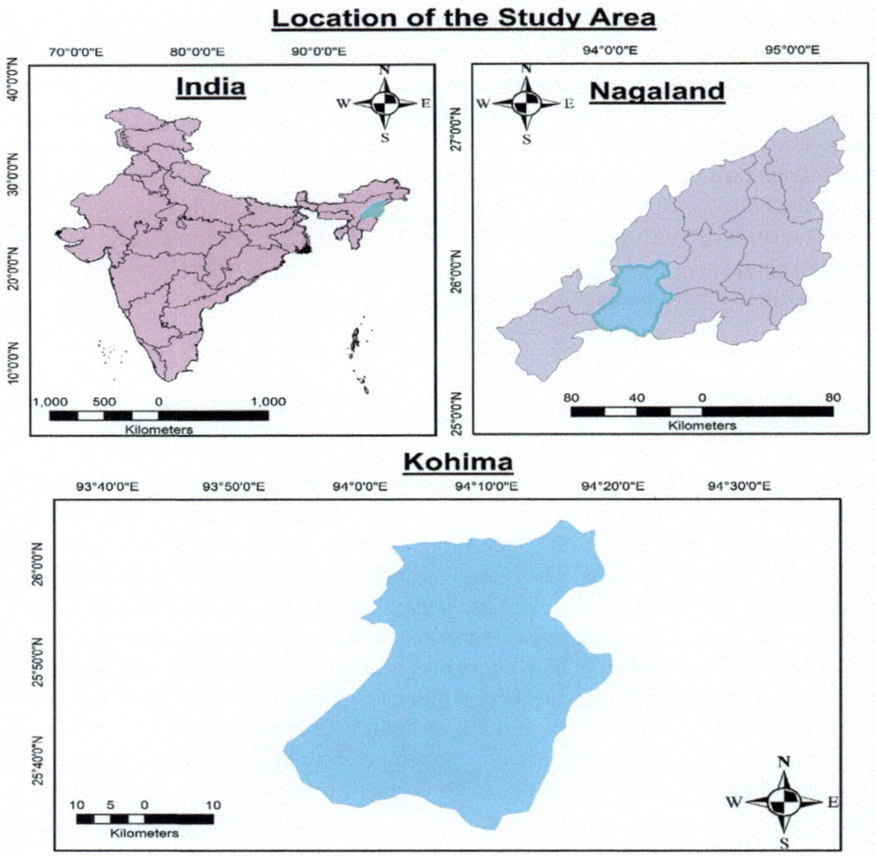

Fig. 13.1 Location map of the study area

Table 13.1 Specifications of datasets used in this study

Satellite-sensor	Path/Row	Date of acquisition	Spatial resolution (m)	Number of spectral bands
Landsat 8-OLI/ TIRS	135/42	25 November 2022	30	11
Landsat 4/5-TM	135/42	28 November 2000	30	7

13.3.2 LULC Classification of Images

Firstly, the images of different years are imported into ERDAS imagine software, and layer-stacking of all the bands has been performed. The co-registration of both the images is performed by applying image-to-image registration method to a common projection system. Afterward, spatial sub-setting was performed to get the area of interest. In addition to it, visual interpretation is also carried out to identify five major LULC classes, namely, forest area, agricultural lands, grasslands/vegetation, built-up and waterbodies for the images of years 2000 and 2022 respectively. An appropriate band combination is chosen to produce the false-colour composites (FCCs) of both the images followed by the creation of training samples for identified LULC classes in the study area. The training and testing pixels including various LULC categories are collected using random sampling approach.

The LULC maps are derived for the years 2000 and 2022 through classification of the images of individual years. The classification of individual images is performed by using supervised maximum likelihood classifier. It is a commonly used classification method based on the probability of a cell supposed to be assigned as a certain class (Jensen 2005; Mishra et al. 2019).

13.3.3 Assessment of LULC Classification Accuracy

The LULC classification results obtained for different years are assessed by utilising error matrix approach. The accuracy is measured with regard to the overall accuracy (OA), producer's accuracy (PA), user's accuracy (UA) and kappa coefficient (Kc) (Congalton and Green 1999).

13.3.4 Detection of LULC Changes

The change detection is a fundamental task focusing on the identification and analysis of the differences or changes that have occurred between thematic spatial maps captured at different time periods. It plays a crucial role in monitoring and understanding various dynamic phenomena, such as landscape changes, urban growth and deforestation. Detection and analysis of the changes is based on the classified LULC maps of the years 2000 and 2022 respectively. To assess the LULC changes during a particular period of 22 years, a post-classification approach is used.

13.3.5 Forest Fragmentation Analysis

Forest fragmentation analysis allows a comprehensive understanding of the spatial patterns and ecological consequences of forest cover. It supports conservation efforts, landscape planning and design of land management strategies sustainably to mitigate the adverse impacts of fragmentation on forest ecosystems. In this study, a qualitative and quantitative analysis of forest fragmentation is conducted by using the landscape fragmentation tool (LFT v2.0). This tool is designed to map different forms of fragmentation that occur in a certain landscape type. Here forest area and grasslands/vegetation's classes are selected for fragmentation analysis. It requires raster LULC data in raster format that has been categorised into forest and non-forest types.

13.4 Results and Discussion

13.4.1 LULC Classification and Accuracy Assessment Results

LULC classification corresponding to multi-date Landsat images of 2000 and 2022 is carried out into five classes based on supervised MLC algorithm. The LULC data collected for the years 2000 and 2022 provide an overview of its changing pattern in the study area.

Accuracy assessment is one of the most vital stages in the classification procedure. It evaluates the accuracy with which the pixels are sampled into suitable LULC classes. Accuracy assessment was performed using the error matrix method for the LULC classified results of years 2000 and 2022. Individual LULC results was compared to the ground reference data randomly collected through field visit and Google earth. The OA for the classified results of 2000 and 2022 are found to be 80.50% and 81.31%, respectively. The Kc values were found to be 0.748 and 0.754 for the classified results of years 2000 and 2022 respectively.

13.4.2 LULC Status and Change Detection

Spatial pattern of the LULC categories in Nagaland district for the years 2000 and 2022 is given in Fig. 13.2. The class-wise area distribution is given in Table 13.2. It revealed that in year 2000, 9.16% (135.07 km^2) area was under agriculture land, 1.90% (28.08 km^2) under built up, 0.31% (4.53 km^2) under grassland/vegetation, 88.46% (1304.94 km^2) under forest area, 0.18% (2.61 km^2) under water bodies. In 2022, the area under these classes was found to be 7.42% (109.48 km^2) for the agriculture land, 6.24% (92.02 km^2) for built up, 1.01% (14.86 km^2) for grassland/vegetation, 85.06% (1254.91 km^2) under forest area, 0.27% (3.97 km^2) for water bodies.

The changes in LULC occurred during the period (2000–2022) are assessed by implementing post-classification method. It evidently shows the significant changes witnessed by LULC classes in the study area over the specific period. During the period of investigation, both the positive and negative changes were witnessed by LULC classes. The LULC change detection including the amount of area changes (km^2) is given in Table 13.2.

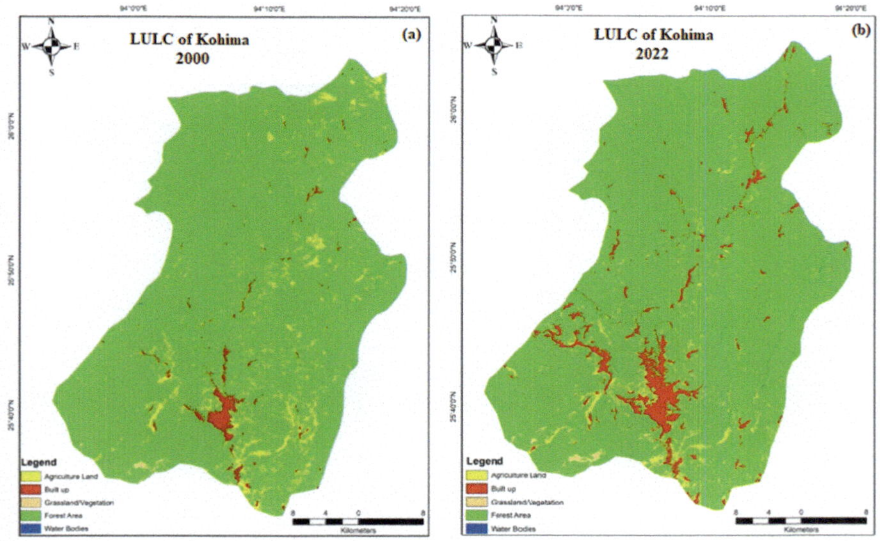

Fig. 13.2 LULC status of Kohima district for the years **a** 2000 and **b** 2022

Table 13.2 Area distribution of LULC and changes between 2000 and 2022

LULC class/year	2000		2022		2000–2022	
	Area (km^2)	Area (%)	Area (km^2)	Area (%)	Change (km^2)	Change (%)
Agriculture land	135.07	9.16	109.48	7.42	−25.59	−1.73
Built up	28.08	1.90	92.02	6.24	63.94	4.33
Grassland/ vegetation	4.53	0.31	14.86	1.01	10.33	0.70
Forest area	1304.95	88.46	1254.91	85.06	−50.04	−3.39
Water bodies	2.61	0.18	3.97	0.27	1.36	0.09

During 2000–2022, the agriculture land decreased by 25.59 km^2 which accounts for 1.73% of the entire study region. The built-up land increased by 63.94 km^2 which credits for 4.33% of the total study region. The grassland/vegetation class increased by 10.33 km^2 which accounts for 0.70%. There was a reduction of 50.04 km^2 area in the forest area of the study region. The water bodies increased from 2.61 to 3.97 km^2 in 2020, accounting for 0.09% of the total region.

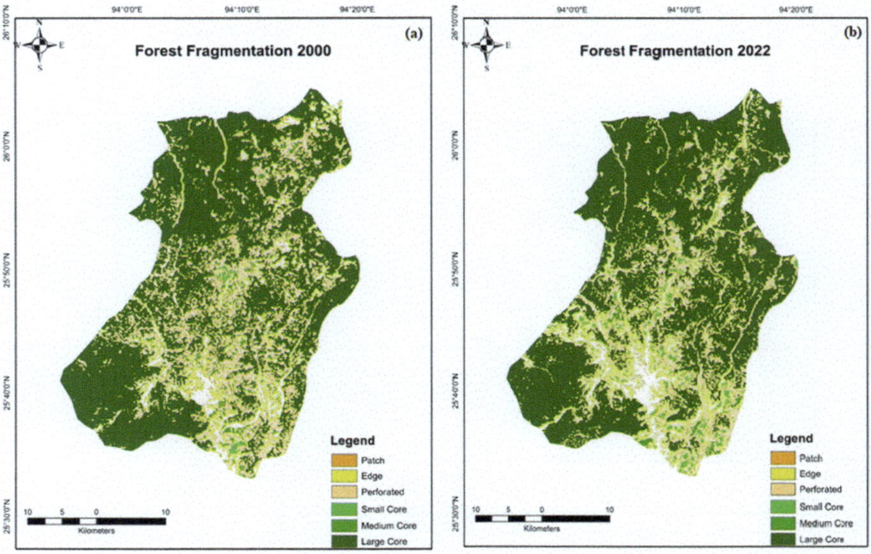

Fig. 13.3 Forest fragmentation of Kohima district for the years **a** 2000 and **b** 2022

13.4.3 Assessment of Forest Fragmentation

The classified LULC maps of different dates are categorised into two distinct classes (forest and non-forest) followed by forest fragmentation using LFT tool. Forest fragmentation of the years 2000 and 2022 are given in Fig. 13.3.

13.4.3.1 Patch Forest

It is scattered, entirely eroded by the "edge effect" pieces. Patch forests are completely engulfed by the edge effect and do not include any forest pixel that are more than 100 m from non-forest regions. Forest pixels are little patches of forest encircled by non-forested terrain. Core forest pixels and patch forest pixels must be totally separated. The process of forest fragmentation only starts when a land cover feature of non-forest must be 300 feet or less from a patch of forest. In the year 2000, patch forests covered an area of 5.72 km^2 and by 2022, that area had grown to 8.51 km^2 as shown in Figure. This small increase in patch forest area is due to a faster rate of forest conversion to non-forested regions. Outside of protected areas, most of the patch formations are visible, primarily in transition zones, followed by buffer zones and core regions. Native peoples' reliance on timber, forestry activities, forest fires and other infrastructural improvements as sources of income are the driving forces behind this. It is the most disturbed area of all the forest classifications.

13.4.3.2 Edge Forest

The forest pixels separate the main forest from expansive non-forested terrain features. The edge forest must be next to core forest and within 300 feet of a significant non-forest land cover feature. The area of Edge Forest declined from 78.14 sq. km in 2000 to 73.58 sq. km in 2022. The creation of farmland, roads and urban settlements are the main causes behind the formation of edge forests.

13.4.3.3 Perforated Forest

The interior border of small non-forest portions within a core forest is made up of perforated pixels, which belong to the least-disturbed group. The forest pixels separate the wooded landscape's main forest from its comparatively minute clearings (perforations). A somewhat modest forest clearing and core forest must be nearby and within 300 feet of the perforated forest. The perforated woods increased from 341 km^2 in 2000 to 370 km^2 in year 2022. The spatial extent of perforation depends on the overall extent of artificial forests in private enclosures with their related agricultural and plantation operations. Patch forests begin to emerge as the perforation grows worse, followed by edge forests, which are further broken up into non-forest regions.

13.4.3.4 Core Forest

These are the pixels of forest located at significant distance from the border between a forest and non-forest area. These are essentially the forest areas surrounded by other forests. To demonstrate the viability of the core patches in relation to patch size, the core forest was divided into three groups. Total forest cover within a landscape is divided into three categories such as of *small, medium and large cores*. It plays a superior role in preserving biodiversity than forest patch size. In the forest fragmentation map of the year 2000, the small core covered an area of 26.60 km^2 that grew to 32.69 km^2 in the year 2022. In the year 2000, medium core covered an area of 3.89 that increased up to 7.20 in the year 2022. The large core area decreased from 854 km^2 in 2000 to 827 km^2 in 2022. There is decline in core forest due to built-up expansion, construction of roads and agroforestry practices. The development of patch and edge forest speeds up the edge effect, putting core regions at risk of increased fragmentation. Table 13.3 shows the changes in different classes and patterns of forest in Kohima District during 2000–2022.

Table 13.3 reveals the forest fragmentation over the period of 22 year in Kohima district of Nagaland, India. Small and medium core forest areas grew by 0.45% and 0.25%, respectively, while the area covered by large core forests decreased by 2.35% during the period of study. It shows decline in dense forest areas. The edge forest decreased by 0.38% and patch forest zones grew by 0.21%, while the perforated forest increased by 2.12%. A widespread degradation of the forest cover is shown by the increasing perforation in the inner forest regions. Most of the forest has

Table 13.3 Changes in different forest classes in Kohima district during 2000–2022 based on forest fragmentation

Fragmentation class/year	2000		2022		2000–2022
	Area (km^2)	Area (%)	Area (km^2)	Area (%)	Change (%)
Patch	5.72	0.44	8.51	0.65	0.21
Edge	78.14	5.97	73.58	5.59	−0.38
Perforated	341	26.03	370.43	28.15	2.12
Small core	26.60	2.03	32.69	2.48	0.45
Medium core	3.89	0.30	7.20	0.55	0.25
Large core	854	65.23	827	62.88	−2.35

been fragmented along the highways, connecting towns, and villages. Therefore, the existing roads in the forest region of Kohima district provide easy access to the forest, which causes the fragmentation of forest land.

Furthermore, the characteristics of satellite data and classification methods may lead to uncertainties in this work. Inappropriate description of LULC categories may be considered as an uncertainty during the classification process. Interpretation expertise and shadows caused by hilly regions may also affect classification accuracy. Consequently, the LULC change assessment may vary with different scales and spatial resolution of the datasets.

13.5 Conclusion

The study is carried out to assess LULC changes followed by forest fragmentation between 2000 and 2022 in Kohima District, Nagaland. It witnessed decrease in forest area and agriculture land by 3.39% and 1.73% respectively. On the other hand, built up, grassland/vegetation and water bodies increased by 4.33%, 0.70% and 0.09%, respectively. Analysis of forest fragmentation showed that the area occupied by small and medium core forests increased by 0.45% and 0.25%, respectively, whereas the area occupied by large core forests decreased by 2.35%, confirming the reduction of dense forest from 2000 to 2022. Perforated forest has risen by 2.12%, while edge forest has decreased by −0.38% and patch forest zones have grown by 0.21%. Analyses of forest fragmentation, in addition to social and economic considerations, offer crucial insights into the possible repercussions of human disturbances. It is essential to develop effective management plans by analysing spatial patterns of forest change to preserving these imperilled ecosystems. As a result of the availability of temporal remote sensing data, analyses of forest changes are available. In a fragmentation study, the size of the core and the spatial resolution of images are closely associated. By considering nearby smaller patches, smaller core sizes have the effect of lowering the average inter-patch distance, which also lowers the edge effect. The study of ongoing LULC changes and forest fragmentation will help to contribute toward the conservation of forest and other natural resources in hilly terrain.

References

Arora A, Pandey M, Mishra VN, Kumar R, Rai PK et al (2021) Comparative evaluation of geospatial scenario-based land change simulation models using landscape metrics. Ecol Indic 128:107810

Batar A, Watanabe T, Kumar A (2017) Assessment of land-use/land-cover change and forest fragmentation in the Garhwal Himalayan region of India. Environ 4:34

Congalton RG, Green K (1999) Assessing the accuracy of remotely sensed data: principles and practices. CRC/Lewis Press, Boca Raton, FL

Devi AR, Shimrah T (2022) Assessment of land use and land cover and forest fragmentation in traditional landscape in Manipur, Northeast India. Int J Environ Sci Technol 19:10291–10306

Dutta S, Dutta I, Das A, Guchhait SK (2020) Quantification and mapping of fragmented forest landscape in dry deciduous forest of Burdwan forest division, West Bengal, India. Trees, Forests and People 2:100012

Guidigan MLG, Sanou CL, Ragatoa DS, Fafa CO, Mishra VN (2019) Assessing land Use/Land cover dynamic and its impact in Benin republic using land change model and CCI-LC products. Earth Syst Environ 3(1):127–137

Gupta R, Sharma LK (2020) Efficacy of spatial land change modeler as a forecasting indicator for anthropogenic change dynamics over five decades: a case study of Shoolpaneshwar Wildlife Sanctuary, Gujarat, India. Ecol Indic 112:106171

Jensen JR (2005) Introductory digital image processing: a remote sensing perspective, 3rd edn. Prentice Hall, Upper Saddle River

Li C, Roy SS, Grant R, Rhode-Barbarigos L (2023) Analysis of the spatial and temporal patterns in land use land cover in Belize city from 1991 to 2021. Remote Sens Appl: Soc Environ 32:101000

Mishra VN, Rai PK (2016) A remote sensing aided multi-layer perceptron-Markov chain analysis for land use and land cover change prediction in Patna district (Bihar), India. Arab J Geosci 9:249

Mishra VN, Rai PK, Prasad R, Punia M, Nistor MM (2018) Prediction of spatio-temporal land use/land cover dynamics in rapidly developing Varanasi district of Uttar Pradesh, India, using geospatial approach: a comparison of hybrid models. Appl Geomat 10:257–276

Mishra VN, Prasad R, Kumar P et al (2019) Assessment of spatio-temporal changes in land use/ land cover over a decade (2000–2014) using earth observation datasets: a case study of Varanasi district, India. Iran J Sci Technol Trans Civ Eng 43:383–401

Mishra VN (2023) Investigation of land use/land cover changes in Alaknanda river Basin, Himalaya during 1976–2020. In: Pandey M, Pandey C, Ray Y et al (eds) Advances in remote sensing technology and the three poles. John Wiley & Sons Ltd.

Mmbaga NE, Munishi LK, Treydte AC (2017) How dynamics and drivers of land use/ land cover change impact elephant conservation and agricultural livelihood development in Rombo, Tanzania. J Land Use Sci 12(2–3):168–181

Negassa MD, Mallie DT, Gemeda DO (2020) Forest cover change detection using geographic information systems and remote sensing techniques: a spatio-temporal study on Komto protected forest priority area, East Wollega zone, Ethiopia. Environ Syst Res 9:1

Ramchandra TVR, Setturu B, Chandran S (2016) Geospatial analysis of forest fragmentation in Uttara Kannada district, India. For Ecosyst 3:10

Reddy CS, Dutta K, Jha CS (2013) Analysing the gross and net deforestation rates in India. Curr Sci 105(11):1492–1500

Sajan B, Mishra VN, Kanga S, Meraj G, Singh SK, Kumar P (2022) Cellular automata-based artificial neural network model for assessing past, present, and future land use/land cover dynamics. Agronomy 12:2772

Shastri S, Singh P, Verma P, Rai PK, Singh AP (2020) Assessment of spatial changes of land use/land cover dynamics, using multi-temporal landsat data in Dadri block, Gautam Buddh Nagar, India. Forum Geogr 19(1) (June 2020):72–79. https://doi.org/10.5775/fg.2020.063.i

Singh A, Rai PK, Deka G, Biswas B, Prasad D, Rai VK (2021) Management of natural resources through integrated watershed management in Nana Kosi micro watershed, district Almora, India. Ecol Environ & Conserv 27(February Suppl. Issue):(S260−S268)

Wu J, Hobbs R (2002) Key issues and research priorities in landscape ecology: an idiosyncratic synthesis. Landsc Ecol 17:355–365

Zhang C, Wei S, Ji S, Lu M (2019) Detecting large-scale urban land cover changes from very high-resolution remote sensing images using CNN-based classification. ISPRS Int J Geo-Inf 8:189

Chapter 14
Overview of Sustainable Development Goals

Surendra Kumar Sharma, Deeksha, Praveen Kumar Rai, Satyavati Shukla, and Anoop Kumar Shukla

Abstract The adoption of the 17 Sustainable Development Goals (SDGs) was a significant step forward in humanity's pursuit for a more sustainable future. The world's advancement toward SDG accomplishment is tracked by 231 distinct social-ecological metrics distributed over 169 objectives, and remote sensing (RS) supplies Earth observation data for 30 of these metrics, directly or indirectly. Nonetheless, a number of projects have been launched in response to the demand for free access data. RS data may also be utilized to create potentially useful extra measures or sub-indicators. By doing so, they can help to overcome one of the primary challenges with SDG tracking: how to execute the SDG indicators effectively. The study looks at the durability of Earth Observation data in terms of SDG ongoing planning, monitoring, and assessment. The scientific community has made tremendous progress in producing geographic information at numerous spatial, spectral, radiometric, and temporal resolutions, allowing the data to be used for a variety of reasons. It looks at papers from a variety of scientific fields that employ geospatial data for mapping, monitoring, and analyzing the earth's properties and processes, as well as laying the groundwork for its application in accomplishing the SDGs.

S. K. Sharma
Indian Institute of Remote Sensing, Dehradun, India
e-mail: ssharma3@ce.iitr.ac.in

Deeksha · A. K. Shukla (✉)
Manipal School of Architecture and Planning, Manipal Academy of Higher Education, Manipal, Karnataka, India
e-mail: anoop.shukla@manipal.edu

Deeksha
e-mail: deeksha.msapma2@learner.manipal.edu

P. K. Rai
Department of Geography, Khwaja Moinuddin Chishti Language University, Lucknow, India

S. Shukla
Key Laboratory of Geospatial Informatics, Guilin University of Technology, Guilin, PR China

P. K. Rai et al. (eds.), *Natural Resources Management and Sustainable Livelihoods in the Mountainous Region*, Advances in Geographical and Environmental Sciences,
https://doi.org/10.1007/978-981-97-2100-9_14

Keywords Sustainable development goals · Remote sensing · Geospatial technology · GIS

14.1 Introduction

In an era marked by unprecedented global challenges, the Sustainable Development Goals (SDGs) have emerged as a beacon of hope and a roadmap for creating a more equitable, prosperous, and sustainable world. Encompassing a wide spectrum of interconnected objectives, the SDGs were established by the United Nations in 2015 as a comprehensive framework to address the most pressing economic, social, and environmental issues facing humanity.

The SDGs, consisting of 17 distinct but interlinked goals and 169 targets, serve as a blueprint for governments, organizations, and individuals to work collectively toward creating a sustainable future for all. These goals recognize the interdependence of economic, social, and environmental dimensions of development, and underscore the need for holistic and integrated approaches to problem-solving. The 2030 Agenda for Sustainable Development, which was approved by 193 member nations at the UN General Assembly Summit in September 2015, includes the 17 SDGs and 169 objectives. This became effective on January 1, 2016. These goals are the result of an exceptional consultation process that brought together national governments and millions of people from across the world to debate and support the global path to sustainable development over the next 15 years.

Issues such as poverty, hunger, well-being, education, gender parity, water and sanitation, energy access, economic progress and meaningful employment, infrastructure development, industrial innovation, inequality mitigation, sustainable urbanization, responsible resource use, climate change mitigation, ecosystems preservation, peace establishment, and collaborative partnerships should all serve as catalysts for action. This all-encompassing agenda underscores that merely prioritizing economic growth is inadequate; instead, it highlights the necessity of fostering a fairer and more equitable society, along with ensuring a safer and more prosperous planet.

Sustainable development is the practice of developing land and building projects in a way that minimizes their environmental effect by allowing them to generate energy-efficient self-sufficiency models. In its 1987 report "Our Common Future," the Brundtland Commission stated that "sustainable development is development that meets the needs of the present without jeopardizing future generations' ability to meet their own needs."

More than only the environment is at stake in sustainable development. It has a far broader reach than that. To ensure a vibrant and healthy society, it all comes down to addressing the diverse needs of individuals in many sectors, social cohesion, and equal opportunity. Sustainable development is also focused on finding better ways to accomplish things while maintaining our standard of living.

The establishment of the Millennium Development Goals (MDGs) in 2000 marked a significant change in bringing together international political determination for alleviating poverty and improving population health and well-being. As a result, the work begun with the MDGs must be completed. Many new issues have surfaced since the MDGs were established, while current ones have been aggravated. Conflicts and political instability have slowed or reversed progress in many countries, while inequality has increased, ecological degradation has gotten worse, migration issues and youth unemployment have gotten worse, and organized crime, including drug and human trafficking, violates human rights and stunts development. People demand for better governance, more receptive governments, and universal human rights everywhere in the world. The increasing interconnectedness of people's and countries' lives necessitates a global agenda tackling the world's most urgent concerns and exploiting opportunities. In light of these serious and unprecedented issues, there is a need for a shift in the process of developing global goals.

The MDGs were adopted by the United Nations member states in 2000. They set forth an ambitious vision for the future and acknowledged the need to coordinate efforts to advance development, peace and security, and human rights on three fronts at once. The MDGs fostered global awareness, political responsibility, improved measurements, social feedback, and public pressure by defining measurable and time-bound targets. There has been significant progress toward fulfilling the MDGs, as well as many achievements in meeting particular objectives both internationally and in individual nations. Since 2012, the United Nations has conducted large discussions on the shifting from MDGs to SDGs. More than a million individuals have contributed their views on the post-2015 development agenda through a series of worldwide, local and national forums in nearly 100 countries, as well as a social media platform. Reports from the Sustainable Development Solutions Network, the Global Compact Office, the United Nations System Task Team on the Post-2015 United Nations Development Agenda, regional commissions, civil society organizations, and accrediting bodies are among a prominent group of informing reports on the matter. The High-Level Panel of Eminent Persons on the Post-2015 Development Agenda, which was chaired by the Presidents of Indonesia, Liberia, and the United Kingdom, also released a report on the topic.

14.2 UN's SDGs

The UN's announcement of 17 Sustainable Development Goals and 169 objectives demonstrates the scope and ambition of the new global Agenda. These objectives aim to expand on the MDGs and finish what they left unfinished. They aspire to achieve all people's human rights, as well as gender equality and the empowerment of all women and girls. They are indivisible and interconnected, balancing the three pillars of sustainable development: economic, social, and environmental (Fig. 14.1).

Fig. 14.1 UN's 17 SDGs (*Source* https://sdgs.un.org/goals)

14.3 Role of Remote Sensing and GIS in Achieving SDGs

Remote sensing and Geographic Information Systems (GIS) are critical to accomplishing the Sustainable Development objectives by providing critical data, information, and tools for informed decision-making and tracking progress toward these objectives. The following are some major ways in which remote sensing and GIS contribute to the SDGs.

SDG 2: Zero Hunger

Remote sensing technology can assist to increase food security and alleviate hunger by monitoring agricultural operations, assessing crop health, and predicting production estimates. GIS supports sustainable farming practices by assisting in the effective management and optimization of agricultural resources (FAO 2016; Mishra et al. 2021; Rai et al. 2021, 2022; Shukla et al. 2023).

As food safety becoming an extensively discussed and contentious issue of discussion throughout the world, having accurate and strong food crop production projections becomes critical. Soil is likely a highly important element of nature that sustains every living thing on Earth, either directly or indirectly. It is becoming more crucial to constantly check soil health because healthy soil supports a strong crop yield. Tripathi et al. (2022) used remote sensing techniques to understand the soil health parameters to increase the crop production rate. The effects of climate change are perhaps the most serious hazard to the world food supply. The undeniable effects of warming temperatures on food systems have lately been recognized. As a result, considerable research attempts are being made across the world to investigate and evaluate both the immediate and future effects of the changing climate on the evolution of world food networks. Jabal et al. (2022) discussed about the impact of climatic condition on the crop production.

Determining agricultural field and produce constitutes one of among the most essential challenges that policymakers and decision makers need to address in order to estimate yearly crop production and food supply. Meraj et al. (2022) carried out the study on various crop types with respect to the plot area to get precise amount of crop production to facilitate food security. If a major climatic catastrophe happens, it is critical that the nations immediately affected as well as other players in the worldwide food industry pay heed. When a nation has massive natural disasters and bad food production, citizens can lose faith in the government's ability to provide safety. Research also focuses on examining the impact of natural disasters on the food crops and their mitigation strategies (Wang et al. 2022; Zhang et al. 2022a, b; Rai 2023a, b). Soil is an important ecosystem service, which plays a vital role in food production. Zhang et al. (2022a, b) discussed about the regulation of land use land cover to improve the crop cultivation in the cities.

SDG 6: Safe Drinking Water and Sanitation

Remote sensing can track water quality and detect pollution sources, while GIS allows for optimal water resource management and sanitation infrastructure development (Singh and Rai 2017; Shukla et al. 2018; Singh et al. 2021; UNESCO 2021).

The natural land use and land cover (LULC) of our planet has changed from prehistoric times and is anticipated to do so in the years to come. There has been a major effect on the ecosystem as a consequence of this transformation, which was brought about by both natural and human sources. A vital supply of freshwater that is required for maintaining life is groundwater. For the extraction, utilization, and management of water resources, it is necessary to identify potential groundwater zones. Due to the interdependence of water, food, and energy and the effects these components have on human lifestyles, national economies, and the global economy, water shortage is a worldwide concern. Mechal et al. (2022) analyzed the effects on the elements of the water cycle and stream flow. Li et al. (2023a, b) discussed about the delineation and identification of potential groundwater resource locations using remote sensing data. Dasgupta and Sanyal (2022) determined hydrological and geological factors impacting groundwater resources, forecast shallow and deep groundwater change, evaluated the drivers based on their influence on groundwater distribution, and comprehended groundwater distribution as a result of LULC shift (Rai 2023a, b). The recent study focuses on water as ecosystem services where, people and civilizations benefit from ecosystem services such as food, water, pollination, nutrient cycling, and many others. Hydrological ecosystem services (HESS), also known as water-related ecosystem services, connect these services to the hydrological cycle, indicating that the amount of the ecosystem service is dependent on water availability, i.e., quantity and quality (Gupta et al. 2023; Ha et al. 2023; Nayak and Shukla 2022; Mohan et al. 2011).

Deeksha and Shukla (2022) presented a systematic review and future directions about freshwater Ecosystems. Patel and Nihalani (2023), presented a review related to the management of urban water networks using GIS.

SDG 11: Sustainable Cities and Communities

Remote sensing and GIS aid in urban planning, monitoring urban expansion, and enhancing infrastructure management, resulting in more sustainable and resilient cities (Mishra et al. 2018; SDNS 2019).

Currently, the world's ecosystems are experiencing massive human disruptions, which are causing the natural environment to change faster than at any time prior. Increasing degrees and intensities of human activity have caused disruptions in habitats on multiple stages. There is a rising demand for effective tools for quantifying human disturbances and detecting the consequent geographical and temporal alterations in ecological state (Yu et al. 2023).

Remote sensing images is critical in urban studies for extracting and tracking urban built-up zones. Several researchers have proposed techniques for calculating regional, national, and global built-up areas. The bulk of them, however, employed restrictive criteria and a manual sample selection process, resulting in long and useless algorithms. By integrating Night Time Lights (NTL), Sentinel-2 multispectral data, Sentinel-1 Radar imaging, and SRTM elevation data with Google Earth Engine, researchers can detect built-up regions in real time (Samadzadegan et al. 2023). Green space within towns is critical for reducing extreme temperatures, pollution, and floods (Tillihal and Shukla 2023). In addition, it is becoming obvious that urban greenery is crucial for both the physical and mental well-being of city dwellers, as well as that urban green area may contain distinctive species. The research works focus on national capital cities, researchers measured the influence of environmental and socioeconomic variables (temperature, rainfall, growth in society, and people concentration) on urban green space internationally (Bille et al. 2023). Adding on the viability of utilizing satellite data to calculate the Cooling Establishment Time for various Blue-Green Infrastructures within the city is calculated (Gobatti et al. 2023). The new focus of these days concentrated on evaluating the impact of LULC change on the ecosystem service (Geng et al. 2022).

Zheng et al. (2023) discussed about various urban-related applications and challenges of night time lights.

SDG 13: Climate Action

Remote sensing aids in the monitoring of climate change consequences such as deforestation, carbon emissions, and land use changes, whereas GIS aids in climate modeling and adaptation measures (Sharma et al. 2021; UNEP 2017).

The impact of climate change is now an international issue as a result of carbon emissions caused by modernization and urbanization. Carbon dioxide (CO_2) is a major greenhouse gas released mostly as a result of the usage of fossil fuels and deforestation across the world. Carbon storage entails creating a carbon sink through plant photosynthesis, and the most common form of area utilized for carbon storage is green space, which is structured with a variety of plant species to absorb CO_2. Therefore, works establish a comprehensive strategy to assess the balance among green space availability and demand based on carbon-related ecosystem services

(Xu et al. 2023). There are works which also give insight on remote sensing tools to investigate spatiotemporal patterns changes in carbon storage (Kafy et al. 2023).

Regardless of a regional, national, or global scale, environmental pollution is a critical issue owing to its impact on the well-being of humans and the surroundings. The study of the problem of atmospheric air pollution employs a variety of methodologies. The employment of monitoring devices, such as ground-based air quality monitors and satellite equipment, is one of the primary techniques for the investigation of atmospheric air pollution. The application of computational models to examine the responses of contaminants in the environment has risen over the past decade owing to the tight connection between the field and distant research methodologies. Tabunschik et al. (2023) presented the evaluation of air pollution concentrations using cutting-edge geoinformation research methodologies. Human progress has resulted in the loss of species hundreds of times quicker than the normal rate over the last 10 million years, with over a million species being threatened with extinction. Ma et al. (2022) discussed about determining the factors influencing plant diversity in urban parks and investigated the role of the park's internal patch variables and exterior landscape qualities in species richness, and also investigated the quantitative correlations between these driving factors and species diversity.

SDG 15: Life on Land

Remote sensing aids in the monitoring of biodiversity, ecosystem health, and deforestation, while GIS aid in the mapping and conservation of natural resources and protected areas (IPBES 2019).

Human productivity and existence are strongly reliant on ecosystem structures, functions, and processes. These ecosystems offer critical biological functions that support people's well-being and the production of ecological goods required for human life. Costanza introduced a technique based on the link among ecosystem services and functions for assessing ecosystem functions in an organized manner in 1984, which was later expanded with Kellert's loss-benefit method for a standardized quantitative evaluation of biosphere materials. Lot of studies related to the work pertaining to evaluation of ecosystem services, functions, etc., have been discussed by (Akbar et al. 2023; Yao et al. 2023). Vatitsi et al. (2023), described the work related to the eco-environmental health and ecosystem services.

Forest biomass is critical in understanding the state of ecosystem functioning and terrestrial carbon stock. Satellite remote sensors for effective and spatiotemporally consistent forest monitoring have become a basic technology for above ground biomass estimation. Chen et al. (2023) presented the study related to the use of satellite data for object-based mapping.

River corridors have a significant role in pollution removal, biological habitat improvement, biodiversity conservation, and landscape enhancement. The urban river is a significant component of the urban ecosystem, serving as a transporter of critical environmental components as well as a distinctive landscape that influences urban regulation and cultural services. Li et al. (2023a, b) presented the study related to assessing the ecological health of urban riverscape corridors.

Efficient environmental evaluation is a serious task that necessitates an integrated strategy. To capture these multidimensional approaches, researchers are working on the influence of urban development on the biological environment in hilly places (Zhao et al. 2023).

SDG 17: Partnerships for the Goals

Remote sensing and GIS promote cooperation among multiple stakeholders by offering a centralized platform for data exchange, analysis, and decision-making (World Bank 2018).

SDGs and India

India, like many other countries, has agreed to the SDGs of the United Nations as the basis for tackling its social, economic, and environmental challenges. The SDGs have been incorporated into the Indian government's development agenda, policies, and programs, with the goal of achieving sustainable development by 2030. India has made extensive use of remote sensing technologies to track and analyze progress toward the SDGs. The following are some significant areas that India is working to align with the SDGs:

SDG 1: No Poverty

To enhance financial inclusion and create work chances for the poor, India has established numerous poverty alleviation programs, such as the Mahatma Gandhi National Rural Employment Guarantee Act (MGNREGA) and the Pradhan Mantri Jan Dhan Yojana (PMJDY). Furthermore, programs such as the National Social Assistance Program (NSAP) target social protection for disadvantaged communities.

SDG 2: Zero Hunger

The Indian government is promoting agricultural productivity and food security through programs such as the Pradhan Mantri Fasal Bima Yojana (PMFBY) and the National Food Security Act (NFSA). Post-harvest losses are being reduced, and food distribution networks are being improved. The Indian Space Applications Centre (SAC) and the Indian Council of Agricultural Research (ICAR) have teamed to use remote sensing data to monitor crop health and predict production. The data is combined with on-the-ground observations to estimate food security and develop suitable agricultural policy.

SDG 3: Good Health and Well-Being

During the COVID-19 pandemic, India employed remote sensing and GIS to track people's movements and detect possible disease breakout regions. This information aided health officials in developing targeted containment measures and allocating resources.

SDG 6: Clean Water and Sanitation

India has taken initiatives to improve accessibility to clean water and sanitary facilities, especially in rural areas. The Swachh Bharat Mission seeks universal sanitation coverage, whereas the Jal Jeevan Mission seeks to provide piped water to every family. The National Remote Sensing Centre (NRSC) of India has been monitoring water bodies, groundwater levels, and water quality using remote sensing data. The data aids in the development and implementation of water resource management plans, particularly in water-stressed areas (NRSC—ISRO). Janardhanan et al. (2023) investigated the long-term groundwater storage losses in the Ganga Basin regional aquifer in India. Shukla et al. (2020), investigated the influence of spatial urbanization on water basins.

SDG 7: Affordable and Clean Energy

To satisfy its energy demands while lowering carbon emissions, India is pushing renewable energy sources such as solar and wind. Through numerous projects, the nation is also attempting to electrify distant and rural regions.

SDG 11: Sustainable Cities and Communities

To construct sustainable and inclusive cities, India is focused on smart city development, urban redevelopment initiatives, and upgrading public transit. Initiatives such as the Atal Mission for Rejuvenation and Urban Transformation (AMRUT) and the Pradhan Mantri Awas Yojana (PMAY) aim to develop urban infrastructure and offer affordable homes. The Smart Cities Mission of the Government of India focuses on making cities and human settlements inclusive, safe, resilient, and sustainable. The Smart Cities Mission aims to transform selected cities across India into smart and sustainable urban centers by leveraging technology, data, and innovative solutions.

The National Remote Sensing Centre and India's urban development authorities have been using remote sensing technology to track urban expansion, monitor encroachments, and plan sustainable urban development. This information contributes to the development of smart cities and inclusive urban planning (Smart Cities Mission, Government of India). Remote sensing serves as one of the most effective and affordable methods for evaluating LULC shifts on the Earth's surface using modern computer techniques like categorization and change detection (Dahiya et al. 2023; Sahoo et al. 2022). To estimate the effect of LULC change on the temperature, researchers study surface temperature and urban heat islands (UHIs) fluctuations during two seasons, winter and summer (Sahoo et al. 2022).

SDG 13: Climate Action

India has pledged to reduce the intensity of its greenhouse gas emissions while expanding the percentage of non-fossil fuel-based energy sources. The government is also actively involved in global climate discussions and climate change projects.

Monitoring Forest Carbon and REDD+ Implementation

The Forest Survey of India (FSI) assesses forest cover changes and monitors carbon sequestration activities using remote sensing data. This information is critical for India's participation in REDD+ (Reducing Emissions from Deforestation and Forest Degradation) efforts (Forest Survey of India).

Wildfires in forests are growing into a severe environmental issue not exclusively in India, but in several other countries throughout the world, as a result of climate change and accompanying regional and national warming. Furthermore, forest fires are increasingly a major source of forest destruction in India, particularly in the dry deciduous wooded zone. To access this situation researchers analyzed the effects of wildfires on the efficiency of terrestrial ecosystems in India (Sannigrahi et al. 2020a, b). Few other researchers estimated and anticipated the aboveground biomass, carbon stock, and carbon sequestration capacity of a region's various land uses and connected these projections to LULC changes (Bordoloi et al. 2022). Indian mangrove forests occupy an estimated 4662.56 km2 and are found in nine coastal states and three union territories. Bindu et al. (2020) focused on the mangrove vegetation health and its potential to hold the carbon stock (Bindu et al. 2020).

SDG 15: Life on Land

India is striving to conserve biodiversity-rich regions through afforestation. Initiatives such as the Green India Mission and Project Tiger are devoted to the preservation of forests and animal habitats.

Remote sensing data is used by India's Ministry of Environment, Forests, and Climate Change (MoEFCC) to monitor protected areas, follow wildlife habitats, and analyze deforestation. The information aids conservation efforts and aids in the preservation of biodiversity-rich areas (Source: MoEFCC—Government of India).

The expansion of LULC changes during the last three decades has resulted in a massive shift of global topography characteristics. Landscape shifts are caused by a variety of variables, including abiotic variation such as climate, terrain, and soils, biotic interactions that induce spatial patterning even in uniform environmental circumstances, and historical and current urbanization trends (Shamsudeen et al. 2022). These impacts are found on ecosystem services, which can be monitored using earth observation satellite data. In a few cases, geographically explicit Ecosystem Service Value (ESV) of six eco-regions in the Sundarbans Biosphere Reserve (SBR), India, was calculated employing time-series satellite data sets and multiple unit values (Sannigrahi et al. 2020a, b).

Given its large population and complex socioeconomic situation, India confronts considerable obstacles in reaching all of the SDGs. The government, in collaboration with civil society and private sector stakeholders, is putting policies and programs in place to foster sustainable development across the country.

14.4 Conclusions

The 17 SDGs were established to increase human well-being, safeguard natural assets, and reduce the effect of human activity on the world for subsequent generations to come. In contrast to the preceding MDGs, the SDGs are intended for countries that are developing as well as developed. Given the SDGs' many themes and sectors, monitoring is crucial to meeting the targets by 2030, as well as modifying current methods for enhanced operational and precise targeting. Spatial data may be used to observe geographic variability. Consequently, it is advantageous to identify social and economic inequalities at both the national and local levels. Numerous studies have demonstrated the value of using geospatial data to monitor SDG progress and growth and plan future strategies. It is, however, not adequately utilized in the monitoring and assessment of global issues and objectives. Given the diversity of SDG objectives, geospatial information is one of the most crucial instruments for tracking their success. It will also pave the road for the successful implementation of the SDGs. Based on this finding, geospatial strategies for the implementation and monitoring of SDGs 5, 8, 10, and 17 remain vital, despite the fact that relatively little research has been conducted. Despite the notion of sustainability, and the notion that today's population is responsible for the next generation, the world isn't on pace to meet the majority of the SDGs by the goal date of 2030.

Achieving the SDGs would surely need significant worldwide collaboration to make optimal use of data sharing, processing, and aggregation in a highly interdisciplinary environment. To provide reliable and precise data that can be utilized to create comprehensive sustainable development strategies, national geospatial information centers will need to collaborate with experts in national statistics and earth observation. Nonetheless, a number of projects have been launched to meet the demand for free access data. One of the primary challenges in SDG monitoring is how to put the SDG indicators into practice. RS data can help with the formulation of possibly important additional indicators or sub-indicators.

The evidence-driven goals serve as a framework for action, highlighting the need of complete biodiversity surveys, sound natural resource management techniques, and inclusive community development. There is a request for improved monitoring systems, improved data collecting in remote places, and a clearer knowledge of the complicated interactions between different aspects of the ecosystem to overcome existing gaps. Once filled, these gaps will pave the way for more effective tactics and treatments.

Future solutions must be dynamic and flexible, taking into account the changing problems provided by climate change, socioeconomic dynamics, and global interdependence. A forward-thinking strategy includes sustainable livelihood programs, climate-smart behaviors, and participatory governance approaches. Furthermore, education and awareness campaigns will be critical in building a sense of communal responsibility for sustainable mountain development, ensuring that communities are not only aware of the difficulties, but also equipped with the knowledge to actively contribute to solutions. Continuous research and innovation should be pursued, with an emphasis on new technologies and interdisciplinary studies that combine traditional knowledge with current science. The purpose of policy integration is to match regional and national policies with the imperatives of sustainable mountain development. Infrastructure development, like all other aspects of development, must be addressed with forethought, with an emphasis on reducing environmental effect and guaranteeing long-term sustainability of projects.

In essence, the data, gap analysis, and future plans provided here constitute a roadmap—a complete guide for policymakers, scholars, and communities devoted to mountainous area sustainability. The problems are significant, but so are the potential for constructive change. By accepting these aims and tactics, we begin on a path toward a more sustainable and peaceful cohabitation with the unique and sensitive ecosystems present in our alpine areas. The road ahead will need dedication, teamwork, and a common vision for a future in which both environment and communities thrive.

References

Akbar A, Abulizi A, Abliz A, Abliz A, Jiang J, Yu T, Ou B (2023) Land use and land cover change effects on the value of ecosystem services in the Konqi River Basin, China, under Ecological Water Conveyance Conditions. Forests 14(5). https://doi.org/10.3390/f14051028

Bille RA, Jensen KE, Buitenwerf R (2023) Global patterns in urban green space are strongly linked to human development and population density. Urban Forest Urban Greening 86. https://doi.org/10.1016/j.ufug.2023.127980

Bindu G, Rajan P, Jishnu ES, Ajith Joseph K (2020) Carbon stock assessment of mangroves using remote sensing and geographic information system. Egypt J Remote Sens Space Sci 23(1):1–9. https://doi.org/10.1016/j.ejrs.2018.04.006

Bordoloi R, Das B, Tripathi OP, Sahoo UK, Nath AJ, Deb S, Das DJ, Gupta A, Devi NB, Charturvedi SS, Tiwari BK, Paul A, Tajo L (2022) Satellite based integrated approaches to modelling spatial carbon stock and carbon sequestration potential of different land uses of Northeast India. Environ Sustain Ind 13. https://doi.org/10.1016/j.indic.2021.100166

Chen L, Ren C, Zhang B, Wang Z, Man W, Liu M (2023) Improved object-based mapping of aboveground biomass using geographic stratification with GEDI data and multi-sensor imagery. Remote Sens 15(10). https://doi.org/10.3390/rs15102625

Dahiya N, Singh S, Gupta S, Rajab A, Hamdi M, Elmagzoub MA, Sulaiman A, Shaikh A (2023) Detection of multitemporal changes with artificial neural network-based change detection algorithm using hyperspectral dataset. Remote Sens 15(5). https://doi.org/10.3390/rs15051326

Dasgupta B, Sanyal P (2022) Linking land use land cover change to global groundwater storage. Sci Total Environ 853. https://doi.org/10.1016/j.scitotenv.2022.158618

Deeksha, Shukla AK (2022) Ecosystem services: a systematic literature review and future dimension in freshwater ecosystems. Appl Sci (Switzerland) 12(17). https://doi.org/10.3390/app12178518

FAO (2016) The state of food and agriculture, 2016. Climate change, agriculture and food security

Geng J, Yuan M, Xu S, Bai T, Xiao Y, Li X, Xu D (2022) Urban expansion was the main driving force for the decline in ecosystem Services in Hainan Island during 1980–2015. Int J Environ Res Pub Health 19(23). https://doi.org/10.3390/ijerph192315665

Gobatti L, Bach PM, Scheidegger A, Leitão JP (2023) Using satellite imagery to investigate Blue–Green infrastructure establishment time for urban cooling. Sustain Cities Soc 104768. https://doi.org/10.1016/j.scs.2023.104768

Gupta LK, Pandey M, Raj PA, Shukla AK (2023) Fine sediment intrusion and its consequences for river ecosystems: a review. J Hazard Toxic Radioactive Waste 27(1):04022036. https://doi.org/10.1061/(ASCE)HZ.2153-5515.0000729

Ha LT, Bastiaanssen WGM, Simons GWH, Poortinga A (2023) A new framework of 17 hydrological ecosystem services (HESS17) for supporting river basin planning and environmental monitoring. Sustainability (Switzerland) 15(7). https://doi.org/10.3390/su15076182

IPBES (2019) The global assessment report of the intergovernmental science-policy platform on biodiversity and ecosystem services

Jabal ZK, Khayyun TS, Alwan IA (2022) Impact of climate change on crops productivity using MODIS-NDVI time series. Civil Eng J (Iran) 8(6):1136–1156. https://doi.org/10.28991/CEJ-2022-08-06-04

Janardhanan S, Nair AS, Indu J, Pagendam D, Kaushika GS (2023) Estimation of groundwater storage loss for the Indian Ganga Basin using multiple lines of evidence. Sci Rep 13(1). https://doi.org/10.1038/s41598-023-28615-y

Kafy AAl, Saha M, Fattah MA, Rahman MT, Duti BM, Rahaman ZA, Bakshi A, Kalaivani S, Nafiz Rahaman S, Sattar GS (2023) Integrating forest cover change and carbon storage dynamics: leveraging Google Earth Engine and InVEST model to inform conservation in hilly regions. Ecol Ind 152. https://doi.org/10.1016/j.ecolind.2023.110374

Li J, Huang L, Zhu K (2023a) Ecological health assessment of an urban river: the case study of Zhengzhou City, China. Sustainability (Switzerland) 15(10). https://doi.org/10.3390/su15108288

Li Y, Abdelkareem M, Al-Arifi N (2023b) Mapping potential water resource areas using GIS-based frequency ratio and evidential belief function. Water (Switzerland) 15(3). https://doi.org/10.3390/w15030480

Ma Z, Zhang P, Hu N, Wang G, Dong Y, Guo Y, Wang C, Fu Y, Ren Z (2022) Understanding the drivers of woody plant diversity in urban parks in a snow climate city of China. J Forest Res. https://doi.org/10.1007/s11676-022-01535-9

Mechal A, Takele T, Meten M, Deyassa G, Degu Y (2022) A modeling approach for evaluating the impacts of land use/land cover change for Ziway Lake Watershed hydrology in the Ethiopian Rift. Model Earth Syst Environ 8(4):4793–4813. https://doi.org/10.1007/s40808-022-01472-w

Meraj G, Kanga S, Ambadkar A, Kumar P, Singh SK, Farooq M, Johnson BA, Rai A, Sahu N (2022) Assessing the yield of wheat using satellite remote sensing-based machine learning algorithms and simulation modeling. Remote Sens 14(13). https://doi.org/10.3390/rs14133005

Mishra VN, Rai PK, Singh P (2021) Geo-information technology in earth resources monitoring and management (edited book). Nova Science Publishers, USA, pp 1–395. ISBN: 978-1-53619-669-6

Mishra VN, Rai PK, Rajendra P, Puniya M, Nistor MM (2018) Prediction of spatio-temporal land use/land cover dynamics in rapidly developing Varanasi district of Uttar Pradesh, India, using geospatial approach: a comparison of hybrid models. Appl Geomat (Springer). https://doi.org/10.1007/s12518-018-0223-5

Mohan K, Shrivastava A, Rai PK (2011) Ground water in the City of Varanasi, India: present status and prospects. Quaestiones Geographicae 30(3):47–60. https://doi.org/10.2478/v10117-011-0026-9

Nayak D, Shukla AK (2022) Review of state-of-the-art research on river hydrological hazards, restoration, and management. In: River dynamics and flood hazards: studies on risk and mitigation). Springer Nature Singapore, pp 463–482. https://doi.org/10.1007/978-981-19-7100-6_25

Patel K, Nihalani S (2023) A review of _urban water networks management using GIS. MethodsX 11:102261. https://doi.org/10.1016/j.mex.2023.102261

Rai PK (2023a) River conservation and water resource management. Springer Nature, Singapore, pp 1–352. ISBN No.: 978-981-99-2604-6. eBook ISBN: 978-981-99-2605-3

Rai PK (2023b) Advances in water resource planning & sustainability. Springer Nature, Singapore, pp 1–322. ISBN No. 978-981-99-3659-5

Rai PK, Mishra VN, Singh P (2021) Recent technologies for disaster management & risk reduction-sustainable community resilience & responses (edited book). Springer Nature, Switzerland, pp 1–471. ISBN: 978-3-030-76116-5. https://doi.org/10.1007/978-3-030-76116-5.

Rai PK, Mishra VN, Singh P (2022) Geospatial technology for landscape and environment management: sustainable assessment & planning (edited book). Springer Nature, Singapore, pp 1–431. ISBN: 978-981-16-7373-3. https://doi.org/10.1007/978-981-16-7373-3

Sahoo S, Majumder A, Swain S, Gareema, Pateriya B, Al-Ansari N (2022) Analysis of decadal land use changes and its impacts on Urban Heat Island (UHI) using remote sensing-based approach:

a smart city perspective. Sustainability (Switzerland) 14(19). https://doi.org/10.3390/su141911892
Samadzadegan F, Toosi A, Dadrass Javan F (2023) Automatic built-up area extraction by feature-level fusion of Luojia 1–01 nighttime light and Sentinel satellite imageries in Google Earth Engine. Adv Space Res. https://doi.org/10.1016/j.asr.2023.05.015
Sannigrahi S, Chakraborti S, Banerjee A, Rahmat S, Bhatt S, Jha S, Singh LK, Paul SK, Sen S (2020a) Ecosystem service valuation of a natural reserve region for sustainable management of natural resources. Environ Sustain Ind 5. https://doi.org/10.1016/j.indic.2019.100014
Sannigrahi S, Pilla F, Basu B, Basu AS, Sarkar K, Chakraborti S, Joshi PK, Zhang Q, Wang Y, Bhatt S, Bhatt A, Jha S, Keesstra S, Roy PS (2020b) Examining the effects of forest fire on terrestrial carbon emission and ecosystem production in India using remote sensing approaches. Sci Total Environ 725. https://doi.org/10.1016/j.scitotenv.2020.138331
SDNS (2019) Sustainable development solutions network. www.pica-publishing.com
Shamsudeen M, Padmanaban R, Cabral P, Morgado P (2022) Spatio-temporal analysis of the impact of landscape changes on vegetation and land surface temperature over Tamil Nadu. Earth (Switzerland) 3(2):614–638. https://doi.org/10.3390/earth3020036
Sharma C, Shukla AK, Zhang Y (2021) Climate change detection and attribution in the Ganga-Brahmaputra-Meghna river basins. Geosci Front 12(5):101186. https://doi.org/10.1016/j.gsf.2021.101186
Shukla AK, Ojha CSP, Garg RD, Shukla S, Pal L (2020) Influence of spatial urbanization on hydrological components of the Upper Ganga River Basin, India. J Hazard Toxic Radioactive Waste 24(4). https://doi.org/10.1061/(asce)hz.2153-5515.0000508
Shukla AK, Shukla S, Surampalli RY, Zhang TC, Yu YL, Kao CM (2023) Modeling microconstituents based on remote sensing and GIS Techniques. In: Microconstituents in the environment: occurrence, fate, removal and management, pp 227–246. https://doi.org/10.1002/9781119825289.ch10
Shukla AK, Ojha CSP, Mijic A, Buytaert W, Pathak S, Garg RD, Shukla S (2018) Population growth, land use and land cover transformations, and water quality nexus in the Upper Ganga River basin. Hydrol Earth Syst Sci 22(9):4745–4770. https://doi.org/10.5194/hess-22-4745-2018
Singh S, Rai PK (2017) Application of earth observation data for estimation of changes in land trajectories in Varanasi District, India. J Landsc Ecol 11(1):5–18. https://doi.org/10.1515/jlecol-2017-0017
Singh A, Rai PK, Deka G, Biswas B, Prasad D, Rai VK (2021) Management of natural resources through integrated watershed management in Nana Kosi micro watershed, District Almora, India. Ecol Environ Conserv 27(February Suppl. Issue):S260–S268
Tabunschik V, Gorbunov R, Gorbunova T (2023) Unveiling air pollution in Crimean Mountain Rivers: analysis of Sentinel-5 satellite images using Google Earth Engine (GEE). Remote Sens 15(13):3364. https://doi.org/10.3390/rs15133364
Tillihal SB, Shukla AK (2023) River flood monitoring and management: a review. Adv Water Resour Plan Sustain 223–238. https://doi.org/10.1007/978-981-99-3660-1_13
Tripathi A, Tiwari RK, Tiwari SP (2022) A deep learning multi-layer perceptron and remote sensing approach for soil health based crop yield estimation. Int J Appl Earth Observ Geoinform 113. https://doi.org/10.1016/j.jag.2022.102959
UNEP (2017) Environmental sustainability is essential for global equity.
UNESCO (2021) Valuing water. World Water Assessment Programme (United Nations)
Vatitsi K, Ioannidou N, Mirli A, Siachalou S, Kagalou I, Latinopoulos D, Mallinis G (2023) LULC change effects on environmental quality and ecosystem services using EO data in two Rural River Basins in Thrace, Greece. Land 12(6):1140. https://doi.org/10.3390/land12061140
Wang LC, Hoang DV, Liou YA (2022) Quantifying the impacts of the 2020 flood on crop production and food security in the middle reaches of the Yangtze River, China. Remote Sens 14(13). https://doi.org/10.3390/rs14133140
World Bank (2018) Ending poverty investing in opportunity

Xu L, Fang K, Huang Y, Xu S (2023) Demand priority of green space from the perspective of carbon emissions and storage. Sustainability (Switzerland) 15(14). https://doi.org/10.3390/su151411199

Yao L, Zhang X, Luo J, Li X (2023) Identification of ecological management zoning on arid region from the perspective of risk assessment. Sustainability (Switzerland) 15(11). https://doi.org/10.3390/su15119046

Yu T, Abulizi A, Xu Z, Jiang J, Akbar A, Ou B, Xu F (2023) Evolution of environmental quality and its response to human disturbances of the urban agglomeration in the northern slope of the Tianshan Mountains. Ecol Ind 153. https://doi.org/10.1016/j.ecolind.2023.110481

Zhang D, Jing P, Sun P, Ren H, Ai Z (2022a) The non-significant correlation between landscape ecological risk and ecosystem services in Xi'an Metropolitan Area, China. Ecol Ind 141. https://doi.org/10.1016/j.ecolind.2022.109118

Zhang M, Liu D, Wang S, Xiang H, Zhang W (2022) Multisource remote sensing data-based flood monitoring and crop damage assessment: a case study on the 20 July 2021 extraordinary rainfall event in Henan, China. Remote Sens 14(22). https://doi.org/10.3390/rs14225771

Zhao B, Han J, Li P, Li H, Feng Y, Hu B, Zhang G, Li J (2023) Evidence for urbanization effects on eco-environmental quality: a case study of Guyuan City, China. Sustainability (Switzerland) 15(11). https://doi.org/10.3390/su15118629

Zheng Q, Seto KC, Zhou Y, You S, Weng Q (2023) Nighttime light remote sensing for urban applications: Progress, challenges, and prospects. ISPRS J Photogram Remote Sens 202:125–141. Elsevier B.V. https://doi.org/10.1016/j.isprsjprs.2023.05.028

Chapter 15
Effectiveness of Drainage Morphmetric Analysis for Water Resource Management in Changing Climate Scenarios

Prafull Singh and Praveen Kumar Rai

Abstract The present work represented an integrated methodology for the assessment of drainage morphometric components of Kali river basin for water resource management using satellite data and field survey. Various linear, aerial and relief aspects of the Kali basin were evaluated and their relationships are analyzed which provides the valuable hydrological and hydrogeological data base for understanding of topographical aspects and complexity of the basin and their inter-relationships for management of water resources at river basin level. The study revealed that the basin possesses a fourth order stream and the length of stream segment is maximum for first order stream and decreases as the stream order increases. The drainage density (Dd) of basin is observed 0.41 km/km^2, which indicates that the area possesses highly permeable soils and low relief. The mean bifurcation ratio of the Kali river basin is 4.38 which indicate that the basin is not much influenced by geological structures. The basin also have lower form factor (0.22), representing elongated shape and signifying a flat hydrograph peak for longer time period. Low value of length of overland flow (1.22) indicates steep slopes and shorter flow paths. The present work shows that basin slope, drainage density (D), length of overland flow (Lo) and basin relief (H) are closely connected and expressively related with a large number of morphometric variables. The results observed from the morphometric analysis of drainage basin provide the very informative hydrological information for water resource management program of the basin and flood management.

Keywords River morphometry · ASTER DEM · ARC GIS · Upper Ganga plain · India

P. Singh (✉)
Remote Sensing and Groundwater Modeling Lab, Department of Geology, Central University of South Bihar, Gaya 824236, India
e-mail: prafullsingh@cusb.ac.in; pks.jiwaji@gmail.com

P. K. Rai
Department of Geography, Khwaja Moinuddin Chishti Language University (U.P. State Govt. University), Lucknow, India

P. K. Rai et al. (eds.), *Natural Resources Management and Sustainable Livelihoods in the Mountainous Region*, Advances in Geographical and Environmental Sciences,
https://doi.org/10.1007/978-981-97-2100-9_15

15.1 Introduction

The analysis of the drainage does not appear to be complete if it lacks the systematic approach towards the development of drainage basins in an area. River characteristics are sensibly implicit by the morphometric analysis of the specific river basin. It helps to elaborate a primary hydrological analysis in order to envisage estimated behaviour of a watershed directly combined with geomorphology and ecology (Esper 2008).

River basins encompass a distinctive morphologic region and are particularly applicable to drainage patterns and geomorphology (Strahler 1957; Dornkamp and King 1971; Rai et al. 2017, 2018, 2019; Mishra et al. 2021). Morphometric analysis of drainage basins thus provides not only a well-designed explanation of the landscape, but also helps as an influential means of associating the form and process of drainage basins that may be extensively unglued in space and time (Easthernbrook 1993). Assessing morphometric factors needs preparation of a drainage map; ordering the numerous streams; and calculating the basin area and perimeter, drainage length, drainage density and frequency, and bifurcation ratio etc. These parameters comprehend the nature of river basins (Krishnamurthy et al. 1996; Kumar et al. 2000; Reddy et al. 2002, 2004; Nag and Chakraborty 2003; Ratnam et al. 2005; Vijith and Satheesh 2006). The hydrological processes in river basins are generally determined by climate and physiographic aspects (Rai et al. 2021, 2022). Geology, relief and climate of any river basin are also the key elements of a running water ecosystem effective at the basin scale. Therefore, it is very significant to understand the basin morphology and their hydrological consequence, as both are indissolubly related through geomorphic process of soil development, erosion and deposition (Beven et al. 1988; Samal et al. 2015).

The channel patterns are the surface expressions of the variables influencing river dynamics (Zwnnitz 1932; Lahiri 1996; Casteltort and Simpson 2006; Raj 2007; Rai 2023a, b). The nature and characteristics of a river is changed, when the character of one or more reliant variables is transformed and hence the river geomorphology essentially associated with the basin geometry and its channel network (Baiju et al. 2015; Reddy et al. 2004). The morphometric characteristics at the watershed scale may comprise a vital information concerning its development because all hydrologic and geomorphic progressions arise within the watershed (Bardossy and Schmidt 2002; Mahtab et al. 2003; Pareta and Pareta 2012; Romshoo et al. 2012; Babu et al. 2014; Vandana 2013; Kumar et al. 2015; Pike 2000; Sujatha et al. 2015).

An extensively recognized principle of morphometry is that river basin morphology reflects geological and geomorphic processes over time, as specified by several morphometric studies and the inventive study on the river basin morphometry has been completed by Horton (1932, 1945), Miller (1953), Smith (1950), Strahler, Bhatt and Ahmed (2014) and others by conventional methods. In recent years, the use of Geographic Information System (GIS) technology assisted much to the work of hydrologists in the scientific study and management of water resources. GIS-based drainage morphometry evolution analyses are not only faster but also computationally competent, which have been used for understanding basin characteristics, hydrological response and watershed prioritization studies (Abdel-Lattif and Sherief 2012; Patel et al. 2012; Romshoo et al. 2012; Tamma Rao et al. 2012; Thomas et al. 2012; Wakode et al. 2013). In recent years, morphometric factors are also very helpful for watershed prioritization which is the process of ranking different sub-watersheds of a large watershed according to the order in which they have to be taken for treatment with respect to soil and water conservation actions (Chowdary et al. 2013). Furthermore, morphometric analysis was employed for characterizing watersheds (Nag 1998; Sreenivasa et al. 2004), prioritization of watersheds and for the development of ground water resources (Sreedevi et al. 2005, 2009). Vijith and Satheesh (2006), Manu and Anirudhan (2008), Rai et al. (2014) and Singh et al. (2013, 2014) evaluated the drainage characteristics using satellite data. Magesh et al. (2011) calculated numerous morphometric parameters from Survey of India (SOI) topographical maps and shuttle radar topography mission (SRTM) DEM in GIS platforms.

The objective of this paper is to build composite morphometric indices of the Kali river basin to understand the hydrological behaviour and there prospect. In the present study, the GIS analysis techniques were used to evaluate linear, aerial and relief aspect of Kali river basin based on Morphometric analysis.

15.2 Geographic Setting of Study Area

Kali River basin originates from the Doon valley in the western part of Uttrakhand and passes through Saharanpur, Muzaffarnagar and Bagpat districts of Uttar Pradesh and lastly merged in Hindon river which is Major River in the western U.P. The Kali River, basin which is important tributary of Hindon river of Indo-Gangetic plain and contributes important source of water resource of the area for surface and ground water resources. The basin forms a part of Middle Ganga alluvial aquifer of North India with flat topography under the geographic latitudes 29° 10′–29° 30′ N and longitudes 77° 30′–78° 00′ E (Fig. 15.1). The normal average annual rainfall of the basin is about 750 mm. May is the hottest month in the basin and maximum temperature reached upto 46 °C and lowest temperature reached upto 6 °C in the month of January.

It is well known that the Indo-Gangetic basin is one of the best aquifer systems of the world in term of availability of groundwater. In last two decades the groundwater resources of the Indo-Gangetic basin are severely depleted in terms of quantity and quality due to the high population density and rising trend of sectorial demand. The reports and research papers published on this areas shown that the water quality and quantity depleted in fast rate particularly in urban area where groundwater withdrawal is very high and recharge is very low due conversion of open land into the impervious surfaces. The purpose of present research work to evaluate the availability of groundwater resources of the Kali River basin based on quantitative assessment of the basin for generation of data base for management of the water resources using satellite data and field survey.

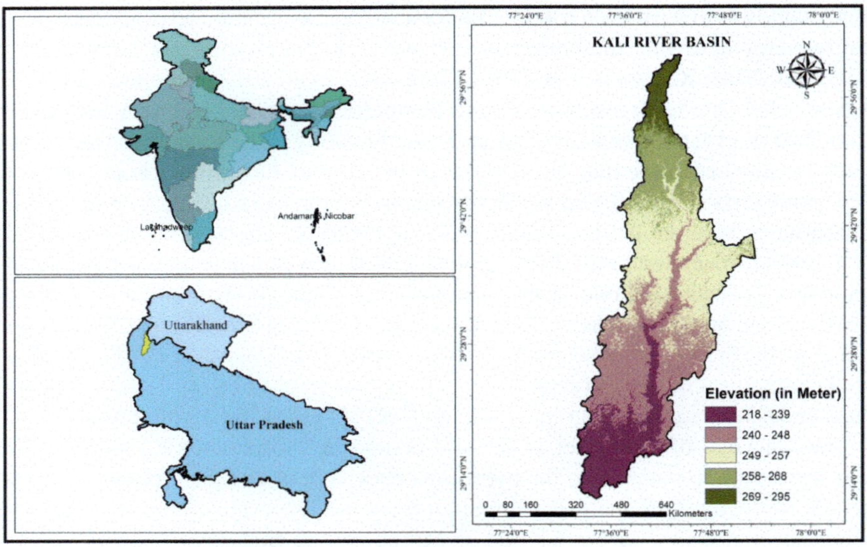

Fig. 15.1 Location map of the study area

The Kali River originated from hilly terrain of Doon valley and mostly flow in the plan area of Middle Ganga Plain. Geomorphologically they make older alluvial plan and younger alluvial plains with diversified geomorphic landforms along river course and surrounding areas such as flood plains, sand bar, scattered ravenous, ox-bow lakes, Paleo-channels etc. The geomorphic setup of the basin have good prospect for groundwater management plans. Hydrogeologically the entire basin is comes under the quaternary alluvial deposits by Ganga river system and geologically composed with varying amount of alluvial sediments of sand, silt, clay and Kankar. The groundwater occurs in the unconsolidated alluvial sediments under phreatic to semi-confined conditions at upper level.

15.3 Data Used and Methodology

Data used in this study contains the digital data of Advanced Space-borne Thermal Emission (ASTER) DEM of 15 m spatial resolution acquired on 28th September 2012 by the TERRA satellite downloaded from the USGS (http://www.earthexplorer.usgs.gov). The basin area and stream network were demarcated using automated processes (Maidment 2002; Ozdemir and Bird 2009; Ahmed et al. 2010; Singh et al. 2014; Rai et al. 2014; Pande and Moharir 2015) in the Arc GIS version-10.3 software using Arc hydro tool in Spatial Analyst Extension.

Flow direction is a significant factor in hydrologic modelling to identify where landforms drains, it is essential to determine the flow direction for each cell in the landscape. Flow direction was deliberated for each pixel using the filled DEM (Fairfiled and Leymarie 1991; Bhatt and Ahmed 2014). Arc Hydro in Arc GIS-10.3 platforms, permits water from a given cell to flow into only one adjoining cell. Flow accumulation is the subsequent step in hydrologic modelling. Flow accumulation was used to produce a drainage network, dependent on the flow direction of each cell (Mark 1983; Bhatt and Ahmed 2014). A threshold may also be computed in that case only sinks, whose depth is lower than the threshold, will be filled. Likewise, river basin can be delineated by giving pour points where water flows out of an area in the Kali basin; this is the lowest point along the Kali river basin boundary. Basically, the pour point is a user-provided point to the cells of highest flow accumulation in any basin (Magesh et al. 2013, Magesh and Chandrasekar 2014; Bhatt and Ahmed 2014; Rai et al. 2014).

The morphometric parameters were separated into three categories: linear, areal and relief aspects. These parameters were calculated by means of many mathematical formulas mentioned in Tables 15.1 and 15.2. The morphometric parameters for the delineated basin were calculated based on normal processes (Horton 1945; Miller 1953, Schumm 1956, Strahler 1964). The extracted drainage networks are ordered based on Strahler (1952) scheme of classification. The highest stream order in the Kali basin is computed as fourth order, which is the main Kali River. The methodology of stream extraction of the Kali basin is shown in the Fig. 15.2. The process of extraction of stream network of the Kali basin from ASTER DEM is shown in the Fig. 15.3. Finally, thematic database layers were brought to common spatial reference system (WGS-84/UTM) for better analysis in GIS platform.

15.4 Result and Discussion

Quantitative morphometric assessment provides very reliable information to evaluate and understand the hydrological behaviour of the rocks and their hydraulic characteristics. Hydrogeological observations, integrated with drainage analysis, provide useful clues regarding broad relationships among the geological framework of the basin (Singh et al. 2014). Various important linear, areal and relief aspect of Kali river basin and their hydrological inferences were discussed in detail in the below paragraphs and also shown in the Tables 15.1 and 15.2.

15.4.1 Linear Aspects

Linear aspects such as stream order, stream number for various orders, bifurcation ratio, mean bifurcation ratio, stream length for various stream orders, total stream length, stream length ratio and Rho coefficient were calculated (Table 15.1).

Table 15.1 Drainage parameters of Kali river basin

Stream order	Stream number (Nu)	Total stream numbers (Num)	Stream length (Lu) (km)	Total stream length (km)	Mean stream length (km) (Lsm)	Stream length ratio (Lur)	Bifurcation ratio (Rb)	Mean bifurcation ratio (Rbm)	Rho coefficient (ρ)
I	81		233.208546		2.879118	–	–	–	–
II	15		122.621489		8.174766	0.52	5.40	–	0.096
III	4	101	103.15639	506.22	25.789098	0.84	3.75	4.38	0.224
IV	1		47.227752		47.227752	0.46	4	–	0.115

Table 15.2 Linear, areal and relief parameter of the Kali basin

S. no.	Parameter	Formula	References	Results
A	*Drainage network*			
1	Stream order (U)	Hierarchical rank	Strahler (1964)	1–4
2	Stream length (Lu)	Length of the stream	Horton (1945)	506.22 km
3	Mean stream length (Lsm)	Lsm = Lu/Nu	Strahler (1964)	84.04 km
4	Stream length ratio (RL)	RL = Lu/(Lu − 1)	Horton (1945)	0.46–0.34
5	Bifurcation ratio (Rb)	Rb = Nu/Nu + 1	Schumm (1956)	3.75–5 40
6	Mean bifurcation ratio (Rbm)	Rbm = average of bifurcation ratios of all order	Strahler (1957)	4.38
7	Rho coefficient (ρ)	Lur/Rb	Horton (1945)	0.145
B	*Basin geometry*			
8	Bain length (Lb)	Longest diagonal across the basin	Schumn (1956)	74.33 km
9	Basin width (W)	Longest vertical distance across the basin	Schumn (1956)	
10	Mean basin width (Wb)	Calculated from average vertical distance across the basin	Horton (1932)	128 65 km
11	Basin perimeter (P)	Calculated from Kali basin boundary in GIS	Schumn (1956)	298.22 km
11	Basin area (A)	Calculated from Kali basin boundary in GIS	Schumn (1956)	1221.02 km^2
12	Relative parameter (Pr)	Pr = A/P	Schumn (1956)	4.09 km
13	Length area relation (Lar)	$Lar = 1.4 * A^{0.6}$	Hack (1957)	99.58
14	Lemniscate's (k)	$k = Lb^2/A$	Chorley et al. (1957)	4.52
15	Form factor (Ff)	$Ff = A/L^2$	Horton (1945)	0.22
16	Shape factor ratio	$Sf = Lb^2/A$	Horton (1945)	4.52
17	Elongation ratio (Re)	$Re = D/L = 1.128\sqrt{A}/L$	Schumm (1956)	0.53
18	Circulatory ratio	$Rc = 4\pi A/P^2$	Strahler (1964)	0.17

(continued)

Table 15.2 (continued)

S. no.	Parameter	Formula	References	Results
19	Compactness coefficient (Cc)	$Cc = 0.2841 * P/A^{0.5}$	Gravelus (1914)	2.42
20	Fitness ratio (Rf)	Rf = Cl/P (main channel length/perimeter)	Melton (1957)	0.25
21	Texture ratio (Rt)	T = Nu/P	Schumm (1956)	0.34
C	*Drainage texture analysis*			
23	Drainage density (Dd)	Dd = Lu/A	Horton (1945)	0.41
24	Drainage intensity (Di)	Dt = Fs/Dd	Horton (1945)	0.20
25	Stream frequency (Fs)	Fs = Nu/A	Horton (1945)	0.08
26	Length of overland flow (Lg)	$Lg = 1/D \times 2$	Horton (1945)	1.22
27	Infiltration Number (If)	If = Fs * Dd	Faniran (1968)	0.032
28	Constant of channel maintenance (km^2/km)	C = 1/Dd	Schumn (1956)	2.44
29	Drainage texture (Dt)	Dt = Dd * Fs	Smith (1950)	0.34
D	*Relief characteristics*			
29	Total basin relief (R)	R = H – h (height of the mouth)	Strahler (1952)	77
30	Relief ratio (Rr)	Rr = R/L	Schumm (1956)	1.035
31	Maximum relief (H)	Calculated from DEM data in GIS	–	295
32	Minimum relief (h)	Calculated from DEM data in GIS	–	218
33	Relative relief ratio (Rhp)	H * 100/P	Melton (1957)	22.87
34	Gradient ratio (Rg)	H – h/Lb	Sreedevi et al. (2005)	1.035
35	Watershed slope (Sw)	Sw = H/Lb	Sreedevi et al. (2005)	3.96
36	Ruggedness number (Rn)	Dd * (H/100)	Patton and Baker (1976)	1.20
37	Melton ruggedness number (MRn)	H/A0.5	Melton (1957)	2.2

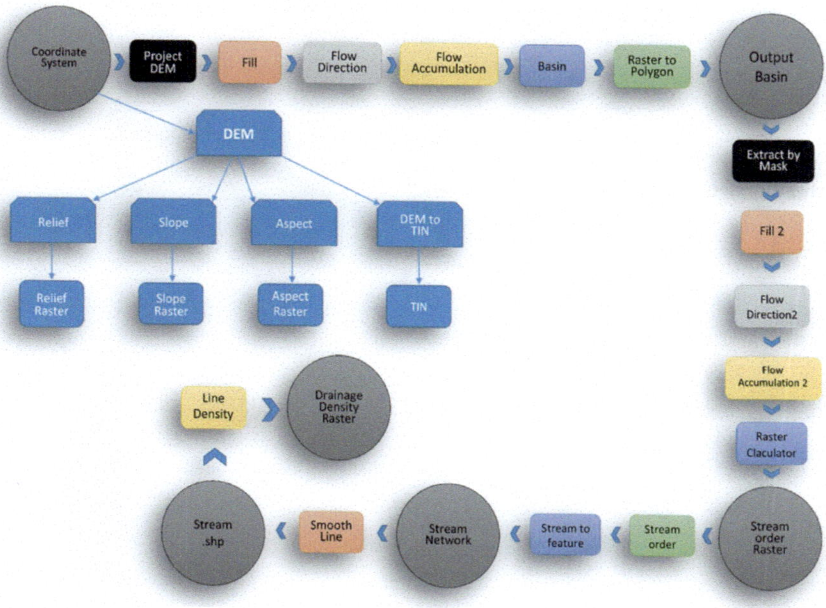

Fig. 15.2 The methodology of stream extraction of the Kali basin

Stream number (Nu)

Number of streams in each order was calculated and total numbers of streams were computed. It is noticeable that the total number of streams progressively decreases as the stream order increases (Table 15.1). High values of first-order streams designate that there is a chance of unpredicted flash flood after heavy rain in the down streams (Chitra et al. 2011).

Stream order (U)

Stream ordering is the hierarchical ranking of stream networks. Strahler's method of stream ordering was followed in the study. It is a useful parameter of stream dimension, discharge and drainage area (Strahler 1964). The number of streams (Nu) in each order is shown in Table 15.1. Kali basin is designated as fourth-order basin (Fig. 15.4). The lower number of streams of a basin designates the maturity of landscape; however the higher number of streams (first- and second-orders) specifies that the region is prone to erosion. In the present study, maximum stream frequency is detected in the first-order streams (Table 15.1). The first-order streams are most sensitive to tectonics and thus are the best indicators of areas with high rates of uplift (Keller and Pinter 2002). More number of first-order streams (81) is perceived in the mountainous area of the Kali basin, which point towards topography density and compacted nature of the bedrock lithology. Besides, the presence of large number of streams in the river basin specifies that the landscape is still experiencing erosion, and

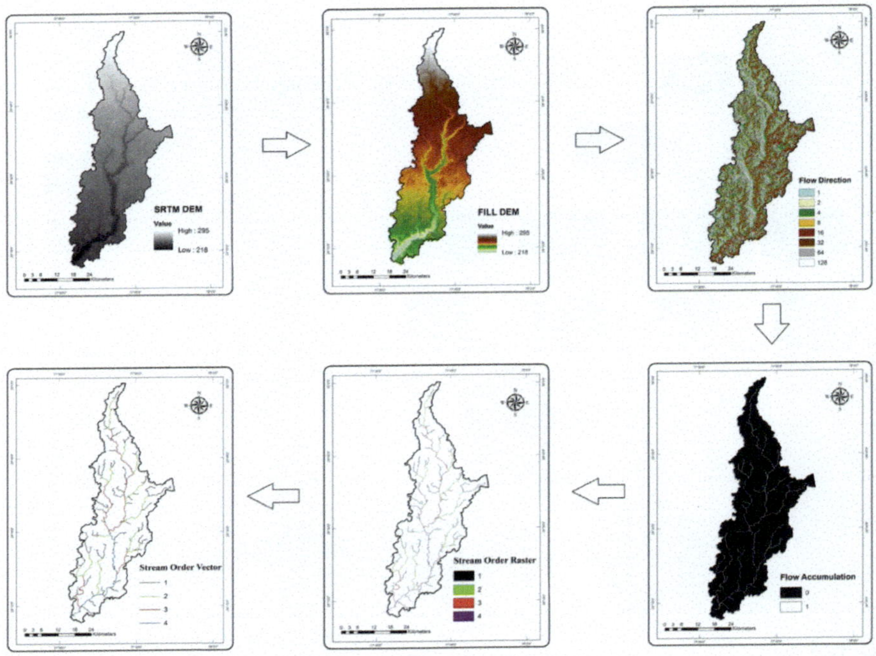

Fig. 15.3 Process of stream extraction of the Kali basin

at the same time, less number of streams in the basin designates mature topography (Strahler 1964).

Stream length (Lu)

Stream length (Lu) is a dimensional property used in understanding the drainage system components which clearly shows the hydrological characteristics of the underlying rock surfaces over the regions of successive stream orders. Drainages with comparatively small lengths are characteristic of areas with steep gradients and finer texture, however lengthier lengths of stream are normally symptomatic of low slopes (Strahler 1964).

The mean and total stream length of each stream order is presented in Table 15.1. The mean length of channel segments of a given order is more than that of the next lower order, but less than the next higher order, representing that the basin development follows erosion process acting on geologic material with similar weathering, erosion characteristics (Nag and Chakraborty 2003). The Lsm values vary with respect to different basins, as they are directly proportional to the size and landscape of the basin. Total stream length is 506.22 km in the Kali basin. Further, it is also distinguished that the Lu is maximum (233.20 km) in the case of first-order streams of the basin, as geometrical resemblance is conserved in the basins of increasing stream order (Strahler 1964). The Lu decreases subsequently with an increase in

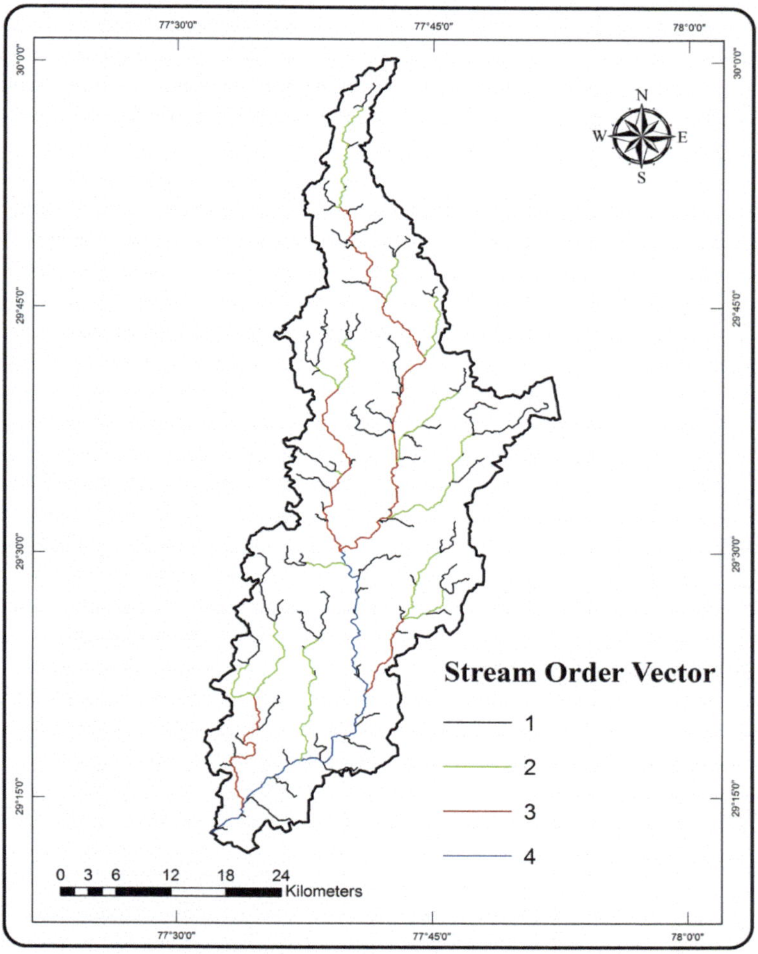

Fig. 15.4 Drainage map of Kali basin with stream orders

stream order, representing constant variation in the elevation over which the streams occur (Subba Rao 2009).

Stream length ratio (Lur)

The mean Lur of Kali basin is 0.60 and varies from 0.46 to 0.84 in each successive stream order (Table 15.1). The variability in 'Lur', among successive stream orders, is a reflection of variances between gradient and landscape and hence it has a vital control on discharge and erosional stage of the basin (Sreedevi et al. 2005).

The 'Lur' of the third order streams shows a high value of 0.84 compared to the 'Lur' the second and forth fourth-order streams within the basin, i.e. 0.52 and

0.46 respectively. This designates that the rock formations in the area, drained by the third-order streams, are more permeable and the gradients are quite gentler than those formations drained by the lower order streams. Changes of 'Lur' from one order to another order specify their late youth stage of geomorphic development.

Bifurcation ratio (Rb)

Bifurcation ratio is the average number of specific stream order segments meet together to form a next higher order drainage segment. It is calculated as the number of stream segments of any given order to the number of streams of the next higher order (Table 15.1). It is a dimensionless measure stating the arrangement of drainage basin. The value of Rb is highly stable and displays a small range of difference from region to region or environment to environment, except where influential structural controls govern (Strahler 1957).

During the flood time, the 'Rb' shows a significant role in control over the peakedness of the runoff hydrograph (Jain and Sinha 2003). The 'Rb' of 2 is obtained for plane topography and 3–4 for hilly or highly dissected topography (Horton 1945). In similar, 'Rb' value characteristically ranges between 3.0 and 5.0 for basin in which the geological structure do not distort the drainage pattern and its system (Strahler 1964).

In the study area, the 'Rb' values of the Kali basin are not same from one order to its next order (Table 15.1), the variation in 'Rb' values clearly displays that the differences in stage of geomorphic development and topographic disparities (Abdel-Lattif and Sherief 2012). Gieusti and Schneider's (1965) identified that the bifurcation ratio within a river basin decreases with increasing stream order and that bifurcation ratio increases with increasing catchment area originally and then tends to become constant do not hold well here. Average bifurcation ratio for Kali basin is 4.38. Higher Rb value (>4.0) quantitatively indicated that river basins are hilly origin.

Rho coefficient (ρ)

Rho coefficient relates the drainage density and physiographic change of the basin, permitting the assessment of the storage capability of the drainage system (Horton 1945). It is mainly prejudiced by climatic factors, geological factors, biological parameters, geomorphic and anthropogenic aspects. The 'Rho' value of the Kali river basin is 0.145 as shown in Table 15.2. This is clearly shows higher hydrologic storage through decrease of impacts of erosion during elevated discharge.

15.4.2 Areal Aspects

Area of a basin (A) and perimeter (P) are the key parameters in quantitative morphology. The Kali basin was calculated by converting the map of the basin into polygon form. The total basin area is found 1221 km^2, and perimeter is 298.22 km. River basin area is hydrological significant because it directly affects the size of the storm hydrograph and the extents of peak and mean runoff. It is stimulating that the maximum flood discharge per unit area is inversely associated to size (Chorley et al. 1957). The aerial parameters of the Kali river basin are described below.

Drainage density (Dd), stream frequency (Fs) and drainage intensity (Di)

Drainage density (Dd) is a key factor assessing the degree of fluvial dissection and delivers substantial evidence concerning the infiltration ability of the topography, climate and run-off potential and vegetation cover of the river basin (Horton 1945; Strahler 1964). It is vital to signify the stages of the morphologic evolution of fluvial landscape and calculated as total length of stream of all order upon the total basin area and its measurement reduced to L-1. It specifies the nearness of channel spacing and offers a mathematical dimension of topography dissection and runoff potential (Reddy et al. 2004). The 'Dd' of Kali basin is 0.41 km/km^2. Dd clearly reflects an idea about the physical properties of the underlying rocks in the Kali basin. According to Horton (1945), low 'Dd' is a sign of the occurrence of highly permeable rock and soil formation and low relief while, high 'Dd' prevails in the impervious/impermeable rocks under hilly relief. Slope gradient and relative relief is the key morphological parameters controlling drainage density. Strahler (1964) distinguished that low 'Dd' is seen where basin relief is low, while high 'Dd' is identified where basin relief is high.

The stream frequency (Fs) is also a function of the underlying lithology and topographic complexity of the drainage basin (Horton 1945). It is the number of stream segments per unit area. During this study, it is detected that the 'Fs' value show a positive correlation with drainage density (Dd) of the Kali basin representing the increase in stream networks with respect to increase in drainage density. The calculated 'Fs' value of Kali basin range is 0.08. Faniran expresses the drainage intensity, as the ratio of the stream frequency to the drainage density. During the calculation, it is found that the Kali river basin shows a low drainage intensity of 0.20. The low 'Di' value suggests that drainage density and stream frequency have slight consequence on the extent to which the surface has been lowered by denudation agents. With these low 'Dd', 'Sf' and 'Di' values of the Kali basin, surface runoff is not speedily removed from the area, making it vastly susceptible to flooding, gully erosion and landslides.

Constant of channel maintenance (C)

The reverse of drainage density is called as the constant of channel maintenance and it says the number of sq. unit of river basin surface essential to sustain one linear unit of drainage channel (Samal et al. 2015). Constant of channel maintenance (C) mainly depends on the rock type, permeability, climatic system, vegetal cover and elevation as well as period of erosion (Schumm 1956). Kali basin having the highest 'C' value of 2.44 is under very high structural controls and high runoff situations. Higher values of 'C' suggest more area is essential to develop surface flow, which recommends that part of water, may get lost by evaporation, percolation, etc. whereas lower value specifies less chance of infiltration and henceforth more surface runoff (Bhagwat et al. 2011). It also shows the relative size of topography units in a river basin and has a precise genetic. Alluvial basin of plain and piedmont zone displays maximum value, as the porousness in this zone is high.

Drainage texture (Dt)

The value of the difference of drainage texture mainly depends on different type of natural aspects i.e. precipitation and other climatic characteristics, rock, soil type, vegetation characteristics, permeability, elevation, infiltration capability within the basin. Smith (1939) has categorized drainage texture into five different textures i.e. very coarse (2), coarse (2–4), moderate (4–6), fine (6–8) and very fine (8). The drainage texture of the Kali basin is 0.34, indicates the highly permeable rock (very coarse texture) with low infiltration rate and high elevation. The weak rocks unprotected by vegetation produce a fine texture, whereas resistant rocks cause coarse texture. Scant foliage of dry climate causes finer textures than those developed on similar rocks in a moist climate. The rock texture is usually reliant upon foliage type and climate (Dornkamp and King 1971). Presence of vegetal growth reduces the 'Dt' value and increases deep water percolation which additional depends upon geology and soil characteristics (Samal et al. 2015).

Length of overland flow (Lo)

Length of overland flow (Lo) is a key morphometric factor accompanying both hydrologic and physiographic development of the river basin (Horton 1945). 'Lo' is distinct as non-channel flow from a point on drainage divide to the adjacent stream channel and approximately equal to the one half of the reciprocal of the drainage density (Strahler 1964). During overland flow, the infiltration process takes place for entire basin. The Kali river basin accounts 'Lo' value of 1.22. The higher value of 'Lo' of the river basin or sub-watersheds specifies geomorphic maturity whereas lower value characterized by late youth stage of geomorphic development (Prakash et al. 2016). River basins with lower 'Lo' values also specify that in such basins low rainfall is adequate to contribute a substantial amount of surface run-off to stream discharge. This parameter relays inversely to the average slope of the channel and is fairly identical with the length of sheet flow to a large degree. It allows an active association with the drainage density and constant channel maintenance.

Form factor (Ff)

Form Factor (Ff) is defined as the ratio of the basin area to the square of the basin length. The index of Ff displays the inverse relationship with the square of the axial length and as a direct correlation with highest discharge (Gregory and Walling 1973). The calculated value of Ff is 0.22 of the Kali basin signifies the elongated shape of the basin (Table 15.2). So, the lower values and elongated shaped basin shows that the basin fits to the flatter peak flow of shorter period.

Infiltration number (if)

The infiltration number of a basin or a watershed is demarcated as the outcome of drainage density and stream frequency and given knowledge about the infiltration physiognomies of the watershed. The higher the infiltration number, the lower will be the infiltration and the higher run-off. It is calculated that the 'If' value of the Kali basin is 0.032. During the study, it is very well identified that the 'If' value plays a key role in distinguishing the infiltration characteristics of the river basin and it is inversely proportional to the infiltration capacity of the river basin.

Elongation ratio (Re)

Elongation ratio (Re) is very well known as the ratio between the diameter of a circle with the same area as that of the basin (A) and the maximum length (L) of the river basin. The values of elongation ratio can be categorized into three groups, namely less than 0.7 'Re' designates elongated shape and 0.9–0.8 values signify the oval shape, and more than 0.9 values characterize the circular shape of the basin. The 'Re' value for the Kali basin is assessed as 0.53 (Table 15.2) that proves that the basin shape belongs to the much more elongated type. The elongated shapes of river sub-watersheds with the larger basin area, are averring the role of head-ward erosion in growth of long channels (Prakash et al. 2016). The difference of the elongated shapes of the basins is due to the guiding consequence of thrusting and faulting in the basin (Sreedevi et al. 2009).

Circulatory ratio (Rc)

Circularity ratio is very much inclined by geological structure, elevation, gradient differences, climatic parameters, stream frequency and stream lengths and land use land cover within the river basin. The basin 'Rc' is 0.17 which signify that the basin is strongly elongated and belongs to the heterogeneous geological structure and materials. The Rc value of the Kali river basin is 0.17 (Table 15.2). If the value of 'Rc' is accurately 1.0, the river basin is set to be a perfectly circular shape, in which the discharge is greater in quantity (Miller 1953). The calculated 'Rc' values also indicate the high runoff capability and low permeable ability of subsoil and sub-surface soil along the Kali basin. It is a vital ratio, which designates the dendritic stage of a river basin (Sreedevi et al. 2009). The stream length, stream frequency and stream slope of the many orders play an important role on the 'Rc' rather than the slope settings and drainage pattern of a river basin.

Texture ratio (Rt)

Texture ratio is demarcated as the ratio between the first order streams and perimeter of the river basin. 'Rt' is a significant feature in the river morphometric evolution which depends on the underlying structure of geology, infiltration ability of bedrock and relief aspects of the sub-basins or watersheds. Rt value of the Kali basin is 0.34.

Compactness coefficient (Cc)

Compactness coefficient (Cc) is the correlation of the shape of the river basin to a circle and stated as a ratio between the length of the river basin boundary and the perimeter of a circle with the same area. If the river basin is a perfect circle, then 'Cc' value would be equal to 1 (Gravelus 1914). A circular river basin is most perilous from a drainage point stand because it will produce the shortest time of concentration before peak flow arises in the river basin. The ‘Cc’ value of the Kali basin is 2.42 and it is independent of size of watershed and dependent only on the slope.

Fitness ratio (Fr)

Melton stated that the fitness ratio (Fr) is the ratio of main channel length to the length of the basin perimeter, which is a significant measure of topographic fitness. The fitness ratio for Kali river basin is 0.25.

Length area relation (Lar)

Hack (1957) identified that for a large number of river basins, the stream length and basin area is associated by a simple power function as follows: Lar = 1.4 * A0.6. The Lar value of the Kali basin is 99.58.

Lemniscate’s (k)

Chorely et al. (1957), express the lemniscate’s value to define the slope of the basin. The lemniscate (k) value for the Kali river basin is 4.52, which displays that the watershed occupies the maximum area in its regions of inception with large number of streams of higher order.

15.4.3 Relief Aspects

The relief aspects are also significant for proper water resources management in the river basin. The relief parameters in morphometric study bring into consideration the impact of aspect and elevation over a large river basin. The parameter of the distribution of slope angles sampled over the whole river basin depends on the elevation distribution within it. It is a vital attribute of topography in general and the drainage basin is in particular. According to Strahler (1957), relief parameters are indicative of the potential energy of the drainage system because of its elevation above mean sea level. Assessment of some of the relief aspects of the basin is deliberated below.

Maximum and minimum heights (H, h)

The maximum and minimum height corresponds to the highest and lowest point of the river basin. The minimum height is 218 m and maximum of 295 m detected in the Kali river basin. Total basin relief (R) calculated in the Kali basin is 77 m (Table 15.2).

Relief ratio (Rh) and relative relief ratio (Rhp)

An elevation difference between the highest point of a river basin and the lowest point of the river valley is recognized as the total relief of the river basin while relief ratio (Rh) is demarcated as the ratio between the total relief of the river basin and the lengthiest measurement of the basin parallel to the main drainage line (Schumm 1956). The ‘Rh’ of the Kali basin is 1.035. During the calculation, it is detected that areas with low to moderate relief and slope are considered by moderate value of relief ratios. Low value of relief ratios are generally due to the impervious basement rocks of the river basin and low slope gradient. The 'Rhp' value of the Kali basin is 22.87.

Gradient ratio (Rg)

Gradient ratio (Rg) is a significant parameter of channel slope, which allows calculation of the runoff volume (Sreedevi et al. 2005). It is a vital indicator of channel gradient, which permits assessment of the surface runoff volume. The 'Rg' value of the Kali basin is 1.035, which reflects the mountainous nature of the terrain.

Ruggedness number (Rn) and Melton ruggedness number (MRn)

According to Strahler's (1957), the ruggedness number (Rn) is the results of the basin relief and the drainage density and practically combines slope gradient with its length. The 'Rn' value of the Kali basin is 1.20. The low ruggedness value of basin indicates that river basin area is less prone to soil erosion and have intrinsic structural intricacy in connotation with relief and drainage density.

The 'MRn' is a slope index that provides specialized representation of relief ruggedness within the watershed. The 'MRn' is a slope index that offers specific view of relief ruggedness within the drainage basin. The 'MRn' value of the Kali basin is 2.2. Incidence of debris flows or fluvial sediment transport clearly depends on availability of debris and yet low ruggedness of landforms designates the availability of loci suitable to trap debris from upstream areas and out of tributaries conquered by debris flow, instigating a transition from debris flows to bed load transport (Marchi and Fontana 2005).

15.5 Conclusion

The results observed in this study from various linear, aerial and linear aspects of the basin revealed that basin have good prospect and potential for groundwater management activities. The assessment of drainage characteristics of Kali basin and its fourth order streams revealed the significance of morphometric studies in terrain characterization and basin evolution studies. Morphometric analysis of Kali basin in India offers not only an well-designed description of the basin landforms, but also helps as an influential means of associating the form and process of river basins that may be extensively separated in space and time and its outcomes did show that rainfall has a substantial part in the drainage development whereas structure and relief of rocks dictate the drainage pattern. Lower order streams generally govern the basin. The difference in stream length ratio might be due to change in gradient and topography. The study areas experience intermediate to fine drainage textures. The low bifurcation ratio and elongation ratio of the Kali river indicate the elongated shape of the basin which develop due to head-ward erosion leading to expansion of channels. The Kali basin is moderate to well-drained and revealed a geomorphic maturity in its physiographic development. The relief ratio of the Kali basin specifies that the discharge ability of the basin is low to moderate and the groundwater potential resource may adequate in the valley side. The development of stream segments in the Kali basin area is more or less affected by rain fall. The present study is very beneficial for watershed harvesting structures and watershed management plans. First- and second-order streams are not valuable for building check dams in the basin area because the streams are located on high elevated area with steep gradient. The outcome thus produced delivers satisfactory knowledge base required for decision making during strategic planning and description of prioritized hazard management zones in hilly areas and also valuable for watershed planners while executing soil and water conservation measures in the basin area.

References

Abdel-Lattif A, Sherief Y (2012) Morphometric analysis and flashfloods of Wadi Sudr and Wadi Wardan, Gulf of Suez, Egypt: using digital elevation model. Arab J Geosci 5:181–195. https://doi.org/10.1007/s12517-010-0156-8

Ahmed SA, Chandrashekarappa KN, Raj SK, Nischitha V, Kavitha G (2010) Evaluation of morphometric parameters derived from ASTER and SRTM DEM—a study on Bandihole sub-watershed basin in Karnataka. J Indian Soc Remote Sens 38:227–238

Babu KJ, Sreekuma S, Aslam A (2014) Implication of drainage basin parameters of a tropical river basin of South India. Appl Water Sci. https://doi.org/10.1007/s13201-014-0212-8

Baiju CK, Kumar AB, Gopinath G, Chandran MSS, Ramkumar T (2015) GIS based morphometric analysis of the Kochara sub-watershed of Greater Periyar Plateau (GPP) of South Western Ghats, India. Int J Adv Remote Sens GIS 4(1):895–910

Bardossy A, Schmidt F (2002) GIS approach to scale issues of perimeter-based shape indices for drainage basins. Hydrol Sci J 47:931–942

Beven KJ, Wood EF, Sivapalan M (1988) On hydrological heterogeneity-catchment morphology and catchment response. J Hydrol 100:353–375

Bhagwat TN, Shetty A, Hegde VS (2011) Spatial variation in drainage characteristics and geomorphic instantaneous unit hydrograph (GIUH); implications for watershed management—a case study of the Varada river basin, Northern Karnataka. Catena 87:52–59

Bhatt S, Ahmed SA (2014) Morphometric analysis to determine floods in the Upper Krishna basin using Cartosat DEM. Geocarto Int 29(8):878–894. https://doi.org/10.1080/10106049.2013.868042

Casteltort S, Simpson G (2006) River spacing and drainage network growth in widening mountain ranges. Basin Res 18(3):267–276

Chitra C, Alaguraja P, Ganeshkumari K, Yuvaraj D, Manivel M (2011) Watershed characteristics of Kundah sub-basin using remote sensing and GIS techniques. Int J Geomat Geosci 311–335

Chorley RJ, Donald EG, Malm PHA (1957) A new standard for estimating drainage basin shape. Am J Sci 255:138–141

Chowdary VM, Chakraborthy D, Jeyaram A, Krishna Murthy YVN, Sharma JR, Dadhwal VK (2013) Multi-criteria decision making approach for watershed prioritization using analytic hierarchy process technique and GIS. Water Resour Manage 27:3555–3571

Dornkamp JC, King CAM (1971) Numerical analyses in geomorphology, an introduction. St. Martins Press, New York, p 372

Easthernbrook, 1993Easthernbrook DJ (1993) Surface processes and landforms. Macmillan Publishing Co., New York, 325 pp

Esper AMY (2008) Morphometric analysis of Colanguil river basin and flash flood hazard, San Juan, Argentina. Environ Geol 55:107–111

Fairfiled J, Leymarie P (1991) Drainage network from grid digital elevation models. Water Resour Res 30:1681–1692
Gieusti EV, Schneide WJ (1965) The distribution of branches in river networks. US geological survey, 422G
Gravelus H (1914) Grundrifi der gesamten Gewcisserkunde. Band I: Flufikunde (Compendium of Hydrology, vol I. Rivers, in German). Goschen, Berlin, Germany
Gregory KJ, Walling DE (1973) Drainage basin form and processes: a geomorphological approach. In: Arnold E (ed) Halsted Press, New York, p 456
Hack J (1957) Studies of longitudinal stream profiles in Virginia and Maryland, U.S. geological survey professional paper, 294-B
Horton RE (1932) Drainage-basin characteristics. Trans Am Geophys Union 13:350–361
Horton RE (1945) Erosional development of streams and their drainage basins: hydrophysical approach to quantitative morphology. Geol Soc Am Bull 56:275–370
Jain V, Sinha R (2003) Evaluation of geomorphic control on flood hazard through geomorphic instantaneous unit hydrograph. Curr Sci 85:1596–1600
Keller EA, Pinter N (2002) Active tectonics. Earthquakes, uplift and landscape. Prentice Hall, New Jersey
Krishnamurthy J, Srnivas G, Jayaram V, Chandrasekhar MG (1996) Influence of rock types and structures in the development of drainage networks in typical hardrock terrain. ITC J 3–4:252–259
Kumar R, Kumar S, Lohani AK, Nema RK, Singh RD (2000) Evaluation of geomorphological characteristics of a catchment using GIS. GIS India 9(3):13–17
Kumar A, Samuel SK, Vyas V (2015) Morphometric analysis of six sub-watersheds in the central zone of Narmada River. Arab J Geosci 2015(8):5685–5712. https://doi.org/10.1007/s12517-014-1655-9
Lahiri S (1996) Channel pattern as signature of neotectonic movements—a case study from Brahmaputra valley in Assam. J Indian Soc Remote Sens 24(4):265–272
Magesh NS, Chandrasekar N (2014) GIS model-based morphometric evaluation of Tamiraparani sub-basin, Tirunelveli district, Tamil Nadu, India. Arab J Geosci 7:131–141
Magesh NS, Chandrasekar N, Soundranayagam JP (2011) Morphometric evaluation of Papanasam and Manimuthar watersheds, parts of Western Ghats, Tirunelveli district, Tamil Nadu, India: a GIS approach. Environ Earth Sci 64:373–381
Magesh NS, Jitheshlal KV, Chandrasekar N (2013) Geographical information system-based morphometric analysis of Bharathapuzha river basin, Kerala, India. Appl Water Sci 3:467–477.https://doi.org/10.1007/s13201-013-0095-0
Mahtab A, Narender B, Ajai (2003) Satellite derived digital elevation model and terrain parameters—generation, accuracy assessment and validation. J Indian Soc Remote Sens 31(1):19–24
Maidment DR (2002) Arc hydro: GIS for water resources. ESRI Press, Redlands (CA)
Manu MS, Anirudhan S (2008) Drainage characteristics of Achankovil river basin, Kerala. J Geol Soc India 71:841–850
Marchi L, Fontana GD (2005) GIS morphometric indicators for the analysis of sediment dynamics in mountain basins. Environ Geol 48:21–288
Mark DM (1983) Relations between field-surveyed channel networks and map-based geomorphometric measures, Inez Kentucky. Ann Assoc Am Geogr 73:358–372
Miller VC (1953) A quantitative geomorphologic study of drainage basin characteristics in the Clinch Mountain area, Virginia and Tennessee, Project NR 389042, Tech Report 3. Columbia University Department of Geology, ONR Geography Branch, New York
Mishra VN, Rai PK, Singh P (eds) (2021) Geo-information technology in earth resources monitoring and management. Nova Science Publishers, U.S.A., pp 1–395. ISBN: 978-1-53619-669-6
Nag SK (1998) Morphometric analysis using remote sensing techniques in the Chaka sub-basin Purulia district, West Bengal. J Indian Soc Remote Sens 26(1–2):69–76
Nag SK, Chakraborty S (2003) Influence of rock types and structures in the development of drainage network in hard rock area. J Indian Soc Remote Sens 31:25–35

Ozdemir H, Bird D (2009) Evaluation of morphometric parameters of drainage networks derived from topographic maps and DEM in point of floods. Environ Geol 56:1405–1415
Pande CB, Moharir K (2015) GIS based quantitative morphometric analysis and its consequences: a case study from Shanur river basin, Maharashtra India. Appl Water Sci. https://doi.org/10.1007/s13201-015-0298-7
Pareta K, Pareta U (2012) Quantitative geomorphological analysis of a watershed of a Ravi river basin, H.P. India. Int J Remote Sens GIS 1(1):41–56
Patel D, Dholakia M, Naresh N, Srivastava P (2012) Water harvesting structure positioning by using geo-visualization concept and prioritization of mini-watersheds through morphometric analysis in the Lower Tapi basin. J Indian Soc Remote Sens 40:299–312
Pike RJ (2000) Geomorphometry: diversity in quantitative surface analysis. Prog Phys Geogr 24:1–20
Prakash K, Singh S, Shukla UK (2016) Morphometric changes of the Varuna river basin, Varanasi district, Uttar Pradesh. J Geomat 10(1):48–54
Rai PK (2023a) River conservation and water resource management. Springer Nature, Singapore, pp 1–352. ISBN No. 978-981-99-2604-6. eBook ISBN 978-981-99-2605-3
Rai PK (2023b) Advances in water resource planning & sustainability. Springer Nature, Singapore, pp 1–322. ISBN No. 978-981-99-3659-5
Rai PK, Mohan K, Mishra S, Ahmad A, Mishra V (2014) A GIS-based approach in drainage morphometric analysis of Kanhar river basin, India. Appl Water Sci. https://doi.org/10.1007/s13201-014-0238-y
Rai PK, Chaubey PK, Mohan K, Singh P (2017) Geoinformatics for assessing the inferences of quantitative drainage morphometry of the Narmada basin in India. Appl Geomat (Springer) 9:167–189. https://doi.org/10.1007/s12518-0170191-1
Rai PK, Mishra VN, Singh P (2018) Hydrological inferences through morphometric analysis of Lower Kosi river basin of India for water resource management based on remote sensing data. Appl Water Sci (Springer) 8(15):1–16. https://doi.org/10.1007/s13201-018-0660-7
Rai PK, Singh P, Mishra VN, Singh A, Sajan B, Shahi AP (2019) Geospatial approach for quantitative drainage morphometric analysis of Varuna river basin, India. J Landsc Ecol 12(2):1–21
Rai PK, Mishra VN, Singh P (eds) (2021) Recent technologies for disaster management & risk reduction-sustainable community resilience & responses. Springer Nature, Switzerland, pp 1–471. ISBN: 978-3-030-76116-5. https://doi.org/10.1007/978-3-030-76116-5
Rai PK, Mishra VN, Singh P (eds) (2022) Geospatial technology for landscape and environment management: sustainable assessment & planning. Springer Nature, Singapore, pp 1–431. ISBN: 978-981-16-7373-3. https://doi.org/10.1007/978-981-16-7373-3
Raj R (2007) Strike slip faulting inferred from off-setting of drainages: Lower Narmada basin, Western India. J Earth Syst Sci 116(5):413–421
Ratnam N, Srivastav K, Rao Venkateswara YK, Amminedu V, Murthy KSR (2005) Check dam positioning by prioritization of micro-watersheds using SYI model and morphometric analysis: remote sensing and GIS perspective. J Indian Soc Remote Sens 33(1):25–38
Reddy GPO, Maji AK, Gajbhiye KS (2002) GIS for morphometric analysis of drainage basins. GIS India 11(4):9–14
Reddy GPO, Maji AK, Gajbhiye KS (2004) Drainage morphometry and its influence on landform characteristics in basaltic terrain, Central India—a remote sensing and GIS approach. Int J Appl Earth Obs Geoinf 6:1–16
Romshoo SA, Bhat SA, Rashid I (2012) Geoinformatics for assessing the morphometric control on hydrological response at watershed scale in the Upper Indus basin. J Earth Syst Sci 121:659–686
Samal DR, Gedam SS, Nagarajan R (2015) GIS based drainage morphometry and its influence on hydrology in parts of Western Ghats region, Maharashtra, India. Geocarto Int 30(7):755–778
Singh P, Thakur J, Singh UC (2013) Morphometric analysis of Morar river basin, Madhya Pradesh, India, using remote sensing and GIS techniques. Environ Earth Sci 68:1967–1977

Singh P, Gupta A, Singh M (2014) Hydrological inferences from watershed analysis for water resource management using remote sensing and GIS techniques. Egypt J Remote Sens Space Sci 17:111–121
Smith GH (1939) The morphometry of Ohio: the average slope of the land (abstract). Ann Assoc Am Geogr 29:94
Smith KG (1950) Standards for grading texture of erosional topography. Am J Sci 248:655–668
Sreedevi PD, Subrahmanyam K, Shakeel A (2005) The significance of morphometric analysis for obtaining groundwater potential zones in a structurally controlled terrain. Environ Geol 47(3):412–420
Sreedevi PD et al (2009) Morphometric analysis of a watershed of South India using SRTM data and GIS. J Geolo Soc India 73:543–552
Sreenivasa VS, Govindaonah S, Home Gowda H (2004) Morphometric analysis of sub-watersheds in the Pawagada area of Tumkur district, South India using remote sensing and GIS techniques. J Indian Soc Remote Sens 32(4):351–362
Strahler AN (1957) Quantitative analysis of watershed geomorphology. Trans Am Geophys Union 38:913–920
Strahler AN (1952) Hypsometric (area-altitude) analysis of erosional topography. Bull Geol Soc Am 63
Subba R (2009) A numerical scheme for groundwater development in a watershed basin of basement terrain: a case study from India. Hydrogeol J 17:379–396
Sujatha ER, Selvakumar R, Rajasimman UAB, Victor RG (2015) Morphometric analysis of sub-watershed in parts of Western Ghats, South India using ASTER DEM. Geomat Nat Hazards Risk 6(4):326–341. https://doi.org/10.1080/19475705.2013.845114
Tamma Rao G, Gurunadha Rao VVS, Dakate R, Mallikharjuna Rao ST, Raja Rao BT (2012) Remote sensing and GIS based comparative morphometric study of two sub-watershed of different physiographic conditions, West Godavari district, A.P. J Geol Soc India 79:383–390
Thomas J, Joseph S, Thrivikramji KP, Abe G, Kannan N (2012) Morphometrical analysis of two tropical mountain river basins of contrasting environmental settings, the southern Western Ghats, India. Environ Earth Sci 66:2353–2366
Vandana M (2013) Morphometric analysis and watershed prioritization: a case study of Kabani river basin, Wayanad district, Kerala, India. Indian J Geo-Mar Sci 42(2):211–222
Vijith H, Satheesh R (2006) GIS based morphometric analysis of two major upland subwatersheds of Meenachil river in Kerala. J Indian Soc Remote Sens 34(2):181–185
Wakode HB, Dutta D, Desai VR, Baier K, Azzam R (2013) Morphometric analysis of the upper catchment of Kosi river using GIS techniques. Arab J Geosci 6:395–408
Zwnnitz ER (1932) Drainage pattern and their significance. J Geol XL 6:498–521

Chapter 16
Pfafstetter Code Based Watershed Delimitation, Geological Observations, Land Cover, and Population Density Analysis for Mizoram Through Remote Sensing: Aiding and/or Improvising the Digital Watershed Atlas of India

Priyanka Puri

Abstract Integrated Watershed Management cannot proceed without watershed delimitation and this not an easy task. Further, in a landscape dissected and dominated with water in a hill topography, this task becomes challenging yet entrancing. For the state of Mizoram, these aspects are pertinent for the geology marked by uncountable hydrological systems. Remote Sensing technology can prove to be a highly effective tool in this as it can help observe the dynamic landscapes over a period of time, which otherwise is a daunting task to do in the field. The current attempt tends to provide an unfollowed, state of the art method to observe the alternative to the existing planning boundaries for watersheds and similar units. Currently, it is the Digital Watershed Atlas of India that provides a schema for such a planning. With its roots based in spatial scale for watershed delineation, the present attempt tends to create these planning regions by taking natural drainage as the determining factor. The current study is an attempt to provide a perspective to the existing delimitation with the help of remote sensing through the well known Pfafstetter Code of watershed delimitation. This analysis is supplemented with observations on the existing geology of water flow, land cover, and population density of the state. Such an attempt is not observed for the north-east in general and the state of Mizoram in particular. It can help in providing an alternative strategy, aiding or improvising the existing watershed delimitation and also in updating the development measures with regards to existing geology and occupation of space.

Keywords Geology · Hydrology · Mizoram · Pfafstetter code · Watersheds

P. Puri (✉)
Department of Geography, Miranda House, University of Delhi, Delhi 110007, India
e-mail: priyanka.puri@mirandahouse.ac.in

P. K. Rai et al. (eds.), *Natural Resources Management and Sustainable Livelihoods in the Mountainous Region*, Advances in Geographical and Environmental Sciences,
https://doi.org/10.1007/978-981-97-2100-9_16

16.1 Background

Watersheds are the basis of planning in regions where they are a natural component of the landscape. The method of identifying a watershed is basically to observe it as an area where water from precipitation and from different water sources moves downstream from a common channel (Black 1996). Simply put forward, it is a water divide between two regions of drainage (Webster 2023). In terms of definition, a watershed is a natural feature on land which experiences outflows of surface, subsurface and ground water at a common point (Black 1996). Watersheds are of different sizes and have specific characteristics of possessing a delimiting area called drainage basin which lends it its dimensions. Watersheds, can, however, be of different sizes depending upon the geographic scale under consideration (NOAA 2023). These scales lend a systematic classification called Hydrological Unit Codes (HUC) to the watershed in a region which can range at various geographic levels and can also help in the delineation of the watershed at various levels. The levels of scale also lead to nesting of such regions (USGS 2023).

Explaining its characteristics, it can be said that a watershed is an area with drainage at a common point for all sources of water flow. In literature, it is also commonly used as a term with catchment area or drainage basin (USGS 2023). Within discussions in literature and in practice in hydrology, the unit of water shed or drainage divide is considered as a frame of reference of regional planning and serves as a novel method of dealing with such challenges (Randhir 2007) and also as a fundamental of planning in hydrology (Bajjali 2018). Drainage networks are considered to provide base to assessment in geographical analysis. Watershed, in this regard, is taken as the most prominent method of hydrological studies (Ray 2018).

It is also credited to have provided a dependable, one stop solution to complex planning problems by delineating a region through a watershed and then focussing upon the problems existing therein. The enhanced recognition of watershed for a balanced planning purpose has provided it an unmatched popularity as a process for planning as it leads to the provision of a system based approach in which the complexity of environmental, social and economic issues can be resolved (Randhir 2007).

Within this natural system, there are further sub-systems comprising of other natural and human sub-systems which form this unit. These can be identified as mountains, rivers, valleys, forests, and streams on one hand and communities, settlements and ecologies on the other hand (Randhir 2007). Randhir (2007) quotes geographer and scientist Powell to define watershed as an area which has boundaries in hydrology and has the living organisms linked by a common source of water and form a community (Randhir 2007). It is also suggested that the unit is static as a landscape but dynamic as a hydrological one due to the water dynamicity observed in origin and circulation (Black 1996).

Within this background, the next question in consideration is the delineation of this unit. There exist numerous methods to demarcate this region and its boundaries ranging from the traditional to the technologically updated ones (Li et al. 2019) including modelling (Chu et al. 2010). One of the most commonly used methods is the DEM which is intensively used in the traditional and updated methods of watershed delineation and extraction (Li et al. 2019; Liu et al. 2020). Two basic methods which can be identified for the purpose of this delineation are manual and digital (Ray 2018). Remote Sensing and GIS softwares have been found to be particularly useful and effective in this regard (Chu et al. 2010; Sit et al. 2019).

The significance of watershed management has been highlighted in numerous ways in multiple literature. Reducing losses in the form of soil, water and nutrients and ensuring an overall sustainable development are seen as the major objectives of watershed management (Behera and Panda 2003). They are also suggested to provide a 'logical' input to planning for a wholistic planning (Bewket 2003). It has been summed up that watersheds provide the 'best' unit of planning of the precious water resource and that the watershed approach is all inclusive with the incorporation of natural and human elements (Heathcote 2009).

This is also indicative of the fact that this as a planning methodology is a dynamic one and is now beyond the limited realm of resource utilisation to an overall planning for socio-economic dimensions as well. In the context of climate change, this type of planning requires all the more a greater significance (Wang et al. 2016). Current advancements in the techniques of remote sensing and GIS are now able to provide efficient and sufficient information on hydrological parameters and hydrogeological aspects. This has given information for regions and circumstances which are otherwise inaccessible (Duan et al. 2021).

Enhanced techniques of geographical research provided by advances in geospatial technology have been credited to provide unprecedented methods of delineating and analysing a watershed. Remote Sensing and Geographical Information Systems (GIS) assisted analysis specifically comes under this category. The focus of the current research is to provide a delineation methodology of watersheds for the selected region of the state of Mizoram in general and for the specific study area of the district of Lunglei (Mizoram), while comparing the existing watersheds of the region by the Digital Watershed Atlas of India (DWA). The DWA is drawn on a scale of 1:1 million following the hierarchy of streams.

It also tries to explore the advantages and limitations of both the derivations and tries to observe their efficacy as watershed delineation methods. Within this analysis, attempt has also been made to observe watershed delineation in the light of existing hydro-geology of the study area along with land cover and population concentration. This is done for the purpose of identifying priority areas for watershed delineation as well as planning.

16.2 Basis of the Study

Water is the lifeline of a watershed and this is turn is an outcome of the region's geography which is assisted and altered by a combination of natural and anthropogenic factors. Despite the identifiable features, identification and delineation of watershed is evolving in nature. The characteristics of a watershed to be effective include ensuring water supply over a longer period, maintenance of water quality and, sustainable development (Heathcote 2009). It is also suggested that international involvement in the management of local water resource leads to more complications along with other political concerns.

In all such aspects, watershed based planning provides a very viable and practical solution to many issues (Cosgrove and Loucks 2015). While highlighting the need and importance of watershed management, the role of remote sensing and GIS deserves a distinct mention although numerous methods exist for watershed delimitation as mentioned above. In watershed delimitation, the Pfafstetter Coding System takes into consideration the topology of land. Developed in 1989, by Otto Pfafstetter, a Brazilian engineer this system divides the watershed into three categories—basins, interbasins and internal basins. The drawing of these units as per this system is hierarchical in nature and watersheds are drawn from the river junctions.

A Pfafstetter basin has the characteristic that it does not receive drainage from any other drainage area, an interbasin receives water from upstream watersheds, and an internal basin is the one that does not contribute to another watershed or water body. Also an interbasin is a drainage area with no share in flow of any kind (Verdin, n.d.). The classification of watershed regions as per this code provides a global appeal to the method. The main consideration in this system is that it examines the topographic controls of an area drained in a topology of hydrographic network. The details of this method involve the following:

a. This method is believed to stringently apply the rule of drainage area with the main stream and the tributary.
b. The main stream drains the maximum area and the tributary drained area is called a basin.
c. The area between two tributaries is called interbasin.
d. Coding of watersheds begins by identifying four largest tributaries and assigned numbers 2, 4, 6 and 8 in the order encountered going upstream.
e. The interbasins are numbered 1, 3, 5, 7, and 9 going upstream from the mouth of the main stem.
f. A closed basin is numbered as zero-0.
g. A basin can be divided further into basins and interbasins.
h. An interbasin is subdivided by identifying the 4 tributaries with the greatest drainage areas with the same methodology (Verdin, n.d.).

In the data derived through remote sensing here, the Pfafstetter Coding System is used by the database to generate watersheds.

The remote sensing method provides timely and accurate information of features and phenomena for their recognition, categorisation and examination (Choi and

Engel 2003; Rai et al. 2018; Mishra et al. 2021; Rai et al. 2021, 2022) and this was not feasible otherwise with multiple limitations (Wang et al. 2016). Although both methods are not a replacement of any other method but owing to their immense spatial spectrum as compared to others, they tend to make analysis more convenient. They can supplement, complement, improvise and enhance field based data. This has been made particularly possible with the use of GIS (Wang et al. 2016). GIS techniques are not only convenient but also efficacious in dealing with the watershed and its components (Biswas et al. 2022).

Remote sensing techniques have fast gained significance for hydrogeological studies (Rai 2023a, b) and in this the role of vegetation is equally important yet not significant (Rajib et al. 2020). Interestingly, GIS has also made it possible to derive real time watersheds (Choi and Engel 2003). Spatial geometry provided by GIS techniques is one of the best in modelling in hydrological examination and watershed delineation through this method provides much better results than the traditionally applied methods (Liao et al. 2020). Detailing of scales and determining selection criteria for watershed delineation are some exceptional helping aspects of GIS which are not otherwise possible (Liu et al. 2020).

Taking all these aspects into consideration, the present attempt aims at delineating watersheds for the state of Mizoram. That is attempted through the state of the art remote sensing and GIS by taking into account the hydrogeological aspects of the state.

It further tries to examine the derivation with the existing Digital Watershed Atlas for the country at multiple geographic scales. To make the study practically relevant, observations on population and Land Cover and Land Use (LULC) of the state are also depicted in the light of the same. The limitations and advantages of both the methods are outlined. The purpose of the findings is to provide an additional and updated database with a methodological support to the existing database of the DWA for the state. It is also an attempt to observe the advantages and disadvantages of both the databases.

16.3 Study Area

The Study Area comprises of two levels for analysis. The state of Mizoram forms the first level. The second level of spatial data is the district of Lunglei in Mizoram. This is done with an attempt to bring out major observations on watershed basins as delineated from the selected databases at the macro and meso level; the watershed becoming the micro level of study. The state covers a latitudinal extent of 21° 58′ & 24° 35′ N and a longitudinal extent of 92° 15′ & 93° 29′ E. It covers an area of about 21,087 km^2 with international borders with Myanmar and Bangladesh. The total population of the state as per 2011 Census was 10, 91,014 people and population density of 52 persons per km^2. It also has a high literacy rate of about 91.85% as per the Census of 2011. With its location in the far north eastern part of the country, it is also strategically significant.

As per its geography, which is the main focus here, the state is largely mountainous with high variation in topography in the eastern part and marvels in vegetation. The hill ranges are steep with cuts as deep gorges and rivers flowing incessantly north to south in these ranges. The hills have an average height of 900 m; the highest peak being 2,210 m. The natural landscape of the state is observed to be very varied in nature with lush greenery. The state experiences seasons from summer, to rainy, autumn and winter.

Mizoram is defined to have a rare combination of wooded hills, swiftly flowing rivers, streams and still lakes. These physical features lend a distinction to its topography. Physiography in detail highlights the state as having rolling hills, lakes and rivers predominantly with the hills covering all dimensions of the state. This provides a very limited space to plain region and the hills are very rugged and steep (Mizoram G. O.).

Lunglei is the largest district of the state as per its area of 552.61 km^2 (Mizoram G. O.). It had the second largest population in the state according to the Census of India 2011. Lunglei derives its name from a bridge like rock found in the riverine region around Nagasih, which is a tributary of Tlwang river. It has 39 villages with a population density of 36 persons per km^2 (Mizoram 2023). The study area is shown in Fig. 16.1.

The watersheds are derived through remote sensing technique for both the regions viz. Mizoram state and Lunglei district. This is done for the purpose of observing the existing classification with remote sensing based outline of watershed within the administrative divisions which provides a practical base to this delineation.

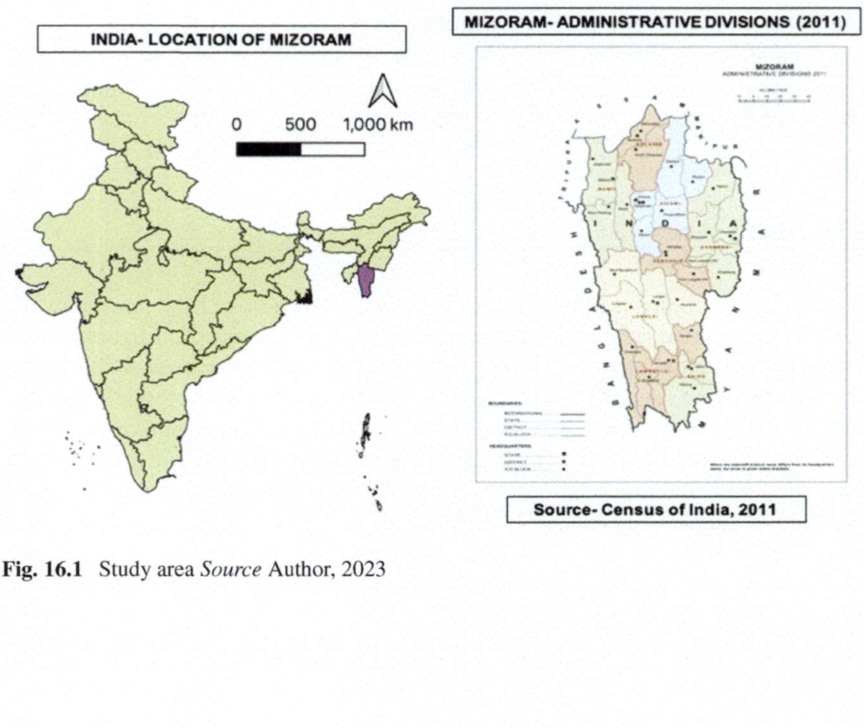

Fig. 16.1 Study area *Source* Author, 2023

16.4 Research Objectives

The study aims at analysing the following:

a. Observing the details of the landscape through landuse, population distribution and hydrogeology of the Study Area.
b. To observe the existing watershed divisions for the study area as per the Digital Watershed Atlas of India.
c. To derive watersheds through the Pfafastetter Coding System through remote sensing.
d. To compare the results and analyse them.

The study tends to derive the hydrological details of the area under study through remote sensing to first observe the geological base of the region. This will form the root of studying the watersheds and can lend insight into the details of geology of the study area. Next attempt is drawn through observations on the watersheds through the DWA. The Atlas of 1990 and 2012 form the basis of watershed delineation in the country.

16.5 Data Source and Methodology

High quality images of the Google Earth Engine are seen as incomparable in the context of watershed delineation (Ray 2018; Ghorbanian et al. 2020). The current study encompasses aspects of relief and drainage and geological characteristics related to orientation, slop, direction of flow, land use and land cover and population density to derive a background of the study area. Watersheds as delineated at various levels through remote sensing. State level analysis has been conducted for land use and land cover, for the parameters of population density and hydrogeological details for deriving the watersheds. The study area at the second level is examined in detail for watershed derivation only. The Google Earth Engine Code Editor has been accessed for multiple data sources to derive information. The first basic information is derived for the land use and land cover of the state through the ESA World Cover 10 m 2020. The European Space Agency World Cover 10 m 2020 product gives data for land use and land cover at 10 m resolution from Sentinel- 1 and 2. This provides data for 11 classes. For population density, GPWv411: Basic Demographic Characteristics (Gridded Population of the World Version 4.11, 2000—2020) has been utilised. This database is derived from NASA SEDAC at the Center for International Earth Science Information Network and models the distribution of world human population for the years 2000, 2005, 2010, 2015, and 2020 on 30 arc-second (around 1 km) grid cells wherein. The population data is put in cells with proportional allocation of population from Census and administrative units. For hydrogeology, multiple examinations are conducted. Hydrological analysis of drainage direction and river channel widths are taken to support this analysis. Further, hydrogeological information is derived from Merit Hydro: Global Hydrography Datasets. MERIT Hydro is defined as a global dataset at about 90 m resolution and is derived from MERIT DEM elevation and water body datasets.

It shows river networks and removes involved noises. Error in this dataset is less than 0.05 for 90% of the databases. It is supposed to improvise upon spatial coverages and that of small streams as well with improved quality. The Global SRTM Landforms data base is used for deriving detailed information on the landscape through this. It combines the SRTM CHILI and SRTMmTPI datasets which are based on 30 m SRTM DEM. To extend the analysis for hydrographic information, WWF HydroSHEDS mapping dataset is used. This is particularly done to observe the flow accumulation of water bodies in the region. It draws details on ecologically relevant information also which is useful in the light of climate change studies. Elevation data from NASA's SRTM. The dataset provides information at various scales for network of rivers, boundaries of watersheds, direction of drainage and accumulation of flow (Google 2023).

The Pfafstetter coding based watersheds are derived from the WWF HydroAtlas for Basins. The flow accumulation is examined here. It tends to show the amount of upstream area draining into each cell. Finally, the watersheds information from the Digital Watershed Atlas of India (2012) is derived for the state and the study area at the second level. The Atlas has been drawn by the Central Water Commission (CWC) and Indian Space Research Organisation (ISRO) in 2012 (Table 16.1).

Dividing the whole country into six River Resource Regions, the Atlas organised drainage at a scale of 1:1 million with 35 river basins, 112 catchments, 550 sub catchments and 3,257 watersheds. In this Atlas, the delineation has been done in seven stages starting with Water Resource Regions and their subsequent division and subdivisions into Basins, Catchments, Sub-catchments, Watershed, Sub watershed and Microwatersheds in decreasing size of the outlined hydrologic unit. Each of the sub-catchment is then divided into watersheds following the lower order streams, a

Table 16.1 Average size and size ranges for each hydrological units—Digital Watershed Atlas of India (2012)

S. no	Category of hydrologic units	Example of code	Size range (ha)	Average size (ha)
1	Water resource region	2	270,00,000–1130,00,000	5,50,00,000
2	Basins	A	30,00,000–300,00,000	95,00,000
3	Catchments	1	10,00,000–50,00,000	30,00,000
4	Subcatchments	A	200,000–10,00,000	7,00,000
5	Watersheds	2	20,000–300,000	1,00,000
6	Subwatersheds	a	5,000–9,000	7,000
7	Microwatersheds	2	500–1,500	1,000

Source India (2019)

group of tributaries or the left and right bank of higher order streams moving downstream upwards using the same base. Watersheds through remote sensing database have been derived from WWF HydroATLAS Basins at Pfafstetter Codes from levels 3–12. With all these datasets, the analysis for both parts of the study area is done or the watersheds.

16.6 Analysis

The analysis begins by depicting the land cover and land use for the state of Mizoram. This is also important for other practical concerns and is shown in Fig. 16.2.

It is visible that tree cover dominates the land cover of the state. This is also an indicator of the fact that the hydrology of the region is rich enough to support this. A visualisation of the region's hydrogeological set up can be seen in the following Figs. 16.3, 16.4 and 16.5.

As is visible from the Figs. 16.3, 16.4 and 16.5, it can be said that the geological and hydrological observations are very clearly discernible and are difficult to access. Population density as is shown in Fig. 16.2, is concentrated along the capital city of Aizwal but otherwise no distinct pattern is indicated. An examination of the above images indicates that the geology of the region is highly complex and variegated in this relief as is shown in Fig. 16.3a.

This complexity is also visible in the contours of the region in Fig. 16.3b. This topography along with the dense vegetation cover also makes it difficult to identify watersheds manually. Remote sensing technique, thus, provides an unprecedented advantage in such a landscape of high complexity. River network as is shown in

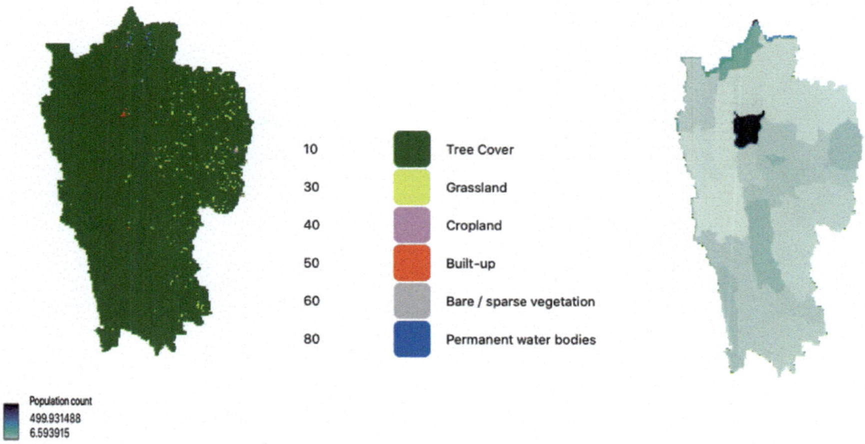

Fig. 16.2 Mizoram—**a** Land Use and Land Cover (2020) and **b** Population density. *Source* Author, 2023 derived from Google Earth Engine Code Editor, ESA World Cover 10 m 2020

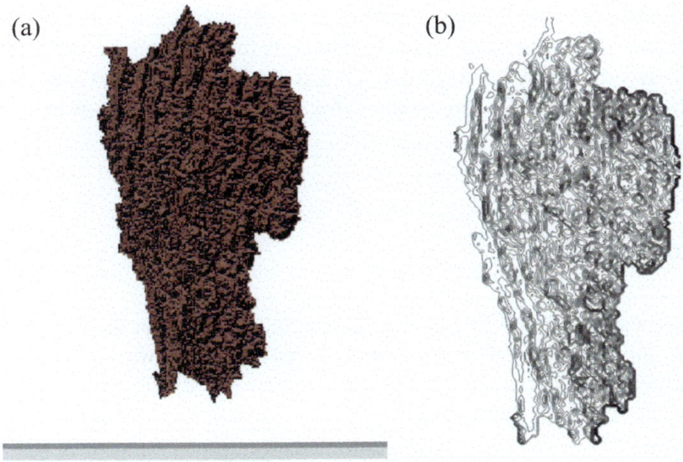

Fig. 16.3 Mizoram—Hydrogeology and contours. *Source* Author, 2023 derived from Google Earth Engine Code Editor from Merit Hydro: Global Hydrography Dataset

Value	Color	Color Value	Description
11		#141414	Peak/ridge (warm)
12		#383838	Peak/ridge
13		#808080	Peak/ridge (cool)
14		#ebeb8f	Mountain/divide
15		#f7d311	Cliff
21		#aa0000	Upper slope (warm)
22		#d89382	Upper slope
23		#ddc9c9	Upper slope (cool)
24		#dccdce	Upper slope (flat)
31		#1c6330	Lower slope (warm)
32		#68aa63	Lower slope
33		#b5c98e	Lower slope (cool)
34		#e1f0e5	Lower slope (flat)
41		#a975ba	Valley
42		#6f198c	Valley (narrow)

Fig. 16.4 Mizoram: SRTM based geology. *Source* Author, 2023 derived from Google Earth Engine Code Editor from Global SRTM database

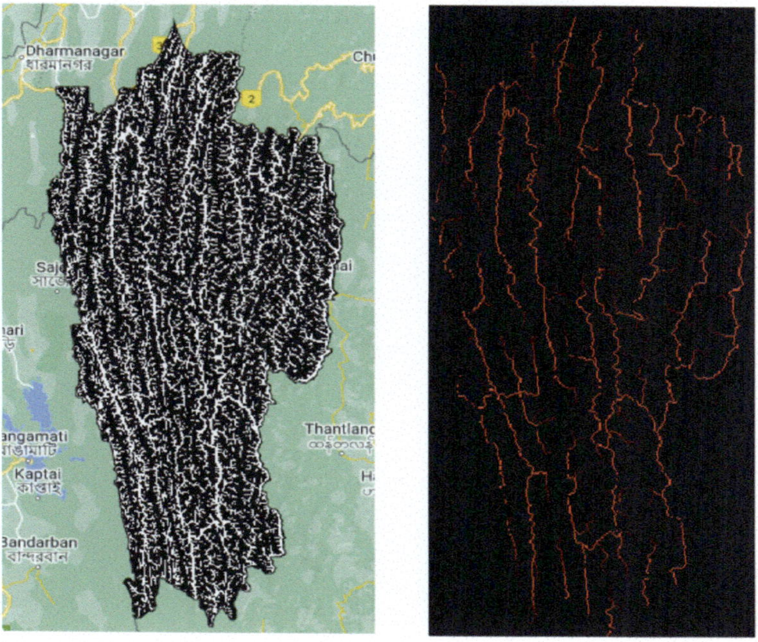

(a) (b)

Fig. 16.5 Mizoram—**a** River network, width and **b** Flow accumulation. *Source* Author, 2023 derived from Google Earth Engine Code Editor from Merit Hydro: Global Hydrography Datasets

Fig. 16.5 is again highly complex and it is very difficult again to identify watersheds. The stream network is very intricate and longitudinal in nature. However, this visualisation also provides an added clarity to the landscape in which the stream network can be seen to flow in an identified geological setting.

This can be taken as the starting point of watershed delineation for the region. The hydrology brings out the geological pattern in a more identifiable manner. The analysis is well supplemented by flow accumulation. Figure 16.4 also highlights the landscape geology. Figure 16.5 is very helpful in the watershed analysis, particularly in the identification of main stream and tributaries. The Pfafstetter identification will follow this natural landscape for identification of watersheds through the remotely sensed data. Within this background an attempt has been made to compare the output generated by the Digital Watershed Atlas and the remotely sensed data. This is done at the 'first' level of both the classifications and is shown in Figs. 16.6 and 16.7.

Upon comparison, it is visible that the two outcomes are different. However, one significant mention in this regard is that the DWA is depicting watersheds as per the political boundaries in its database. However, it is also mentioning that the delineation is on the basis of natural boundaries. But upon the download, the administrative boundaries are being indicated. This is slightly confusing to the reader as is shown in

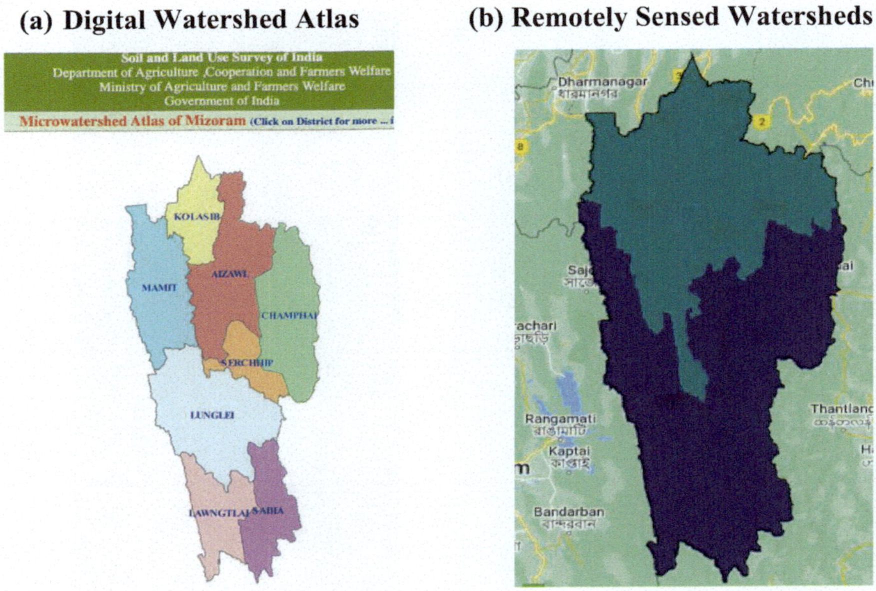

Fig. 16.6 Mizoram—'First' level of watersheds. *Source* Author, 2023 from **a** DWA, 2023 from India, S.o. (2019). **b** WWF HydroSHEDS Basin Pfafstetter Level 03 derived from Google Earth Engine Code Editor

Fig. 16.7 Mizoram—Minutest level of watersheds through upstream area for delineation through remotely sensed data *Source* Author, 2023 derived from derived from Google Earth Engine Code Editor from WWF HydroSHEDS Basin Level 12

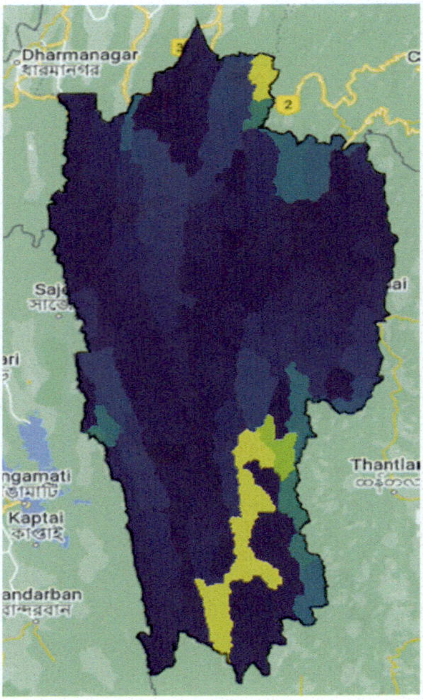

Fig. 16.7. This can also pose a difficulty in understanding the delineation. Further, this aspect is flexible and accurate with the remotely sensed data as it is user defined. So, the minimum qualifying level of the watershed that is visible through remotely sensed data is not visible in DWA initially. This analysis is next extended to the minutest level of information which can be derived from both the datasets. For this purpose, the second category of study area is taken.

This is depicted in Fig. 16.8.

As is visible through Fig. 16.8, the minutest level of water shed is available for the administrative level of the state through remote sensing database. In terms of this detailing at the state level, data is not directly available through DWA, with the exception of administrative boundaries. But remote sensing based datasets can be derived with or without the administrative boundaries. This is a great advantage with the only exception being that of scale based issues. This advantage can be helpful for multiple planning issues.

The study for the second part of the study area is attempted as a detailed sample for the study of watershed delineation. For this purpose, Lunglei district is selected as the representative. For this purpose, the basic watershed delineation is examined for the district at the largest and the smallest available levels. The results are drawn from DWA and remote sensing separately and ae depicted below.

As per Fig. 16.9, the details of the study area can be seen for its hydrogeology. This will form the base for further derivations as has been done for the state. These are visible in Figs. 16.10, 16.11 and 16.12.

Figure 16.10 shows that at the basic level of delineation as per Pfafstetter Coding, there are two basins visible in the upstream area, depicting structural controls as

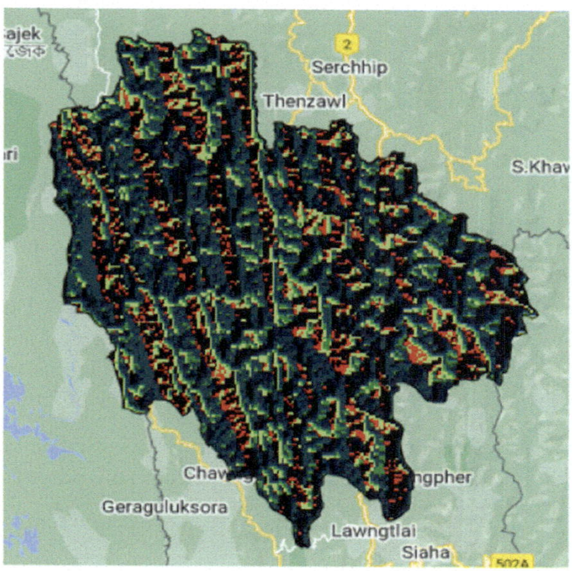

Fig. 16.8 Lunglei district—Hydrogeology. *Source* Author, 2023 derived from Google Earth Engine Code Editor from WWF HydroSHEDS Drainage Direction, 15 Arc-Seconds

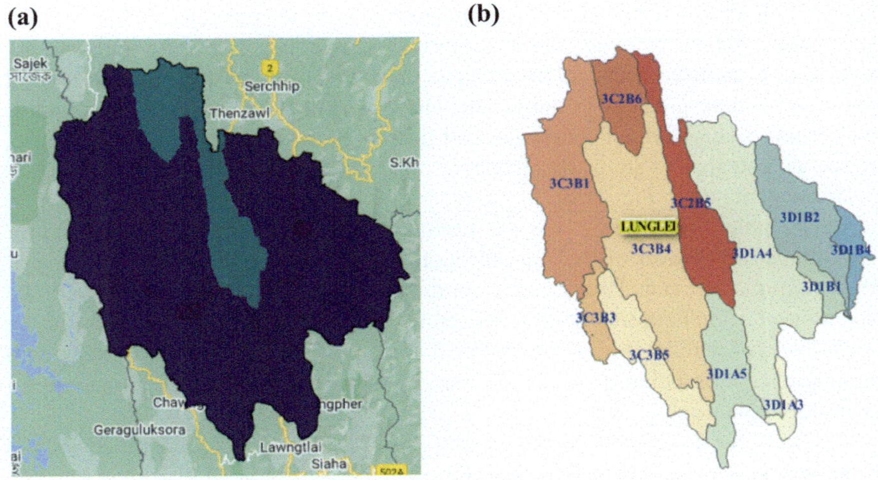

Fig. 16.9 Lunglei—Initial Watershed Basins—Pfafstetter Coding—Level 3 with DWA. *Source* Author (2023), **a** derived from Google Earth Engine Code Editor from WWF HydroSHEDS Basin Level 03 and **b** DWA, 2023 from India. S.o. (2019)

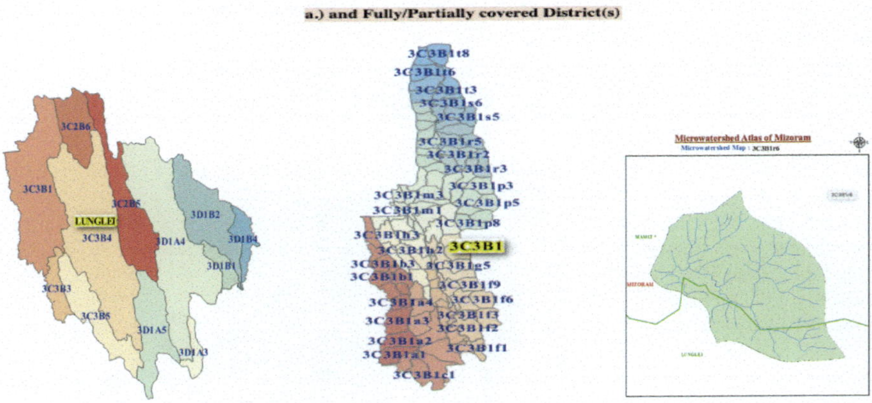

Fig. 16.10 Lunglei—DWA based watersheds at the micro levels. *Source* DWA, 2023 from India. S.o. (2019)

per watershed delineation. The DWA provides this in the form of multiple basins as mentioned and shown in Fig.16.10 earlier. As is observed from Fig. 16.10, the basic level of watersheds for the district are of a varied nature. The DWA provides a more detailed information at this level. For remotely sensed data, the four largest identified tributaries provide two main watershed regions for the state. However, if taken by the individual stream, the DWA representation is shown in Fig. 16.10. But in this regard, the DWA database does not seem to fit in exactly with the geology as can be

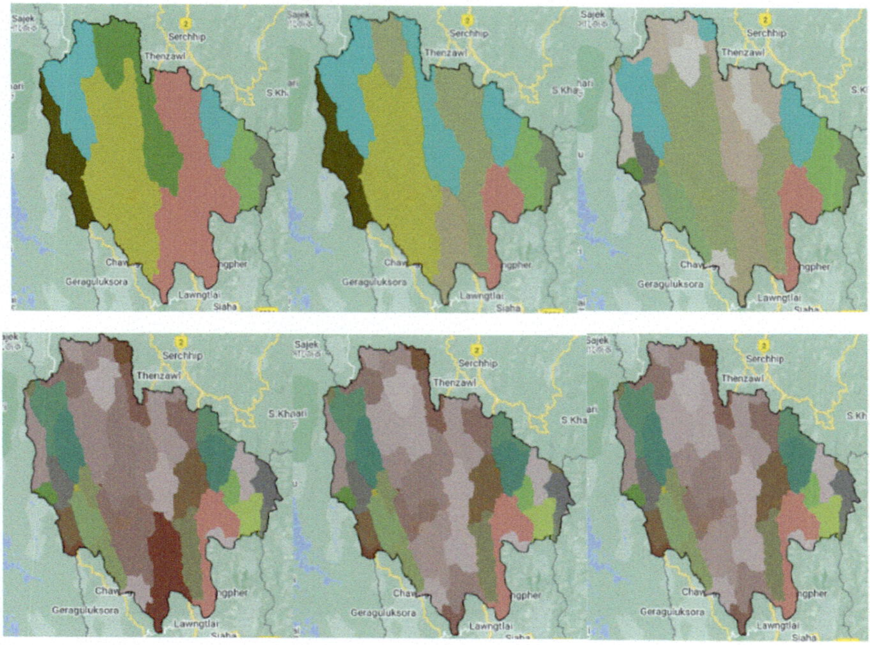

Fig. 16.11 Lunglei—Watersheds through remote sensing Pfafstetter levels 7–12. *Source* Author, 2023 from Google Earth Engine Code Editor WWF HydroATLAS basins

seen from Figs. 16.9 to 16.10. In both the cases, there seem to be some omissions in the stream recognition by the depictions. The reasons for this cannot be deciphered and this appears to be a major flaw in DWA. For remote sensing derived dataset, the methodology is different and the results are also of varied nature but at some levels. This is defined as the strength of the Pfafstetter system that the basins are identified as per the topologies existing; both upstream and downstream of a watershed. The current examination has taken the upstream area as the identifier as the coding begins following the process of upstream identification.

Figures 16.11 and 16.12 are attempts to identify the minimum levels of watersheds through the DWA as well as remote sensing analysis. In this regard, the DWA can be credited for providing a highly detailed classification of watersheds till the last identifiable level. For remote sensing methodology, this cannot be done automatically as the matter of scale is to be manually inserted or the study area identification is required; although this is not impossible.

As is visible from the above diagrams, the levels of Pfafstetter Coding derived through remote sensing show the watersheds delineated towards the lowest possible as per the administrative boundary.

Fig. 16.12 Microwatershed Atlas of (Lunglei District) *Source* (Mizoram G. o., nd.)

16.7 Observations and Conclusions

The above outputs indicate the advantages and disadvantages of the both the datasets in different aspects. The DWA outputs have advantages with regards to the precision with which it is drawn at the micro watershed level. In this context, the scale of calculation of DWA is the benchmark for all its processes of watershed delineation. This scale is further carried over to other levels for drawing of watersheds. However, this classification is static in nature and further information on scale at different levels is not available. This can, however, be problematic as scale based division is not dynamic in observing the existing variations in topography, while topography is a significant component of such methods. The Pfafstetter is one of the methods of watershed classification and is widely used for the purpose. However, a visible disadvantage can be with regards to the choosing of four largest streams which cannot be logically explained as the choice. Besides, the flow accumulation map shows a very different scenario as it indicates majority flow in the big streams. In such a context, the Pfafstetter coding proves to be more beneficial for drawing the watersheds rather than the traditional method of delineation. This is so because it takes into study the 'major' tributaries of the main stream. Besides, the topography of the region is highly complex. In such a scenario, it can be said that the purpose of planning can be the determinant while selecting the methodology for watershed delineation for the region. Both the methods offer comparative advantages and disadvantages. Thus, it become important to examine the requirement of watershed delineation and identifying the purpose of the same. This will prove to be helpful in determining the method which can be more appropriate for the same. Remote sensing methodology is slightly advantageous in dynamic aspects of such a process as it can provide an updated view of the landscape, geology, hydrology and other changes which take place over a larger geographic scale. This updation is not always feasible manually. There is also a possibility of combining remote sensing with other methods but with caution.

References

Bajjali W (2018) Watershed delineation. In: Bajjali W (ed) ArcGIS for environmental and water issues. Springer

Behera SK, Panda RK (2003) Eco-friendly optimum management of land, water and nutrients in a small agricultural watershed. In: Singh VP, Yadava RN (ed) Water and evironment: watershed management. Allied Publishers Pvt. Ltd., New Delhi, pp 114–126

Bewket W (2003) Towards integrated watershed management in highland Ethiopia: the Chemoga watershed case study. Tropical resource management papers. Wageningen University

Biswas S, Sudhakar S, Desai VR (2022) Remote sensing and geographic information system based approach for watershed conservation. J Surv Eng 128(3). https://doi.org/10.1061/(ASCE)0733-9453(2002)128:3(10

Black PE (1996) Watershed hydrology. Lewis Publishers

Choi JY, Engel BA (2003) Real-time watershed delineation system using Web-GIS. J Comput Civil Eng 17(3). https://doi.org/10.1061/(ASCE)0887-3801(2003)17:3(189

Chu X, Zhang J, Chi Y, Yang J (2010) An improved method for watershed delineation and computation of surface depression storage. Watershed Manage. ASCE, pp 1113–1122.https://doi.org/10.1061/41143(394)100

Cosgrove WJ, Loucks DP (2015) Water management: current and future challenges and research directions. Water Resour Res 51(6). https://doi.org/10.1002/2014WR016869

Duan W, Maskey S, Chaffe PB, Luo P, He B, Wu Y, Hou J (2021) Recent advancement in remote sensing technology for hydrology analysis and water resources management. Remote Sens 13(6). https://doi.org/10.3390/rs13061097

Ghorbanian A, Kakooei M, Amnai M, Mahdavi S, Mohammadzadeh A, Hasanlou M (2020) Improved land cover map of Iran using sentinel imagery within google earth engine and a novel automatic workflow for land cover classification using migrated training samples. ISPRS J Photogram Remote Sens 167. https://doi.org/10.1016/j.isprsjprs.2020.07.013

Google (2023) Retrieved from https://code.earthengine.google.com/

Heathcote IW (2009) Integrated watershed management: principles and practice, vol 2. Wiley

India Go (nd) Retrieved from http://slusi.dacnet.nic.in/dwainew.html

India So (2019) Retrieved from http://slusi.dacnet.nic.in/dwainew.html

Li L, Yang J, Wu J (2019) A method of watershed delineation for flat terrain using sentinel-2A imagery and DEM: a case study of the Taihu Basin. Int J Geo-Inform. https://doi.org/10.3390/ijgi8120528

Liao C, Tesfa T, Duan Z, Leung LR (2020) Watershed delineation on a hexagonal mesh grid. Environ Modell Softw 128. https://doi.org/10.1016/j.envsoft.2020.104702

Liu K, Song C, Ke L. Jiang L, Ma R (2020) Automatic watershed delineation in the Tibetan endorheic basin: a lake-oriented approach based on digital elevation models. Geomorphology 358. https://doi.org/10.1016/j.geomorph.2020.107127

Mishra VN, Rai PK, Singh P (2021) Geo-information technology in earth resources monitoring and management (edited book). Nova Science Publishers, USA, pp 1–395. ISBN: 978-1-53619-669-6

Mizoram GO (2023). Retrieved from https://lunglei.nic.in: https://lunglei.nic.in/demography/

Mizoram GO (nd). Retrieved from https://mizoram.nic.in: https://mizoram.nic.in/about/glance.htm

NOAA (2023). Retrieved from https://oceanservice.noaa.gov: https://oceanservice.noaa.gov/facts/watershed.html

Rai PK (2023) River conservation and water resource management. Springer Nature, Singapore, pp 1–352. ISBN No.: 978-981-99-2604-6. eBook ISBN: 978-981-99-2605-3

Rai PK (2023) Advances in water resource planning & sustainability. Springer Nature, Singapore, pp 1–322. ISBN No. 978-981-99-3659-5

Rai PK, Mishra VN, Singh P (2021) Recent technologies for disaster management & risk reduction-sustainable community resilience & responses (edited book). Springer Nature, Switzerland, pp 1–471. ISBN: 978-3-030-76116-5. https://doi.org/10.1007/978-3-030-76116-5

Rai PK, Mishra VN, Singh P (2022) Geospatial technology for landscape and environment management: sustainable assessment & planning (edited book). Springer Nature, Singapore, pp 1–431. ISBN: 978-981-16-7373-3. https://doi.org/10.1007/978-981-16-7373-3

Rai PK, Chandel RS, Mishra VN, Singh P (2018) Hydrological inferences through morphometric analysis of lower Kosi river basin of India for water resource management based on remote sensing data. Appl Water Sci. https://doi.org/10.1007/s13201-018-0660-7

Rajib A, Kim IL, Golden HE, Lane CR, Kumar SV, Yu Z, Jeyalakshmi S (2020) Watershed modeling with remotely sensed big data: MODIS leaf area index improves hydrology and water quality predictions. Remote Sens 12(13). https://doi.org/10.3390/rs12132148

Randhir TO (2007) Watershed management: issues and approaches. IWA

Ray LK (2018) Limitation of automatic watershed delineation tools in coastal region. Ann GIS 1–14

Sit M, Sermet Y, Demit I (2019) Optimized watershed delineation library for server-side and client-side web applications. Open Geospatial Data Softw Stand 4(8). https://doi.org/10.1186/s40965-019-0068-9

USGS (2023). Retrieved from https://water.usgs.gov/: https://water.usgs.gov/GIS/huc.html

Verdin KL (n.d.) Retrieved from https://apo.sdsu.edu: https://apo.sdsu.edu/pfafstetter_system.html

Verdin KL (n.d.). Retrieved from https://proceedings.esri.com/, https://proceedings.esri.com/library/userconf/proc97/proc97/to350/pap311/p311.htm

Wang G, Mang S, Cai H, Liu S, Zhang Z, Wang L, Innes JL (2016) Integrated watershed management: evolution, development and emerging trends. J Forest Res 27. https://doi.org/10.1007/s11676-016-0293-3

Webster M (2023). Retrieved from https://www.merriam-webster.com: https://www.merriam-webster.com/dictionary/watershed

Chapter 17
The Role of Entrepreneurship in Promoting Sustainable Development Through the ODOP Programme in Uttar Pradesh

Zaibun Nisa

Abstract The Uttar Pradesh government's One District, One Product (ODOP) initiative is aimed at promoting local entrepreneurship and empowering rural communities through the development of region-specific products. Pandey (2020) This paper examines the potential of the ODOP programme to contribute to sustainable development in India, with an emphasis on how entrepreneurship plays a part in reaching this objective. The paper reviews the literature on sustainable development, entrepreneurship, and the ODOP programme, and analyzes case studies of successful ODOP initiatives in Uttar Pradesh. The study makes the case that the ODOP program can effectively support entrepreneurship as a means of advancing sustainable development by fostering local innovation, promoting social and environmental responsibility, and generating economic growth. The final section of the study discusses the policy implications of these findings and offers suggestions for further leveraging the ODOP program to support sustainable development in India.

Keywords Sustainable development · Entrepreneurship · ODOP programme · Rural development · India

---

Z. Nisa (✉)
Khwaja Moinuddin Chishti Language University, Uttar Pradesh, Lucknow, India
e-mail: zaibunnisa@kmclu.ac.in

P. K. Rai et al. (eds.), *Natural Resources Management and Sustainable Livelihoods in the Mountainous Region*, Advances in Geographical and Environmental Sciences,
https://doi.org/10.1007/978-981-97-2100-9_17

17.1 Introduction

17.1.1 Background

Entrepreneurship Development is a means for generation of employment and registering growth in the economy (Audretsch 2001). A sustainable approach helps entrepreneurs to bring about innovation, and transit thereby creating a new system (Shepherd). The process of entrepreneurship development contributes in helping new firms and enterprises in achievement of their goals, enhance business activities and upscale the nation's economy. (*(PDF) A Study on Women Entrepreneurship in India*, n.d.) In other words, the process of entrepreneurship development is all about providing necessary support to entrepreneurs in advancement of their skills with the help of training and coaching (Kuratko 2019).This improves their judgments and helps them make considered decisions for all activities of their business.

India being one of the World's largest economies is poised for rapid growth in future. It has in the past decades made huge gains in both the economic and social fields and there exists great potential for increase in productivity by shifting priority of its youth from being job seekers to job providers (Glaeser 2010).

The "One District One Product" Programme was started by the Uttar Pradesh Government to encourage entrepreneurship as a sustainable source of income. The aim of the ODOP program is to give boost to the local traditional industries, expertise in one product, value addition of that product and improvement in the states' GSDP. ODOP focuses on all 75 districts in the state and is for the benefit of the local craftsmen (https://odopup.in/).

Four Pillars of ODOP (Schemes Under ODOP Programme)

Source: Primary

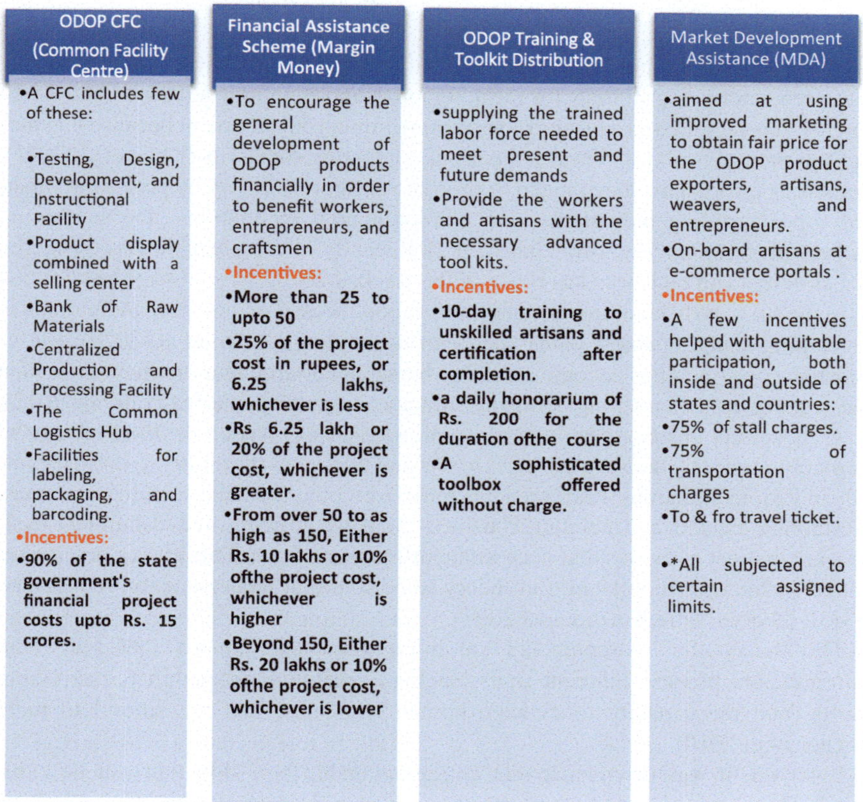

India's Prime Minister, Shri Narendra Modi, has declared his goal for the country's GDP to reach USD 5 trillion by 2024. Uttar Pradesh state has also established a goal to become a USD 1 trillion economy by that time in order to support the PM's agenda. Additionally, in this case, aiming to accomplish this goal also required all of the state's districts. The government of Uttar Pradesh is expected ODOP to play an important role by enabling every district to develop entrepreneurship.

According to Uttar Pradesh Chief Minister Shri Yogi Adityanath, the state's "one district, one product" initiative will be crucial to the state's goal of becoming a USD 1 trillion economy by 2024. (*Time of India, e-newspaper, retrieved on 17/03/2021*). With the introduction of the ODOP program, the export of Bhadohi carpets has increased from Rs 4,000 crore to Rs 8,000 crore. The majority of the state's districts have their own distinctive industries. The districts could help achieve the state's objective of having an economy of $1 trillion USD.

17.2 Literature Review

17.2.1 Entrepreneurship Development

For several decades, authorities and governments worldwide have viewed fostering entrepreneurship as a strategy for local, regional, and economic advancement. Additionally, it is regarded as a means of promoting human development because it generates job prospects and serves as a means of combating social ills (Khan et al. 2021). Numerous efforts have been made to promote entrepreneurship by providing financial support, aiding individuals with entrepreneurial inclinations, and supporting successful entrepreneurs (Brockhaus 1982). Recently entrepreneurship is considered as a skill that can be learned as other skills (Lackeu 2013).

In the past, early business incubation methods focused on assisting individuals in developing sound business plans and securing funding (Milan et al. 2020). But as time has gone on, it has become clear that being an entrepreneur involves more than just coming up with a strong company strategy and getting funding (Cooney 2012).

A successful businessperson is someone who creates wealth for themselves and their community by building business assets and possesses the ability to do so repeatedly in various situations. Entrepreneurs must overcome obstacles to utilize resources to establish, expand, and maintain a business. Starting, growing, and sustaining a business are distinct activities that necessitate different abilities. This idea is best exemplified by the transition from trait theory to behavioural theory and, more recently, to skill theory (Nalia Kutzhonova 2007).

The theory that entrepreneurs are born, not made, asserts that successful entrepreneurs possess inherent traits, such as creativity, adaptability, risk-taking ability, and the capacity to manage human resources, that are natural to them (Lichenstein 2010).

Later on, it was discovered that entrepreneurship is a skill that can be cultivated. This notion has led to efforts to teach entrepreneurial skills to individuals and expand their knowledge in this area, forming the basis for the development of the entrepreneurship theory. Various theories, such as behavioural theory, attribute theory, and cognitive theory, emphasize the essential qualities of an entrepreneur, while the actions taken by entrepreneurs have also played a significant role (Mattare 2010).

India possesses an enormous reserve of natural resources and abundant labour, including a skilled workforce, which has contributed to a surge in unemployment among educated workers in the country (Ku Payal Sharma 2016).

The idea that entrepreneurship promotes economic growth and development is widely accepted, although several empirical studies have produced mixed results regarding the contribution of entrepreneurship to economic expansion due to the presence of various types of entrepreneurship and the different macroeconomic environments in which growth occurs. The literature find out that entrepreneurship creates employment, generates wealth, offers products, and services, and generates revenue for the nation (*GJRA-GLOBAL JOURNAL FOR RESEARCH ANALYSIS X 44*, n.d.).

17.2.2 Sustainable Development

Sustainable development, according to the World Commission on Environment and Development (WCED) is a concept that encompasses economic development, environmental protection, and human development as interconnected elements (HarryHummels 2021; Mishra et al. 2021; Rai et al. 2022).

Entrepreneurship aimed at achieving sustainable development involves a comprehensive approach that links environmental, social, and economic aspects of entrepreneurship, as well as how it changes markets to promote overall societal progress (Entrepreneurship for Sustainable Development 2021).

Small and medium-sized enterprises (SMEs) are crucial drivers of sustainable development and inclusive growth. They generate employment opportunities, promote innovation, and make substantial economic contributions. In numerous developing nations, SMEs account for more than half of the total gross domestic product. Furthermore, in all countries, around 90–95% of registered enterprises fall under the SME category (Guterres 2021).

Entrepreneurship can play a pivotal role in facilitating the growth of villages. This can be achieved by providing entrepreneurship support through Self Help Groups and sustainable financing (Fabien Chidanand Robert 2021).

In present times, entrepreneurs are increasingly exercising caution and assuming greater social responsibility. They have come to appreciate the significance of sustainable entrepreneurship in creating a better future (Manisha Gupta 2022).

Consequently, many governments make concerted efforts to enhance the pool of competent and globally proficient entrepreneurs.

Sustainable growth entails conducting economic activities while ensuring the well-being of future generations. This involves broadening the goal of economic growth to encompass sustainable growth, which encompasses environmental considerations (Milan et al. 2020).

Entrepreneurship aimed at achieving sustainable development seeks to shift economic activities from those that are harmful to the environment to those that are not (Miguel-ÁngelGalindo-Martín, December 2021). Entrepreneurs are discovering fresh business opportunities by distinguishing their products from those of their competitors, crafting new brand images, tapping into untapped markets, and marketing environmentally friendly products. (M´endez-Picazo et al. 2021). The Venture Facilitate entrepreneurship approach emphasizes that sustainable entrepreneurship necessitates government policies, well-functioning markets, supportive banking systems, and other essential components (Miguel-ÁngelGalindo-Martín, December 2021).

17.2.3 One District One Product (ODOP) Programme

The Uttar Pradesh (U.P.) government introduced the One District One Product (ODOP) program in January 2018 with the intention of encouraging traditional industries and crafts unique to each district in the state (Shahi et al. 2021). The program aims to boost rural employment and entrepreneurship and enhance the income of artisans and craftspeople (Singh 2020).

Several studies have been conducted on the ODOP programme, and the following literature review summarizes some of the key findings and insights.

The ODOP programme has helped revive traditional crafts and industries in U.P. districts. The programme has provided market linkages and financial support to artisans, leading to an increase in their income. (*(PDF) The Impact Of The One District One Product (Odop) Programme On Socioeconomic Development Of Kannauj, Uttar Pradesh*, n.d.) The article highlights the success stories of several artisans and entrepreneurs who have benefited from the programme.

An analysis conducted by the National Institute of Panchayati Raj and Rural Development (NIRDPR) suggests that the ODOP programme has the potential to promote rural employment and reduce migration from rural areas. The study also highlights the need for skill development and capacity building among artisans and entrepreneurs to ensure sustainable growth.

An article published in the Economic and Political Weekly (EPW) discusses the challenges faced by the ODOP programme in achieving its objectives. The article argues that the programme is focused mainly on promoting traditional crafts and industries, but does not address the larger issues of infrastructure, market access, and finance that are critical for the growth of these industries.

The effect of the ODOP program on the earnings and employment of craftsmen and craftspeople is examined in a study report that was published in the International Journal of study in Management, Economics, and Commerce (IJRMEC). According to the report, the initiative has significantly increased the income and job options for artists, especially in areas where handicrafts have a long history. An article by the Business Standard highlights the challenges faced by the ODOP programme in terms of implementation and coordination between different government departments. The article argues that there is a need for better coordination and collaboration between the departments of industries, agriculture, and rural development to ensure the success of the programme.

The ODOP programme has received mixed reviews from scholars and experts. While the programme has been successful in promoting traditional crafts and industries, there are challenges related to infrastructure, market access, and coordination between different government departments that need to be addressed to ensure sustainable growth and impact. (*(PDF) Women Entrepreneurship-A Literature Review*, n.d.).

The above literature suggests that the sustainable development includes the following aspects as, local innovation, promoting social and environmental responsibility, and generating economic growth (Fig. 17.1).

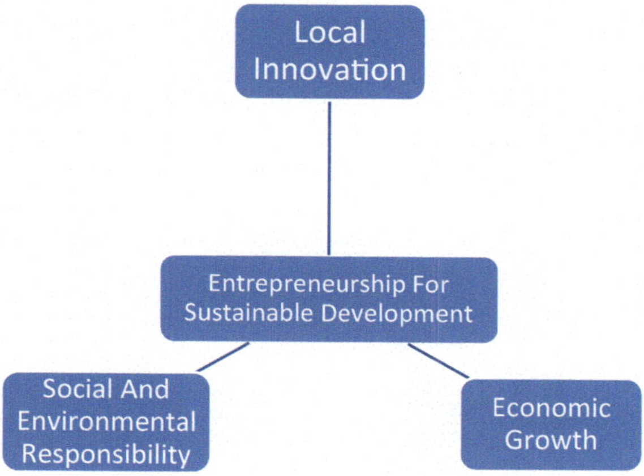

Fig. 17.1 Conceptual framework of study

17.3 Objectives of the Study

The Specific Objective of the Present Study is:

- To evaluate the impact of the ODOP programme on promoting entrepreneurship and sustainable development in Uttar Pradesh.
- To identify the challenges faced by entrepreneurs in the ODOP programme and their impact on sustainable development.

17.4 Research Methodology

The primary and secondary data is utilised in this study. Press releases and official websites are the sources of the secondary data. The primary data is collected from the three districts of Uttar Pradesh i.e., Lucknow, Sitapur and Jhansi.

17.5 Data Analyses and Interpretation

To understand the role of ODOP programme of Uttar Pradesh Government for sustainable development, the parameter selected on the basis of literature review are Innovation, Social and Environmental Responsibility and Economic growth.

17.5.1 *Local Innovation*

One of the objectives of ODOP programme of the government of Uttar Pradesh is to bring innovation in the products of traditional industries of the 75 districts of the state by way of training programme.

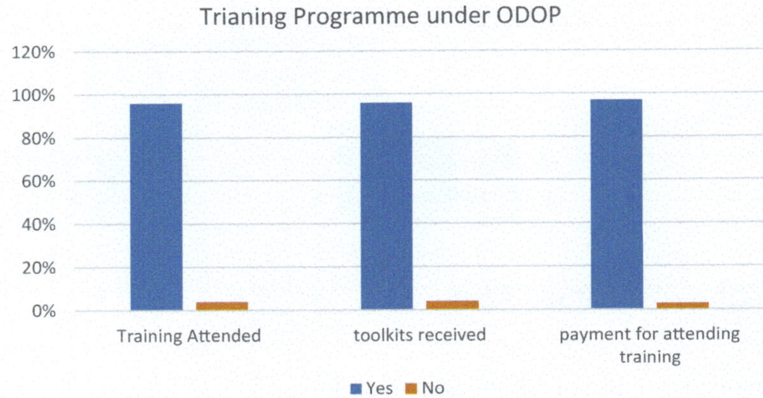

The above figure shows that 96% respondent have attended the training programme under ODOP, 96% received the toolkits and 97% received the payment for the training. The present study finds the training and tool kit distribution scheme of the ODOP programme which is the most popular among beneficiaries.

17.5.1.1 Hypothesis Testing

Ho: There is no dominance of type of training institution being preferred by the beneficiaries.

Ha: There is a dominance of type of training institution being preferred by the beneficiaries.

For testing the above hypothesis a non-parametric chi-square test was used for analysing data. The result is given as follows:

From where they obtained training?			
	Observed N	Expected N	Residual
Recognised institution	79	64	−15
Government training agency	59	64	5
Any other	54	64	10
Total	192		

Test statistics	
A	From where they obtained training?
Chi-Square	4.258[a]
Df	2
Asymp. Sig.	0.119

a. 0 cells (0.0%) have expected frequencies less than 5. The minimum expected cell frequency is 77.7

In the above result, if the p-value is greater than 0.05 (0.119) it means the null hypothesis is accepted. There is an equal dominance of all the institutions working in the field of training. Beneficiaries are equally benefitted from the recognised institutions, government agency and other training institutions.

17.5.2 Areas of Training

Hypothesis Testing:

There is no dominance of Areas of training provided under skill development & tool-kit distribution scheme.

H_1: There is a dominance of areas of training provided under skill development & tool-kit distribution scheme.

For testing the above hypothesis a non-parametric chi-square test was used for analysing data. The result is given as follows:

Areas of training attended			
	Observed N	Expected N	Residual
Skill development &Innovation	96	38.4	57.6
Marketing & selling also	26	38.4	−12.4
Leadership & motivation also	44	38.4	5.6
Social awareness also	3	38.4	−35.4

(continued)

(continued)

Areas of training attended	Observed N	Expected N	Residual
Heath & nutrition also	23	38.4	−15.4
Total	192		

Test Statistics	Areas of training attended
Chi-Square	130.031[a]
Df	4
Asymp. Sig.	0.000

a. 0 cells (0.0%) have expected frequencies less than 5. The minimum expected cell frequency is 38.4

In the above result, if the *p*-value is less than .05 (.000) it means the null hypothesis is rejected. Hence there is a dominance of areas of training attended among all the ODOP beneficiaries. Skill development and innovation has dominance over other areas of training.

17.5.3 Social Responsibility

Hypothesis Testing:

H0: There is no significance relationship between social responsibility and ODOP Programme

The sub-parameters of the social and environment responsibility are:

Binomial Test

		Category	N	Observed Prop	Test Prop	Exact Sig. (2-tailed)
Gender of entrepreneur	Group 1	Male	128	0.85	0.50	0.000
	Group 2	Female	22	0.15		
	Total		150	1.00		
Location of business	Group 1	Urban	59	0.61	0.50	0.011
	Group 2	Rural	91	0.39		
	Total		150	1.00		
Previous experience in business	Group 1	No	70	0.47	0.50	0.463
	Group 2	Yes	80	0.53		

(continued)

(continued)

Binomial Test

		Category	N	Observed Prop	Test Prop	Exact Sig. (2-tailed)
	Total		150	1.00		
Product reserved under scheme	Group 1	No	109	0.73	0.50	0.000
	Group 2	Yes	41	0.27		
	Total		150	1.00		
Type of work force	Group 1	Skilled	129	0.86	0.50	0.000
	Group 2	Unskilled	21	0.14		
	Total		150	1.00		

The observed prop. value of 0.15 is significant at 0.05 level since p value (=0.011) < 0.05. Therefore, Null hypothesis (Ho) is **rejected**. Hence there is a dominance of location in the given sample. Majority of the entrepreneurs taking benefit from the ODOP programme belong to the male gender.

The observed prop. value of 0.85 is significant at 0.05 level since p value (=0.00) < 0.05. Therefore, Null hypothesis (Ho) is **rejected**. Hence there is a dominance of location in the given sample. Majority of the entrepreneurs taking benefit from the ODOP programme belong to urban areas.

The observed prop. value of 0.47 is insignificant at 0.05 level since p value (=0.463) > 0.05. Therefore, Null hypothesis (Ho) is **accepted**. Hence there is a no dominance of previous business experience in the given sample. Fresh as well as experienced entrepreneurs are benefitting from the ODOP programme.

The observed prop. value of 0.73 is significant at 0.05 level since p value (=0.00) < 0.05. Therefore, Null hypothesis (Ho) is **rejected**. Hence there is a dominance of the product reserved in the given sample. Majority of the products taking benefit from the ODOP programme are reserved under ODOP.

The observed prop. value of 0.86 is significant at 0.05 level since p value (=0.00) < 0.05. Therefore, Null hypothesis (Ho) is **rejected**. Hence there is a dominance of type of work force in the given sample. Majority of the ODOP units employ skilled workers.

17.5.4 Economic Growth

Hypothesis Testing:

H_{0}: There is no significant relationship between economic growth and ODOP Programme

The sub-parameters for the economic growth are:

Binomial test

		Category	N	Observed Prop	Test Prop	Exact Sig. (2-tailed)
Nature of the Production Process	Group 1	Continuous	108	0.72	0.50	0.000
	Group 2	Seasonal	35	0.28		
	Total		143	1.00		
Formalities of Raising loan	Group 1	Rigid	45	0.46	0.50	0.238
	Group 2	Flexible	98	0.54		
	Total		143	1.00		
Terms and conditions	Group 1	Unfavourable	38	0.33	0.50	0.000
	Group 2	Favourable	115	0.67		
	Total		143	1.00		
Rate of interest	Group 1	Moderate	89	0.67	0.50	0.000
	Group 2	High	44	0.33		
	Total		143	1.00		

The value of the observed prop is 0.72. Since p value (=0.00) < 0.05, it is significant at the 0.05 level. Consequently, the null hypothesis (Ho) is **not accepted**. Hence there is a dominance of nature of the production process in the given sample. Majority of the ODOP product under study are of continuous nature.

The value of the observed prop is 0.72. Since p value (=0.00) < 0.05, it is significant at the 0.05 level. Consequently, the null hypothesis (Ho) is **not accepted**. Hence, there is a dominance of formalities regarding loan from financial institution. Most of the beneficiaries who are benefitted from the financial assistance consider the process/ formalities of loan sanction as flexible.

The value of the observed prop is 0.67. Since p value (=0.00) < 0.05, it is significant at the 0.05 level. Consequently, the null hypothesis (Ho) is **not accepted**. Hence there is a dominance of attitude regarding the terms and conditions of raising loans amount. Most of the beneficiaries benefitted from the financial assistance scheme consider the terms and conditions of loan as favourable.

The value of the observed prop is 0.67. Since p value (=0.00) < 0.05, is significant at the 0.05 level. Consequently, the null hypothesis (Ho) is **not accepted**. Hence, there is a significant dominance of views regarding rate of interest regarding loan taken from financial institution. Most of the beneficiaries of financial assistance scheme consider rate of interest of loan as moderate.

17.6 Findings of the Study

- **Problems faced in the sphere of production**

  Finance is the most prevalent problem among respondents followed by the problem of demand, logistics, management, power, raw material, labour then government interference.

- **Getting adequate loan amount**

  Most of the respondents (36.9% satisfied & 15% highly satisfied) showed satisfaction towards the loan amount. Whereas 2% are not satisfied the remaining are neither satisfied nor dissatisfied.

- **Timely available of loan**

  Respondents face problem in timely availability of loan as 30% are dissatisfied and 47% are neither satisfied nor dissatisfied whereas only 6% are strongly satisfied, and 10.8% are satisfied.

- **Procedure for financial transaction**

  Only 10.2% respondents are satisfied & 10.2% are highly satisfied whereas 30% are dissatisfied, 2% are highly dissatisfied and 40.8 % are neither satisfied nor dissatisfied.

- **Credit requirements fulfilled**

  Most of the respondent agree that their credit requirement is fulfilled as 46% of the respondent claims that their credit requirement is fulfilled toa great extent, 25% somewhat, 13% claim not at all.

- **Views on State Government's ODOP Scheme in Uttar Pradesh**

  All respondents agree that the ODOP programme is useful for entrepreneurship development as 51.6% claim that the programme is very useful, 43.9% more useful, remaining claim it is useful.

- **Benefits received under govt. scheme**

  Most of the respondents claim that (50.6%) training & stipened is received from the ODOP programme whereas 31.3 claim that they also get other benefits from the programme.

17.7 Conclusion

The government's initiative to boost the MSME sector by the launch of the ODOP programme is contributing to the inclusive and sustainable development of the entire state (*(PDF) The Impact Of The One District One Product (Odop) Programme On Socioeconomic Development Of Kannauj, Uttar Pradesh*, n.d.). The schemes under the programme (common facility centre, skill development and toolkit distribution, financial support, and marketing help), ODOP have improved the state's inclusive development overall. The toolkit distribution scheme is the most popular among the artisans. (Yadav & Tripathi, n.d.).

The current analysis demonstrates the efficacy of the ODOP program with regard to state economic growth, social and environmental responsibility, and local creativity. According to the report, the Uttar Pradesh government's one district, one product programme initiative has been successful in fostering innovation in traditional product development and generating revenue. (Koneru & Head, n.d.) It is helping in capacity building and making entrepreneurs/artisans/weavers self-reliant. But availability of adequate amount of finance and legal formalities required is found to be the major concern especially in rural areas (Khan et al. 2021). This scheme has covered all the districts of Uttar Pradesh. The beneficiaries have got assistance in the form of capital funds with the financial assistance i.e., margin money/subsidy, marketing support, training & skill development as well as common facility centres under the schemes through which many of the owners of small-scale industries have been able to improve their productivity.

The book "Natural Resources Management and Sustainable Livelihoods in the Mountainous Region—Evidence, Gap, and Future Strategies" encapsulates a larger theme that is greatly enhanced by the research on "The Role of Entrepreneurship in Promoting Sustainable Development through the ODOP Programme in Uttar Pradesh." The study clarifies the complex relationship—especially in the local context of Uttar Pradesh—between entrepreneurship, sustainable development, and natural resource management. Through an examination of the ODOP (One District One Product) Program, the study not only clarifies the workings of entrepreneurs but also emphasizes how essential it is for the development of sustainable lifestyles. The study's conclusions are consistent with the book's main idea, which emphasizes the necessity of evidence-based management strategies and strategic interventions in the management of natural resources to guarantee sustainable development in mountainous areas. The connection that this study makes between entrepreneurship and sustainable development is a valuable addition to the collective knowledge not only in the plain area or state like Uttar Pradesh but also in the mountainous region of north-eastern states.

References

Audretsch DB (2001) What's new about the new economy? Sources of growth in the managed and entrepreneurial economies. Ind Corp Chang, pp 10(1), 267–315

Brockhaus HR (1982) The psychology of the entrepreneur. Encycl Entrep, 39–57

Cooney TM (2012) Reducing recidivism through entrepreneurship programmes in prisons. Int J Entrep Innov

Entrepreneurship for sustainable development (pp 364–364). (2021). https://doi.org/10.1007/978-3-319-95867-5_300048

Fabien ChidanandRobert LF (2021). Village development framework through self-help-group entrepreneurship, microcredit, and anchor customers in solar microgrids for cooperative sustainable rural societies. J Rural Stud, 432–440

(Government of Uttar Pradesh. (Official Website): https://odopup.in/)

Gjra-Global Journal For Research Analysis X 44. (n.d.)

Glaeser EL (2010) Urban economics and entrepreneurship. J Urban Econ 67(1):1–14

Guterres A (2021) Investing in sustainable economy. U N Publ, New York

HarryHummels AA (2021) Planetary demands: Redefining sustainable development and sustainable entrepreneurship. Jouurnal Clean Prod

Ku. Payal Sharma TC (2016) Entrepreneurship development in India. Int J Curr Res, 35430–35432

Kuratko DF (2019) Entrepreneurship in India. Int Small Bus J 37(2):99–110

Khan RU, Salamzadeh Y, Shah SZA, Hussain M (2021) Factors affecting women entrepreneurs' success: A study of small- and medium-sized enterprises in emerging market of Pakistan. J Innov Entrep 10(1):1–21. https://doi.org/10.1186/S13731-021-00145-9/TABLES/8

Koneru K, Head (n.d.) Women entrepreneurship in India-problems and prospects

Lackeu M (2013) Developing entrepreneurship competencies. Licent Thesis

Lichenstein L (2010) Investing in entrepreneurs: a strategic approach for strengthening your regional and community economy. Praeger/ABC-CLIO

Manisha Gupta M (2022) Green entrepreneurship and sustainable development: A conceptual framework. In: Materials Today: Proceedings, 3603–3606

Mattare M (2010) Use of self 101: the case for teaching personal development in the entrepreneurship curriculum. N Engl J Entrep, 17–28

Miguel-ÁngelGalindo-Martín M-S-M-T-P (2021) Effects of the pandemic crisis on entrepreneurship and sustainable development. J Bus Res, 345–353

Mishra VN, Rai PK, Singh P (2021) Geo-information technology in earth resources monitoring and management (edit. Book), Nova Science Publishers, U.S.A., pp 1–395. ISBN: 978–1–53619–669–6

Milan R, Hasan N, Nisa Z (2020) Analysis of profile of non-government microfinance organisations: Study at Uttar Pradesh. Int J Sci Technol Res 9(3):3918–3924

Nalia Kutzhonova TS (2007) Skill-Based development of entrepreneurs and the role of personal and peer coaching in entreprise development. Econ Dev

Pandey A (2020) Transforming lives through one district one product (ODOP) scheme: a case of uttar pradesh, India. J Entrep Educ, 23(S2)

(PDF) A study on Women Entrepreneurship in India. (n.d.). Retrieved 14 November 2023, from https://www.researchgate.net/publication/342094983_A_study_on_Women_Entrepreneurship_in_India

(PDF) The Impact Of The One District One Product (Odop) Programme On Socioeconomic Development Of Kannauj, Uttar Pradesh. (n.d.). Retrieved 16 November 2023, from https://www.researchgate.net/publication/372472913_The_Impact_Of_The_One_District_One_Product_Odop_Programme_On_Socioeconomic_Development_Of_Kannauj_Uttar_Pradesh

(PDF) Women Entrepreneurship-A Literature Review. (n.d.). Retrieved 15 November 2023, from https://www.researchgate.net/publication/344350199_Women_Entrepreneurship-A_Literature_Review

Rai PK, Mishra VN, Singh P (2021) Recent technologies for disaster management & risk Reduction-Sustainable Community resilience & responses (edit. Book), Springer Nature, Switzerland, pp 1–471. ISBN: 978–3–030–76116–5. https://doi.org/10.1007/978-3-030-76116-5

Rai PK, Mishra VN, Singh P (2022) Geospatial technology for landscape and environment management: sustainable assessment & planning (edit. Book), Springer Nature, Singapore. pp 1–431. ISBN: 978–981–16–7373–3. https://doi.org/10.1007/978-981-16-7373-3

Shepherd DA (n.d.) The new field of sustainable entrepreneurship: Studying entrepreneurial action linking "what is to be sustained" with "what is to be developed. Entrep Theory Pract,, pp 35(1), 137–163

Shahi S, Singh S, Maurya P, Singh P (2021) True potential of a District—ODOP.

Singh L (2020) ODOP scheme for local craft governance: 'Chikankari' Entrepreneur's Perspective. 2, 129–139. https://doi.org/10.58426/cgi.v2.i1.2020.129-139

Yadav US, Tripathi R (n.d.) Role of one district one product (ODOP) of uttar pradesh: strategies and a new initiative for developing global handicraft index

Chapter 18
Effects of Land Use and Land Cover Change on Physical and Chemical Properties of Soil in Tuirini Watershed, Mizoram, NE India

Vanlaltanpuia

Abstract The present study investigates the effect of land use and land cover (LULC) changes on physical and chemical properties of soil. Studying the dynamics of soil properties in different LULC classes is important for the sustainable management of agricultural resources and mitigation strategies for land degradation. Soil samples were collected and examined from different LULC classes and finds that the dynamics of LULC consequently affect soil properties in the study area. The physical properties of soil like Bulk density and Porosity change significantly. Owing to receiving a good amount of annual rainfall, the study area has slightly acidic soil in nature. The chemical properties like SOC, SOM and available N are tremendously changing their characteristics with the change of LULC. Available K and pH have moderately changed their properties and available P and EC have lower effects on the dynamics of LULC.

Keywords LULC · Physico-chemical properties · Bulk density · Porosity · Soil Organic Carbon (SOC) · Soil Organic Matter (SOM) · Nitrogen (N) · Phosphorous (P) · Potassium (K) · pH · Electrical Conductivity (EC)

18.1 Introduction

Soil is a vital resource that supports life on Earth in several ways. Its physical and chemical properties play a crucial role in agriculture and, by extension, in the economy and human well-being (Vishwakarma et al. 2016; Mishra et al. 2021; Rai et al. 2021, 2022). The fertility and composition of soil influence crop growth, nutrient availability, water retention, and even the health of ecosystems (Singh et al. 2021). Preserving soil health through sustainable agricultural practices is vital for ensuring

Vanlaltanpuia (✉)
Government Aizawl North College, Aizawl, Mizoram, India
e-mail: puiatetea@gmail.com

P. K. Rai et al. (eds.), *Natural Resources Management and Sustainable Livelihoods in the Mountainous Region*, Advances in Geographical and Environmental Sciences,
https://doi.org/10.1007/978-981-97-2100-9_18

food security and maintaining a balance between human needs and environmental conservation. The present study investigates the effect of land use and land cover (LULC) changes on physical and chemical properties of soil. Soil samples were collected and examined from different LULC classes and assessed to find out whether the dynamics of LULC consequently affect or doesn't affect the soil properties in the study area. Manpoong and Tripathi (2019), Sahoo et al. (2019) and Kenye et al. (2019) reported that land use change affects the soil properties in Mizoram. Bier and Singh (2018), Shimrah and Rao (2015) and Saha et al. (2012) also reported that the varying soil property due to land use/land cover change in north east India. Studying the dynamics of soil properties in different LULC classes is important for the sustainable management of soil resources, agricultural resources and mitigation strategies for land degradation (Singh and Rai 2017; Shastri et al. 2020).

18.2 Study Area

The Tuirini watershed, situated between 23° 28′ 40″–23° 53′ 15″ north latitudes and 92° 49′ 15″- 93° 58′ 15″ east longitudes within Mizoram, spans across the central region of the state, encompassing portions of Serchhip and Aizawl districts. Covering an area of approximately 411.38 square kilometers, it represents about 1.94% of Mizoram's total geographical area. This watershed extends in a north–south direction and is contained within the Survey of India topographical maps 84 A/13, 84 A/14, and 84 A/15.

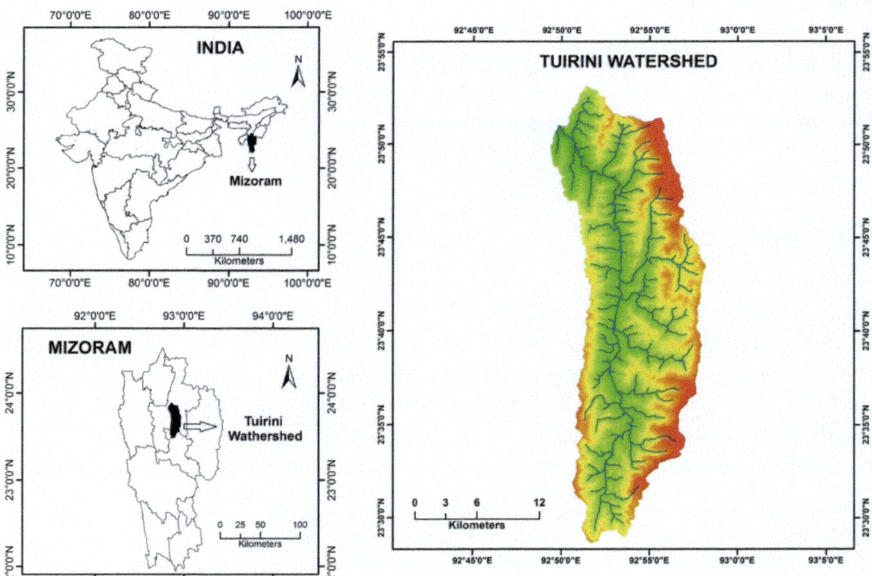

18.3 Objectives

1. To examine the physical and chemical properties of soil in the study area.
2. To assesses the dynamics of physical and chemical properties of soil in different land use and land cover classes in the study area.

18.4 Methodology

The soil samples were collected following the grid based staggered start sampling pattern also known as the diamond sampling method. Within the study area, 73 soil samples were collected from various points with even distribution. In each sample site, the area was divided based on natural slope boundaries into top, medium and low sections. From these sections, three replicate samples were collected along the direction of each slope boundary and then combined or mixed to create one composite sample representing each specific sample site. Chemical properties of the collected soil samples were processed in the laboratory by the Agriculture Department of the Government of Mizoram and the physical properties of the soil samples were processed by the researcher.

For basic statistical analysis, Microsoft Excel has been applied to calculate the Mean and Standard Deviation (SD). Analysis of Variance (ANOVA) in SPSS 21 was employed to evaluate the significant variations in soil physical and chemical properties across different land use or land cover classes within the Tuirini watershed area.

18.5 Result and Discussion

18.5.1 Soil Physical Properties

The two inverse relationship of physical properties of the soil like bulk density (BD) and porosity have been analysed. The result shows that physical properties vary depending on changing land use and land cover changes. The similar results also founded by Reiners et al. (1994), Celik (2005), Abassi et al. (2007) and Kenye et al. (2019). The average bulk density of 1.44 ± 0.12 and porosity of $0.45 \pm 0\ 04$ across the study area provides general condition of the soil's physical characteristics. The lowest BD founded in the very dense forest area has the highest porosity indicating a soil structure of more pore spaces and less compacted soil. Abassi et al. (2007) has reported that the BD value becomes lowest in the forest area at the depth of 0–15 cm. On the other hand, commercial tree plantation area i.e. Teak wood have highest value of BD but have the lowest porosity, indicating a denser soil structure with fewer pore spaces. The variations in BD and porosity among different LULC types highlight the impact of anthropogenic or natural interventions on soil physical characteristics and its suitability for various purposes. The result clearly shows no extreme soil compaction in the study area and is suitable for any agricultural activities.

The mean bulk density of the soil is lowest in a very dense forest area (1.39 ± 0.17) but holds highest porosity value (0.47 ± 0.06). In a moderately dense forest area, BD value increases (1.43 ± 0.09) but soil porosity slightly decreases (0.46 ± 0.04). Open forest area has nearly a similar result with moderately dense forest area having a BD value of 1.43 ± 0.11 and porosity of 0.46 ± 0.03. Saplalrinliana (2016) also found the same sequence record under forest covers in Mizoram. In the current shifting cultivation, the value of bulk density (1.44 ± 0.06) slightly increases than the forest classes with a medium porosity (0.45 ± 0.02). The result indicate that the bulk density and porosity do not change considerably within a short period of one agricultural year and have no negative impact on the jhum cultivation system.

Tree plantation area especially Teak tree (Tectona grandis) has the highest and the lowest value of bulk density and porosity. Sharma (2015) has reported that due to having relatively less moisture content, Teak tree plantation area has incomparable very high bulk density value. Moreover, plantation sites are generally located in the well drain and higher degree of steep slope site. Therefore, the land use/land cover class of the tree plantation area has the highest bulk density and lowest porosity. Owing to land management system practices under fruit plantation area like regular clearance through weeding and fallen leaves removal lead to soil erosion intensity and ranked the second highest in bulk density (1.53 ± 0.07) and porosity of 0.42 ± 0.03 among different LULC classes.

Significant changes (>0.05) of bulk density (0.036) and porosity (0.032) have been found in different land use/land cover classes of the study area and reveal the degradation of soil. Lower bulk density and higher porosity are observed in the very dense forest area. It signifies healthy soil due to better water infiltration and reduced surface runoff. This supports vegetation growth and prevents erosion. Conversely, higher bulk density and lower porosity leads to minimize the pore size of soil resulting runoff intensity and reduced the infiltration rate which increased the surface runoff and intense erosion, limiting plant growth and reducing soil productivity leading to soil degradation (Rai et al. 2023a, b).

The implications on soil productivity and the growth of plants due to changes in bulk density and porosity highlight their significance in land management. Alterations in these soil properties affect surface runoff, infiltration rates, root growth, and overall soil health, emphasizing the importance of maintaining suitable conditions for sustainable land use practices (Table 18.1).

18.5.2 Soil Chemical Properties

The analysis conducted in this study encompasses the examination of soil chemical properties, such as available carbon content like soil organic carbon (OC) and organic

Table 18.1 Changing the soil physical properties in Tuirini watershed

Land use/land cover	BD (g/m^3)	Porosity (in %)
Very dense forest	1.39 ± 0.17	0.47 ± 0.06
Moderately dense forest	1.43 ± 0.09	0.46 ± 0.04
Open forest	1.43 ± 0.11	0.46 ± 0.03
Current shifting cultivation	1.44 ± 0.06	0.45 ± 0.02
Tree plantation	1.57 ± 0.10	0.41 ± 0.04
Fruit plantation	1.53 ± 0.07	0.42 ± 0.03
Overall average	1.44 ± 0.12	0.45 ± 0.04

matters (OM), electrical conductivity (EC), pH (Hydrogen ion concentration), and primary macronutrients viz. nitrogen (N), phosphorus (P), and potassium (K).

18.6 Organic Carbon and Organic Matter

Similar to the physical properties, the chemical properties of the soil also undergo changes with alterations in land use/land cover. Dengiz et al. (2006), Singh et al. (2009), and Sahoo et al. (2019) have emphasized that land use/land cover change stands as a pivotal factor affecting soil organic carbon. This impact stems from the shifts in the input rate (e.g. plant litter) and output rate (e.g. OC mineralization) of soil organic matter due to modifications in plant community and land management practices. Van der Werf et al. (2009), Chen et al. (2010), and Lungmuana et al. (2016) have pointed out that soil organic carbon serves as a fundamental indicator of soil health. It acts as a repository for plant nutrients, maintains a balance among nutrients, regulates the physical condition of soil, and supports soil biota communities.

The average concentration of OC and OM in the study area is 1.36 ± 0.18 and 2.34 ± 0.68, respectively. The very dense forest has the highest concentration of OC and OM (1.56 ± 0.55 & 2.69 ± 0.95), followed by the moderately dense forest (1.42 ± 0.40 & 2.45 ± 0.69) and open forest (1.32 ± 0.35 & 2.28 ± 0.61). These forests classes represent the three highest levels of OC and OM found in the study area. This finding aligns with parallel results identified by Saplalrinliana (2016), Sahoo et al. (2019), and Kenye et al. (2019) in Mizoram, as well as by Bier and Singh (2018) in Nagaland. The forests enhance OC and OM concentration likely linking to increased vegetation, resulting in more litter falls returned to the soil as organic matter. Moreover, the high level of inputs, such as leaf litter and root biomass from the forest trees, along with their resistant nature, prevents microbial decomposition (Sahoo et al. 2019; Kenye et al. 2019).

Following forested areas, current shifting cultivation exhibits a higher amount of OC and OM content (1.29 ± 0.25 & 2.21 ± 0.42) compared to other tree and fruit plantation areas. Sahoo et al. (2019) highlighted that this might be attributed to the inclusion of burnt and partially burnt organic matter in the shifting cultivation areas during slash-and-burn treatments. Similar results were also documented in shifting cultivation systems in Kangchup Hills, Manipur, North-East India (Binarani et al. 2010). However, studies conducted in this area reveal a decline in OC and OM when forests are converted to shifting cultivation. Shifting cultivation, involving the slashing and burning of natural forests, negatively impacts soil organic carbon, diminishes nutrients, and degrades soil quality due to substantial biomass loss (Saha et al. 2012; Osman et al. 2013; Guillaume et al. 2015; van Straaten et al. 2015; Singh et al. 2018; Kenye et al. 2019; Manpoong and Tripathi 2019). Additionally, the high degree steep slope conditions in the study area combined with heavy rainfall contribute to intensify soil erosion, consequently leading to the loss of OC and OM from the top soils.

The study area of the fruit plantation class shows OC and OM contents of 1.13 ± 0.18 and 1.94 ± 0.31, respectively. The ongoing land management practices within these plantations, such as regular clearance of land through weeding and leaf removal, result in a high rate of soil erosion and restrict the time available for decomposition, leading to the accumulation of substantial biomass. Polyakov and Lal (2004) highlighted that continuous cultivation practices cause significant losses of soil organic matter (OM) and other nutrients. This process reduces soil permeability by disrupting soil structure, ultimately impacting soil quality and agricultural productivity (Bhuyan et al. 2013).

Tree plantations have the lowest concentration of OC and OM (1.25 ± 0.15 and 2.15 ± 0.26). They also exhibit the highest bulk density and the lowest porosity values. Analyzing the soil organic carbon stock across different land uses in Mizoram, Kenye et al. (2019) recorded that the OC concentration in teak plantations is lower than that in natural forests and shifting cultivation areas. Tree plantations are predominantly located in well-drained steep slope regions, intensify runoff speed and potentially limit the growth of smaller plants underneath. This limitation reduces litter input, consequently impacting the soil carbon stock.

The observed variations in OC and OM contents across various land use and land cover types indicate soil degradation in the study area. The analysis of variance (ANOVA) shows no significant differences in OC and OM among the different land use/land cover classes (Significance level is 0.168). This lack of variation could stem from the short period of reforestation or limited jhum cycling period and the short-term utilization of land. However, distinct trends are noticeable among different classes, suggesting 84% confidence level in identifying soil degradation.

18.7 PH and EC

The pH levels in the study area exhibit a slightly acidic nature, ranging from 5.61 to 6.07 across various land use/land cover classes. Similar findings have been reported by Lungmuana et al. (2016) and Manpoong and Tripathi (2019) in Mizoram, as well as by Sen et al. (1997) in Northeast India. Among the areas, tree plantations have the highest pH value (6.15 ± 0.54), followed by open forests (6.07 ± 0.60). Mills and Fey (2003) noted that soils beneath plantations tend to become more acidic, often attributed to basic cations being absorbed into forest biomass. The moderately dense forest (5.86 ± 0.45) shows similar pH levels with fruit plantation areas (5.86 ± 0.46), while very dense forests exhibit a pH of 5.74 ± 0.69. In contrast to other reports from tropical countries (Kharal et al. 2018; Islam and Weil 2000; Abassi et al. 2007; Tewari et al. 2016), current shifting cultivation in the study area displays the lowest pH values of 5.61 ± 0.85. This discrepancy might be due to the timing of soil sample collection, steep terrain, and the use of different chemical fertilizers. Moreover, heavy rainfall during the monsoon season may cause surface materials to penetrate and infiltrate into deeper soil layers. However, reports from other tropical countries and Mizoram (Manpoong and Tripathi 2019), thick natural forest regions generally exhibit higher acidity of soil compared to other land use/land cover classes, except for current shifting cultivation. The lower pH in forest soil relates to the relatively higher presence of OC and OM within these soil components and reduced surface evaporation in the study area. The significance test indicates only a 63% (Significance level is 0.38) variation in different LULC.

The average concentration of dissolved salts or alkalinity in the soil measured by EC in the study area is 1.23 ± 0.68, which indicates poor seed emergence (G.O.M. 2009). Saha et al. (2016) also recorded similar findings, noting that Mizoram soils are non-saline with low EC levels. The analysis reveals that current shifting cultivation exhibits the highest EC at 1.58 ± 1.16, followed by very dense forests (1.39 ± 0.91), tree plantations (1.31 ± 0.74), open forests (1.17 ± 0.57), and moderately dense forests (1.15 ± 0.42). Fruit plantations exhibit the lowest EC levels in the study area. While the results indicate that no significant variance in EC, some changes (Significant level is 0.47 or 53%) are observable across different land use/land covers.

18.8 Nitrogen (N)

The average available Nitrogen in the study area measures at 255 ± 93.82. Saha et al. (2016) highlight that Mizoram has generally a high available nitrogen levels, yet the study area falls into a low class as per G.O.M. (2009). Natural forest regions, particularly very dense and moderately dense forests exhibit the highest accumulation of available nitrogen, recording 306.81 ± 126.55 and 279.02 ± 122.18, respectively. Similar high accumulations have been recorded by researchers such as Kharal et al. (2018) in Nepal and Bier and Singh (2018) in Jammu and Kashmir in forested soil areas. However, available nitrogen values notably decrease in open forest areas (226.63 ± 56.89) but rank as the third highest in current shifting cultivation (250.82 ± 80.48) among various land use/land cover classes. The higher level of available nitrogen in current shifting cultivation areas might be attributed to the use of chemical fertilizers, as reported by the farmers. This pattern is also observed in fruit plantation areas (241.40 ± 77.81), whereas the lowest concentration of available nitrogen is found in tree plantation areas (222.59 ± 50.55).

The different land use/land cover classes within the study area show no significant variation in available N, as indicated by the ANOVA results of significance level. However, it can be inferred that approximately 88% (Significance level is 0.122) of the variation among these classes suggests a confidence level in identifying soil degradation resulting from the removal of natural vegetation.

18.9 Phosphorous (P)

According to previous studies by Kumar (2015) and Saha et al. (2016), Mizoram exhibits low available P contents, similar trend also observed in the study area, with an average of 9.18 ± 4.90. This low level of available P might be associated to the region's high rainfall, which leads to the persistence of ion oxides and aluminum fixation, compounded by the low pH values (Tewari et al. 2016). Additionally, the fixation of phosphorus by Fe^{2+}, Mn^{2+}, and Al^{3+} could contribute to this insufficiency (Laskar et al. 1983). Among the various land use/land cover classes, open forests exhibit the highest available P concentration, followed by moderately dense forests and very dense forests (10.11 ± 6.35; 9.79 ± 3.74; 8.85 ± 2.44), indicating that forests possess the highest available P concentrations. This increase might result from the decomposition and recycling of old leaves or vegetation branches, enriching the available P content, a pattern supported by findings from Abassi et al. (2007) and Bier and Singh (2018). Conversely, current shifting cultivation displays the lowest available P concentration in the study area (6.43 ± 2.27). Similar findings by Liu et al. (2002) indicate that cultivated soils have notably less available P compared to forested/vegetated areas, and this variation in P availability correlates with OM content. Following forest classes, fruit plantation areas have higher available P (8.51 ± 6.07), followed by tree plantations (7.57 ± 3.10).

The change in land use/land cover doesn't show significant changes (Significance level is 0.6 or 40%) according to the ANOVA results. However, a noticeable pattern emerges as the vegetation cover decreases, exhibiting a decline that suggests nutrient loss and signifies soil degradation.

18.10 Potassium (K)

The available K concentration in the study area is moderately high at 246 ± 17.39. Saha et al. (2016) also record that Mizoram exhibits moderate to high available K. Remarkably in the study area, current shifting cultivation presents a significantly higher available K concentration (399.94 ± 241.09) compared to all other land use/ land covers. Similar findings have been recorded by Sharma (2015) in Assam, Bier and Singh (2018) in Nagaland, and Kharal et al. (2018) in Nepal. This higher level could be attributed to the use of chemical fertilizers and the adequate retention of available K in the soil through crop residues in agricultural fields. Open forests and moderately dense forests follow the shifting cultivation area with values of 389.24 ± 250.43 and 342.10 ± 252.19 respectively, while the fruit plantation area ranks fourth highest at 316.29 ± 251.15. Very dense forests rank second lowest with 237.00 ± 113.97, and tree plantation areas display the lowest available K concentration at 225.17 ± 68.03. Similar findings recorded that teak plantation areas in Assam also have the lowest available K concentration (Sharma 2015).

Like the result of other primary nutrients, there is no significant variation in available K (Significant level is 0.3 or 60%) across different land use/land cover types in the study area, yet there are dynamics in concentration levels among the different land use/land cover classes (Tables 18.2 and 18.3).

18.11 Conclusion

The present study indicates that the overall natural condition of soil physico-chemical properties is conducive to various vegetation types without fertilizer, yet these properties exhibit dynamics according to land use and land cover (LULC). Chemical properties such as SOC, SOM, and available N undergo significant changes with changing LULC. Available K and pH show moderate alterations, while available P and EC have less pronounced effects on LULC dynamics.

Various anthropogenic activities, such as Jhuming, the use of chemical fertilizers, landscape alterations due to developmental activities and excavations, and soil erosion, contribute to diminishing soil quality and fertility level, thereby expanding the extent of unproductive land.

Management of natural resources in mountainous regions is crucial for fulfilling sustainable livelihoods and to prevent the vulnerability of ecological imbalance. Balancing economic development with environmental conservation is the key to ensure sustainable livelihoods in mountainous regions. It requires a holistic approach that respects the ecological fragility of mountainous region and therefore, valid mitigation strategies is highly recommended to prevent further soil fertility decreases for achieving sustainable management of soil resources. It is suggested that adoption of a permanent land agriculture system to avoid lowering degree of soil fertility caused by jhuming, weeding the fruit and tree plantation area without chemicals, less use chemical fertilizers and replacing forest tree plantation by promoting the conservation of existing natural forest resources *in-situ* through community based forest management.

Table 18.2 Details of the soil chemical properties of different land use/land cover class (LULC)

LULC	SOC (in %)	SOM (in %)	EC (dS m^{-1})	pH	Available N (kg ha^{-1})	Available P (kg ha^{-1})	Available K (kg ha^{-1})
Very dense forest	1.56 ± 0.55	2.69 ± 0.95	1.39 ± 0.91	5.74 ± 0.69	306.81 ± 126.55	8.85 ± 2.44	237.00 ± 113.97
Moderately dense forest	1.42 ± 0.40	2.45 ± 0.69	1.15 ± 0.42	5.86 ± 0.45	279.02 ± 122.18	9.79 ± 3.74	342.10 ± 252.19
Open forest	1.32 ± 0.35	2.28 ± 0.61	1.17 ± 0.57	6.07 ± 0.60	226.63 ± 56.89	10.11 ± 6.35	389.24 ± 250.43
Current shifting cultivation	1.29 ± 0.25	2.21 ± 0.42	1.58 ± 1.16	5.61 ± 0.85	250.82 ± 80.48	6.43 ± 2.27	399.94 ± 241.09
Tree plantation	1.25 ± 0.15	2.15 ± 0.26	1.31 ± 0.74	6.15 ± 0.54	222.59 ± 50.55	7.57 ± 3.10	225.17 ± 68.03
Fruit plantation	1.13 ± 0.18	1.94 ± 0.31	0.91 ± 0.33	5.86 ± 0.46	241.40 ± 77.81	8.51 ± 6.07	316.29 ± 251.15
Overall	1.36 ± 0.18	2.34 ± 0.68	1.23 ± 0.68	5.78 ± 0.61	255 ± 93.82	9.18 ± 4.90	246 ± 17.39

Table 18.3 ANOVA table for significance test

		Sum of squares	df	Mean square	F	Significance
Organic carbon	Between groups	1.202	5	0.24	1.614	0.168
	Within groups	9.98	67	0.149		
	Total	11.182	72			
Organic matter	Between groups	3.555	5	0.711	1.614	0.168
	Within groups	29.524	67	0.441		
	Total	33.079	72			
Electrical conductivity	Between groups	2.143	5	0.429	0.914	0.477
	Within groups	31.417	67	0.469		
	Total	33.559	72			
pH	Between groups	2	5	0.4	1.081	0.379
	Within groups	24.801	67	0.37		
	Total	26.802	72			
Nitrogen	Between groups	75,509.8	5	15,102	1.813	0.122
	Within groups	558,239	67	8331.93		
	Total	633,749	72			
Phosphorous	Between groups	89.485	5	17.897	0.732	0.602
	Within groups	1637.45	67	24.44		
	Total	1726.94	72			
Potassium	Between groups	303,329	5	60,665.7	1.224	0.308
	Within groups	3,320,797	67	49,564.1		
	Total	3,624,125	72			
Bulk density	Between groups	0.161	5	0.032	2.539	0.036
	Within groups	0.852	67	0.013		
	Total	1.014	72			
Porosity	Between groups	0.024	5	0.005	2.619	0.032
	Within groups	0.122	67	0.002		
	Total	0.146	72			

References

Abbasi MK, Zafar M, Khan SR (2007) Influence of different land-cover types on the changes of selected soil properties in the mountain region of Rawalakot Azad Jammu and Kashmir. Nutr Cycl Agroecosyst 78:97–110

Bhuyan SI, Khan ML, Tripathi OP (2013) Soil physico-chemical properties under four agroecosystems in Arunachal Pradesh, Eastern Himalaya. Plant Arch 13:199–206

Bier K, Singh PK (2018) Studies on soil fertility status under different land use systems in Nagaland, J Pharmacogn Phytochem SP1:416–420

Binarani RK, Yadava PS (2010) Effect of shifting cultivation on soil microbial biomass C, N and P under the shifting cultivations systems of Kangchup Hills, Manipur, North-East India. J Exp Sci 1(10):1–14

Celik C (2005) Land-use effects on organic matter and physical properties of soil in a southern Mediterranean highland of Turkey. Soil till Res 83:270–277

Chen XM, Liu JX, Deng Q, Chu GW, Zhou GY, Zhang DQ (2010) Effects of precipitation intensity on soil organic carbon fractions and their distribution under subtropical forests of South China. Chin J Appl Ecol 21(5):1210–1216

Dengiz O, Gol C, Karaca S, Uksel MY (2006) Effects of different landscape position and parent material on soil variability and land use in both sides of Acicay River Çankiri. Int Soil Meet Soil Sustain Life Earth, Manag Soil Technol Proc, Sanlýurfa Turk 2:45–751

G.O.M. (2009) Laboratory testing procedure for soil and water sample analysis. Water Resource Department, Directorate of Irrigation research and Development, Pune, Government of Maharashtra

Guillaume T, Damris M, Kuzyakov Y (2015) Losses of soil carbon by converting tropical forest to plantations: erosion and decomposition estimated by δ (13) C. Glob Chang Biol 21:3548–3560

Islam KR, Weil RR (2000) Land use effects on soil quality in a tropical forest ecosystem of Bangladesh. Agric Ecosyst Environ 79:9–16

Kenye A, Sahoo UK, Singh SL, Gogoi A (2019) Soil organic carbon stock of different land uses of Mizoram, Northeast. AIMS Geosci 5(1):25–40

Kharal S, Khanal BR, Panday D (2018) Assessment of soil fertility under different land-use systems in Dhading district of Nepal. Soil System 2(57):1–8

Kumar M (2015) Phosphate requirement of acidic soils in Northeast India: a reappraisal based on phosphate sorption isotherms. Nat Acad Sci Lett 38:383–386

Laskar S, Dahwal KS, Prasad RN (1983) Soils of Tripura and their fertility management. In: Research bulleting of ICAR research complex for NEH Region, vol 23. Shillong, (Tripura Centre), pp 31–36

Liu SL, Fu BJ, Lu YH, Chen LD (2002) Effects of reforestation and deforestation on soil properties in humid mountainous areas: a case study in Wolong Nature Reserve, Sichuan province, China. Soil Use Mgmt 18:376–380

Lungmuana, Singh SB, Vanthawliana, Saha S (2016) Soil health: importance, options and challenges in Mizoram. Science Vision 16(4):156–159

Manpoong C, Tripathi SK (2019) Soil properties under different land use systems of Mizoram, North East, India. J Appl Nat Sci 11(1):121–125

Mills AJ, Fey MV (2003) Declining soil quality in South Africa: effects of land use on soil organic matter and surface crusting: review article. S Afr J Sci 99(9–10):429–436

Mishra VN, Rai PK, Singh P (2021) Geo-information technology in earth resources monitoring and management (edit. Book). Nova Science Publishers, U.S.A., pp 1–395. ISBN: 978–1–53619–669–6

Osman KS, Jashimuddin M, Haque SMS, Miah S (2013) Effect of shifting cultivation on soil physical and chemical properties in Bandarban hill district, Bangladesh. J for Res 24:791–795

Polyakov V, Lal R (2004) Modeling soil organic matter dynamics as affected by soil water erosion. Environ Int 30:547–556

Rai PK, Mishra VN, Singh P (2021) Recent technologies for disaster management & risk reduction-sustainable community resilience & responses (edit. Book). Springer Nature, Switzerland, pp 1–471. ISBN: 978–3–030–76116–5. https://doi.org/10.1007/978-3-030-76116-5

Rai PK, Mishra VN, Singh P (2022) Geospatial technology for landscape and environment management: sustainable assessment & planning (edit. Book). Springer Nature, Singapore. pp 1–431. ISBN: 978–981–16–7373–3. https://doi.org/10.1007/978-981-16-7373-3

Rai PK (2023a) River conservation and water resource management. Springer Nature, Singapore, pp 1–352. ISBN No.: 978–981–99–2604–6. eBook ISBN: 978–981–99–2605–3

Rai PK (2023b) Advances in water resource planning & sustainability. Springer Nature, Singapore. pp 1–322. ISBN No. 978–981–99–3659–5

Reiners WA, Bouwman AF, Parsons WFJ, Keller M (1994) Tropical rain forest conservation to pasture: changes in vegetation and soil properties. Ecol Appl 4:363–377

Saha R, Chaudhary RS, Somasundaram J (2012) Soil health management under hill agroecosystem of North East India. Appl Environ Soil Sci 2012:1–9

Saha S, Singh SB, Chakrabarty D, Choudhury BU, Das B, Chowdhury S, Syiem EK, Ngachan SV (2016) Analyzing Mizoram weather—A synthesis. ICAR Research Complex for NEH Region Mizoram Centre, Kolasib 796 081, Mizoram

Sahoo UK, Singh SL, Gogoi A, Kenye A, Sahoo SS (2019) Active and passive soil organic carbon pools as affected by different land use types in Mizoram, Northeast India. PLoSONE 14(7)

Saplalrinliana H (2016) Plant litter-soil enzime interactions and biochemical index of soil fertility in jhum agrosystem. Unpublished Thesis of Doctor of Philosophy, School of Natural Resource Management, College of Post Graduate Studies, Central Agricultural University, Umiam, Meghalaya, India

Sen TK, Nayak DC, Singh RS, Dubey PN, Maji AK, Chamuah GS, Sehgal J (1997) Pedology and edaphology of benchmark acid soils of North-Eastern India. J Indian Soc Soil Sci 45:782–790

Sharma C (2015) Physico-chemical properties of soils with special reference to organic carbon stock under different land use systems in Dimoria tribal belt of Assam. IOSR J Agric Vet Sci (IOSR-JAVS) 8(3):32–36

Shastri S, Singh P, Verma P, Rai PK, Singh AP (2020) Assessment of spatial changes of land use/ land cover dynamics, using multi-temporal landsat data in Dadri block, Gautam Buddh nagar, India. Forum Geogr 19(1):72–79. https://doi.org/10.5775/fg.2020.063.i

Shimrah T, Rao KS, Saxena KG (2015) Soil property variations under different land use/cover types in traditional agricultural landscape in Northeast India. J Chem, Environ Sci Its Appl 2(1):73–97

Singh S, Mishra R, Singh A, Ghoshal N, Singh KP (2009) Soil physicochemical properties in a grassland and agro-ecosystem receiving varying organic inputs. Soil Sci Soc Am J 73(5):1530–1538

Singh S, Rai PK (2017) Application of earth observation data for estimation of changes in land trajectories in Varanasi district, India. J Landsc Ecol 11(1):5–18. ISSN: 1805–4196. https://doi.org/10.1515/jlecol-2017-0017

Singh SL, Sahoo UK, Kenye A, Gogoi A (2018) Assessment of growth, carbon stock and sequestration potential of oilpalm plantations in Mizoram, Northeast India. J Environ Prot 9:912–931

Singh A, Rai PK, Deka G, Biswas B, Prasad D, Rai VK (2021) Management of natural resources through integrated watershed management in Nana Kosi Micro Watershed, District Almora, India. Ecol Environ & Conserv 27(February Suppl. Issue):S260−S268

Tewari G, Khati D, Rana L, Yadav P, Pande C, Bhatt S, Kumar V, Joshi N, Joshi PK (2016) Assessment of physiochemical properties of soils from different landuse systems in Uttarakhand, India. J Chem Eng ChemIstry Res 3(11):1114–1118

Van der Werf GR, Morton DC, De Fries RS, Olivier JGJ, Kasibhatla PS, Jackson RB, Collatz GJ, Randerson JT (2009) CO_2 emissions from forest loss. Nat Geosci 2:737–738

Van Straaten O, Corre MD, Wolf K, Tchienkoua M, Cuellar E, Matthews RB, Veldkamp E (2015) Conversion of lowland tropical forests to tree cash crop plantations loses up to one-half of stored soil organic carbon. PNAS 112(32):9956–9960

Vishwakarma CS, Thakur S, Rai PK, Kamal V, Mukharjee S (2016) Changing land trajectories: a case study from India using remote sensing. Eur J Geogr 7(2):63–73

Chapter 19
Natural Disasters in Uttarakhand: Their Root Causes, Effects, and Suggestions

Maisnam Sandhyarani Devi and Shabnam Ara

Abstract One of the world's youngest mountains, the Himalaya is also one of the most active in terms of tectonic activity. Geologically speaking, it is unstable and hazardous. As a part of the Himalayan Mountains, Uttarakhand topography has mountainous in nature. Therefore, it is one of the most disaster-prone states in the nation. Geological instability or hydro-meteorological variability is the main cause of the disasters that has happened in the state. Avalanches, cloudbursts, earthquakes, droughts, and forest fires are all potential threats to the state. The state had its biggest tragedy on June 16, 2013, resulting in significant loss of life and property as well as extensive destruction, leaving a memory to the people that will never be forgotten. The calamity had an impact on the Bageshwar, Pithoragarh, Chamoli, Rudraprayag, and Uttarkashi districts. Due to the disaster's timing, which fell during the busiest travel and pilgrimage season, Rudraprayag the district that was the most severely hit. Compared to other affected areas, the district was experiencing a higher number of casualties. On June 17, major flooding along the banks of the Mandakini and Alaknanda rivers, the two primary tributaries of the Ganga, damaged towns, villages, highways, and bridges for more than 60 km. Cloudbursts and severe rains that struck numerous higher sections of the state were the main contributing factors to this catastrophic calamity. Flash floods and large-scale landslides were brought on by the abrupt rise in water levels brought on by the unprecedented rains in the Mandakini, Alaknanda, Bhagirathi, and other river basins. Floods had affected the pilgrimage sites of Gangotri, Yamunotri, Kedarnath and Badrinath, as well as Hemkunt Sahib. This calamity caused numerous damages in a number of settlements.

Keywords Himalaya · Uttarakhand · Natural disaster · Flashflood · Developmental works

---

M. S. Devi
Independent Scholar and Ex Ph.D. Scholar, Jamia Millia Islamia, New Delhi, Delhi 110025, India

S. Ara (✉)
Department of Geography, HGDC, University of Allahabad, Prayagraj 211002, Uttar Pradesh, India
e-mail: officialshabnam345@gmail.com

P. K. Rai et al. (eds.), *Natural Resources Management and Sustainable Livelihoods in the Mountainous Region*, Advances in Geographical and Environmental Sciences,
https://doi.org/10.1007/978-981-97-2100-9_19

19.1 Introduction

India's natural resources and environment have declined so dramatically in the last many years that it is now ranked among the top ten most disaster-prone nations in the world. With its location, temperature, and geological setting, India has been affected by natural disasters every year (Singh 2006; Rai et al. 2014; Singh et al. 2021; Sur et al. 2021; Singh et al. 2023). A number of natural and man-made reasons, such as unstable geological conditions, topographical features, environmental degradation, population growth, urbanization, and unscientific development practices, are contributing to an increase in disasters across the nation (Mishra et al. 2021; Rai et al. 2021, 2022). Disasters often prove to be a setback to a country's aspirations for prosperity for several decades, destroying the hard-won benefits of challenging developmental efforts and limiting advancement. Certain types of natural hazards are dangerous to humans, while others are damaging to both the land and man-made structures. Natural disaster damage also varies according to the economic development of the region affected; in developing nations, the number of fatalities from natural disasters is rising, whereas in developed nations, the financial losses are usually larger. The population of developing nations has dramatically increased, forcing them to live in dangerous areas near rivers and on steep hillsides. As dangers approach, they are less able to flee. People can relocate more readily, there is more rapid notice, and population growth is lower in developed nations (Donald and David 2010).

A disaster is a phenomenon that is beyond human control that causes significant harm to people, animals, and plants, as well as impairments and damage to property and the environment. Some of the losses are quantified while others are not such as damage to the environment that may not be immediately seen. Human tragedies may be referred to as "disasters" since they can cause the impacted individuals to suffer financially and emotionally, for example, the large scale disasters that are geographically wide spread caused extensive damage and affect large number of lives (Subramanian 2018).

Being the youngest mountain, the Himalaya is highly active in terms of tectonic activity and susceptible to geological dangers. The mountain is also received annually to intense seasonal precipitation during monsoon. Due to its unique geoclimate, the Himalayan range is highly vulnerable to a variety of natural disasters, including earthquakes, landslides, mudflows, debris flows, cloudbursts, and flash floods (Shaw and Nibanupudi 2014). Along with being vulnerable to regular seismic activity, the Himalayan region also faces the risk of river channel siltation, which can lead to frequent floods due to the several large river systems that flow from the Himalaya and the massive amounts of sediment they bring down. Uttarakhand's hilly areas are part of the Himalaya Mountain range. There is a lengthy history of natural disasters in the hilly state. In the Uttarakhand Himalaya, heavy precipitation during the monsoon season, the undermining of the banks by streams, and the destabilisation of the slope material's toe due to road construction and other development activities are all common causes of mass movement (DMMC Report 2012).

The state's hilly areas are particularly unstable, sensitive, fragile, and vulnerable to natural hazards as a result of natural geological forces. The state is also highly vulnerable to natural disasters in the country. Geological instability or hydrometrological variability is the main reason for natural disasters to occur in the region. The state is vulnerable to flashfloods, avalanches, cloudbursts, earthquakes, droughts and forest fires. Uttarakhand has a very common problem of landslides which are quite frequent. The frequency of landslides has been increasing due to extensive construction of roads and haphazard development activities without keeping the environment concerned. Uttarakhand is more sensitive to and probably more affected by climate change. Along with the increase in the global temperatures, changes in the climate at the local level in Uttarakhand also are observed. The hilly terrains of Himalaya are generally very sensitive for landslides, because of combined effect of slope condition, geological structure, nature of rocks, hydrological condition (like spring, underground water, run off etc.), varying temperature and unplanned developments by the people. These are major responsible factors for causing frequent landslides in the hilly terrains of Himalaya. Long and continuous denudation of forest resources by man and livestock make the region prone to massive erosions. During heavy rainfall in monsoon, erosion sheets are frequently formed, which in hilly areas result in massive landslides and flash floods. The Himalaya is a young mountain range that is prone to landslides due to its unstable geology.

Recent studies of remote sensing technique have revealed that the increase in population and settlement in the prone area and climate change are responsible for the disaster in this area. In high mountain areas in Asia recent climate change has causedthe glaciers to reduce and increase the size as well as the number of glacial lakes dammed by debris and when these dams fail GOLF occurs. Such high risk lakes should be identified for action before the disaster take place (Rai 2023a, b). Climatic shifts marked by reduced snowfall and increased rainfall, along with land use changes, lead to frequent disasters like landslides and floods in Kinnaur. Mega hydropower projects, backed by global institutions and local interests, impose unequal risks, disproportionately affecting marginalized communities, where the disaster has impacted the economy of the Kinnaur. The disaster does not happen overnight but it takes time slowly to develop.

19.2 Study Area

The geo-coordinates of Uttarakhand, an Indian state in the Himalayas, are 28°′43′–3127′ N latitude and 77°′34′–83°02′ E longitude. With a total geographic area of 53,483.00 km^2, the state is divided into the regions of Kumaon and Garhwal. In addition, there are 95 developmental blocks and 13 districts. Uttarakhand's capital and largest city is Dehradun. The state has a tropical to temperate climate with 1229 mm of annual rainfall on average. The temperature distribution varies significantly, with Mukteswar at -1.7 °C and Pantnagar at 42 °C (Mishra 2017). The state is home to a wide variety of geomorphic features, including river terraces, major and minor ridges, cliffs, rocky slopes, waterfalls, and highly and moderately dissected denudational hills. Fluvial geomorphic features include point bars, meandering scars, and natural levees. Water logging and floods are common in the Terai and Bhabar regions of the state, which are located in the foothills (NIDM 2015). In the region, the two largest rivers are the Yamuna and the Ganga, which include the Alaknanda and Bhagirathi.

19.3 Causes of June 2013 Devastation

The state of Uttarakhand experienced extremely severe rains from June 14–18, which caused flash floods and landslides in numerous places.

The flood has damaged thirteen districts in the state, with the districts of Uttarkashi, Pithoragarh, Bageshwar, Chamoli, and Rudraprayag suffering the most. The areas around the temple of Kedarnath, the Mandakini, Alaknanda, Pinder, and the banks of the Kali River in the Dharchula suffered the most from the disaster's effects on human settlements (NIDM 2015).The state had an early arrival of the monsoon in 2013. The state as a whole saw heavy rainfall. The reason behind this is the collision between the westerlies and the SW monsoon front. The state received 385.1 mm of rain between June 15 and 18, 2013, which is 440% more than the average of 71.3 mm.

Cloudbursts and heavy to extremely heavy rains that struck various upper regions of the state were the origin and cause of the 2013 tragedy. The situation grew worse when heavy rains in the upper regions caused all of the major rivers' water levels to rise, and the winter snow melted quickly as a result of the impact of falling raindrops. The rainfall also caused extensive landslides in many parts of the state. Sudden increase in water levels by unprecedented rainfall caused flash-floods in the Mandakini, Alanknanda, Bhagirithi and other river basins. On 17 June, towns, villages, roads and bridges for more than 60 miles along the banks of the rivers Mandakini and the Alaknanda were destroyed due to severe flash-floods. Severe flooding affected five major pilgrimage sites—Yamunotri, Gokunt Sahib, Kedarnath, and Badrinath—as well as Gokunt Sahib. Violent flash floods caused by this monsoon's strong rains tore across more than 100 towns, destroying homes, dams, and infrastructure. Toe erosion by swiftly moving mountain rivers full of debris was a major contributor to the state's disaster. Numerous lives have been lost in this calamity, along with a significant amount of property and ecosystem damage. Majority of scientists and academics agreed that the primary cause of this major disaster was the extraordinary amount of rainfall that fell in just two or three days. Even though the damage was widespread throughout the state, the Kedarnath shrine and the area around it saw the worst damage, mostly as a result of a large number of seasonal pilgrims and visitors (NDMA Report 2013).

There were both natural and man-made factors contributing to the horrible Uttarakhand disaster in June 2013. We cannot hold nature alone responsible for this regrettable development. The role of man has also been equal. This flood tragedy was made worse by inadequate disaster management infrastructure, poor building design, extensive deforestation, incorrect agriculture practices on barren hill slopes, and unscientific road and quarry construction. The catastrophic event was made worse by the enormous expansion of hydroelectric plants to satisfy the state's increasing needs and building of highways to handle thousands of tourists visiting Uttarakhand. Since debris was being disposed of in the rivers, the constant construction activity also caused the surface flow to increase and the river bed to rise. To accommodate construction in the name of development, there has been significant deforestation in these areas (Kavita 2017, 494 pp).

19.4 Kedarnath Dham

Kedarnath Dham was one of the severely affected areas by the June 2013 disaster.It is one of India's most well-known and historic pilgrimage sites. Situated in the Mandakini valley of Rudraprayag District, it has a height of 3583 m above the sea level. On the outwash plain of the Chorabari and Companion glaciers is where the town of Kedarnath is situated. The Mandakini and Saraswati river channels cross this outwash and converge near Kedarnath town at the end of the outwash plain. Both glacio-fluvial erosional and depositional processes contributed to the geomorphological formation of the Mandakini Valley. According to India's Seismic Zoning Map, Zone V includes the district of Rudraprayag, which is vulnerable to numerous calamities.

Extremely heavy rains on June 16, 2013, at 5:15 p.m., caused major flooding of the Saraswati River and the Dudh Ganga catchment area, causing significant flow across all of the channels. In every other gulley in the area, active erosion started as an outcome of exceedingly heavy rain. In the major rivers, happened excessive water and sediment accumulation happened (Fig. 19.1). The upper part of Kedarnath town (Sankaracharya Samadhi, Jalnigam Guest House, Bharat Seva Sangh Ashram, etc.) washed away by large amounts of water carrying an enormous amount of debris from the surrounding areas, and glacial moraines had been causing the greatest destruction in the area. The perimeter of the Kedarnath Temple was damaged by landslides, with the base of the temple being flooded with water, mud, and boulders, while the temple itself remains save. There are now completely damaged hotels, rest stops, and stores surrounding the shrine (JRDNA Report 2013, pp. 17).

Over the course of two days on June 15 and 16, 2013, meteorological stations close to the Chorabari glacier recorded 325 mm of rainfall at the glaciers' base. The rain was so heavy that every part of the town of Rambara was washed away. In the second incident, which happened on June 17, 2013, at 6:45 a.m., an enormous quantity of water flowed into Kedarnath town as a result of the overflow and collapse of the Chorabari Lake dam, which was blocked by a moraine. Due to this flash flood, Gaurikund, Sonprayag, Phata, and other areas downstream suffered severe damage (NDMA Report 2013). The heavy rainfall caused the Chorabari glacier to melt quickly, leaving the lake's water with nowhere to go. There were just small

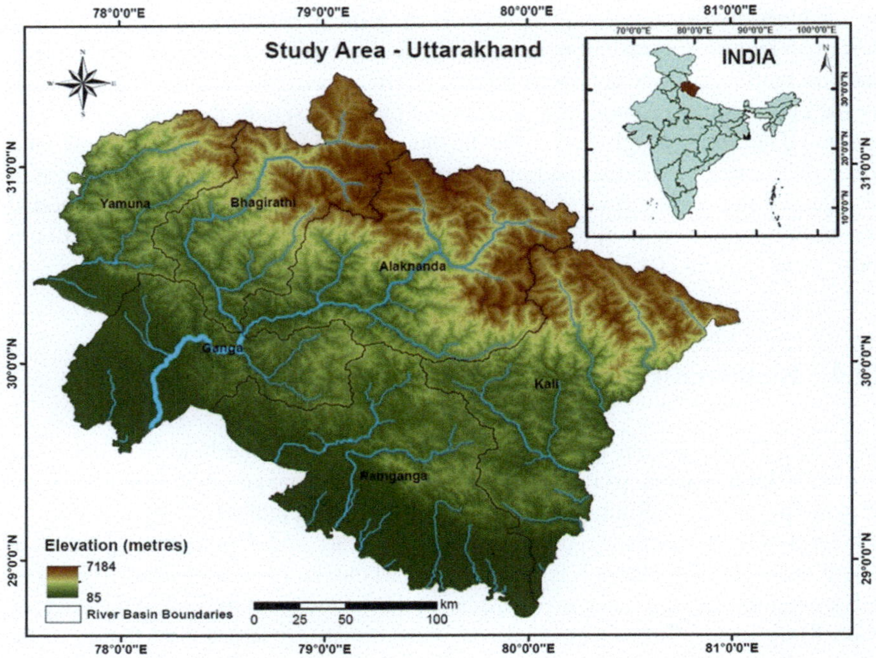

Fig. 19.1 Location map of the study area

openings at the bottom of the lake that allowed the water to be released. Within three days, millions of litres of water unexpectedly gathered in the lake that was blocked by the moraine, increasing their potential energy and weakening the dam's shear strength. In the end, the loose-moraine dam broke, bringing damage throughout the Kedarnath valley. The incident happened at the peak of the pilgrimage season, which caused an enormous death toll in the town of Kedarnath.

The area's recent increase in man-made activity has raised the risk of natural disasters. This pattern is expected to continue growing as more people engage in travel and other activities. When man-made structures were built, they blocked the channels' natural flow routes, causing them to divert from their original direction. Large-scale loss of life and property during the tragedy has been related to the uncontrolled development of hotels, shops, and other structures in the flood plain of the Mandakini and Saraswati rivers to accommodate the demands of an expanding number of tourists (NIDM Report 2013). In Rudraprayag district, there were loss of several livestock and human beings due to this calamity (Table1). Several houses/ lodges/ govt. & private properties were washed away with flash flood.

The tragedy had a particularly bad effect on Uttarakhand state's northern area. The towns of Gaurikund and Rambara, which served as hubs for travel to Kedarnath, were entirely devastated. Severe destruction and casualties also happened in the market town of Sonprayag. In addition to causing widespread flash floods and landslides

Table 19.1 Assessment of loss due to natural disaster in Rudraprayag district till 18 August 2013

Tehsil	No. of villages	Dead/missing person	Injured/affected person	Cattles loss		Building loss			Loss			No. of affected families
				small cattel	Big cattel	Complete	Severe	Semi	Personal commercial shop/lodge	Government buildings	Agriculture land/crop (hec)	
1	2	3	4	5	6	7	8	9	10	11	12	13
Rudraprayag	44	22	1	176	6	106	37	30	34	14	12.416	1146
Ukhimath	117	584	00	8	25	134	20	20	52	10	00	4508
Jakholi	44	44	1	13	11	33	48	77	7	4	7.527	433
Total	**205**	**650**	**2**	**197**	**42**	**273**	**105**	**127**	**93**	**28**	**19.943**	**6087**

Source District Magistrate Office Rudraprayag

in other Uttarakhand districts, the extreme rainfall also severely damaged roads, collapsed buildings, and severely damaged property.

The key factor in killing a great number of people was the powerful water current brought on by the abrupt flash flood. Additionally, landslides killed people by burying them under massive boulders and rubble, and a small number of deaths were caused by a lack of food, water, and medical supplies. The disaster was also attributed to a number of deaths brought on by the intense cold, the ongoing heavy rain, and lack of suitable clothes and shelter (Sharma et al. 2014). Thus, the incident led to one of the biggest tragedies in recent memory, with almost 4000 people losing their lives. Nonetheless, valuable lessons were learnt in handling this crisis, and these will greatly contribute to improving the disaster management system both in Uttarakhand and globally.

The Uttarakhand Disaster of June 2013 resulted in significant losses in terms of human and animal life, infrastructural damage, and damage to both public and private properties. The interruption of the roads and bridges caused this calamity to affect not just visitors and pilgrims but all segments of the local population (Figs. 19.2 and 19.3). The disaster struck during the height of the Char Dham and tourism seasons, when a large number of pilgrims, visitors, and those providing services to them were gathered in the affected locations from all over the country (Fig. 19.3a and b).

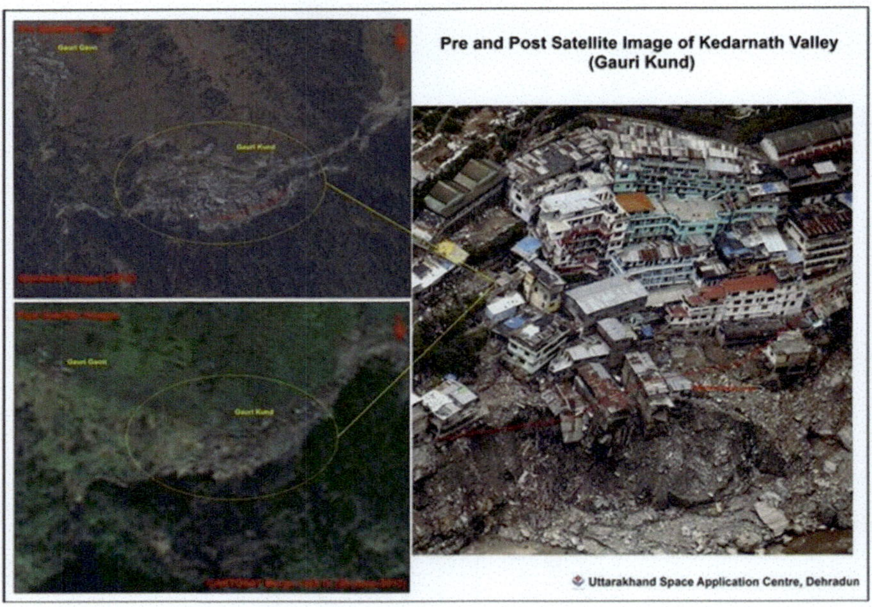

Fig. 19.2 Downstream affected area of Mandakini valley along the river. *Source* USAC

Fig. 19.3 **a** Kedernath before disaster. **b** Kedernath after disaster. *Source* USAC

19.5 Scientific Reason

It has long been known by scientists why such catastrophic weather events occur and the potential impact of climate change on them. As the atmospheric water vapor condenses during cooling, rain and snowfall occur. Global warming is causing an increase in the amount of water vapor in the atmosphere. But the presence of abundant greenhouse gases in the atmosphere makes cooling less effective. There may not be a significant increase in overall rainfall since greenhouse gases in the atmosphere absorb heat and let less of it escape. The accumulation of excess moisture can lead to heavier rainfall in certain areas. The rising frequency and intensity of cyclones, as well as dry weather phenomena like droughts, are connected to the altering patterns of upper air circulation brought on by global warming as floods and landslides happened in Uttarakhand in June 2013. The impacts of the global warming issue, which scientists predict will only become worse as greenhouse gas emissions rise, are already being felt in India.

In Uttarakhand, the 2013 monsoon arrived at least two weeks ahead of schedule. This is a very typical example of how climate change is affecting weather conditions. The Indian Meteorological Department and the Regional and National Weather Forecasting Centers in New Delhi, along with other relevant departments in the state, have released forecasts indicating an increase in extreme weather events and days of heavy rainfall as the monsoon shifts later in the season. Weather events are now more unpredictable due to climate change. The most important thing to remember when it comes to disaster preparedness is to be aware of these general patterns and to be ready for anything from severe storms to heavy rains and the ensuing flooding.

19.6 Environmentalist Views

The history of disaster has been long and continuous. Both manmade and natural disaster are taking human lives and destroying assets/properties. The natural disaster sometimes is predictable and can be controlled. Environmentalists believe that human activity is mostly responsible for a large number of tragedies, including the bombings of Hiroshima and Nagasaki, the Bhopal gas tragedy, the Chernobyl nuclear accident, and others. Besides this many disasters such as land slide, earthquake, construction of dam and tunnel, floods and avalanches are common in hilly areas. Recent decades have seen a significant level of pressure on land and resources arising from unplanned development activities, which has resulted in the loss of life and property. The sensitive ecosystem and nearby populations are negatively affected by the unchecked growth of hydroelectric power plants and the removal of forests in hilly areas. Developmental activities compel the local people to migrate to some other places due to the displacement either choice or by force they have to be shifted from their original places. This not only influence the human beings but the natural habitat and small and macro level ecosystem.

An increasing amount of property and lives lost, according to environmentalists, have come about from reckless developmental activities carried out in recent decades. The immense and uncontrolled growth of massive hydroelectric power plants in the state of Uttarakhand, the continuous building of highways to accommodate the growing number of tourists and urban centers, and the reduction of forest cover have all had an impact on the region's sensitive ecology. The construction of buildings and roads in this area is being done without knowing the local geology. This is a riskier position because blasting is used to move the rocks more often (Nair and Singh 2014). Environmentalists often attribute the tragedy of Uttarakhand disaster 2013 to anthropogenic interventions; particularly hydropower projects. The unscientific ways will never help to deal with any kind of situation. Therefore, a proper River Zoning Regulation, Coastal line Zone Regulation, construction and commercial activities along the rivers banks has been the consideration of scientists and planners.

Mountains and rivers are our natural resources. Mountains give us an aesthetic beauty, variety of flora and fauna. Rivers formed over the mountain region are due to either melting glacier or rainfall, the water is not a basic need for the humans but also for plants and animals on the earth surface as it also helps in the regulation of the bio-geochemical cycle where various nutrients and chemicals flow in various part of our ecosystem and environment. Being a part of the Himalayan mountain region Uttarakhand is vulnerable for disaster when tectonic activities take place or humans induce forces that compel to disturb the region. Water has become precious now–a-days as the ground water level, quality as well as quantity of the water is going down, so it becomes an essential part of our planning and for the existence of life and development, where management of natural resources are important for the sustainable livelihood not only in the mountaineering region but in the downstream area too as the after-effects have been seen there.

19.7 Recommendation

- The state's whole region must be separated into appropriate eco-sensitive zones for development activity; necessitating more thorough planning. At the pilgrim site at least 10 km area should be protected from construction activities.
- In regions that are specifically preserved, like national parks and sanctuaries, there should be minimal human intervention. The geophysical characteristics of the local physical terrain should be taken into consideration when carrying out the tasks.
- If appropriately and successfully maintained, the integrated water management plan, a flagship program of the government, has the ability to preserve the local ecology
- Proper and effective enforcement of environmental laws and policies in conjunction with the relevant department is essential for areas that are delicate and vulnerable.
- To reduce the risk of further landslides massive afforestation with contour wise green planning and well designed drainage system is needed.
- Planned check dams, gully plugging, protection walls, contour bunding, hill sensitive water harvesting structure and large scale preemptive protection works are other measures to minimize landslide.
- Execution of constructional and developmental work should not be done without studying the environmental implication through an Environmental Impact Assessment made mandatory by the extant of government law and decrees.
- There is a need of attention to look in to the matter of illegal mining—both major and minor—in the lower reaches of Himalaya especially for minerals.
- The native communities must be made aware of and sensitive to the relevant regulations and laws related to the development activity that may have an impact on their day-to-day lives.
- A new migration mechanism, one that moves people from vulnerable areas to safe areas within the region, is required in the hilly areas.

References

DMMC Report (2012) State disaster management action plan for the state of Uttarakhand. Disaster Mitigation & Management Centre, Uttarakhand Secretariat, Rajpur Road, Dehradun. http://dmmc.uk.gov.in/files/pdf/complete_sdmap.pdf

Donald H, David H (2010) Natural hazards and disasters, Brooks-Cole Publication, USA, p 4

JRDNA Report (2013) India–Uttarakhand disaster, June 2013 (82643). Joint Rapid Damage and Needs Assessment Report. Jointly submitted by Govt. of Uttarakhand, ADB and the World Bank, August, 2013

Kavita T (2017) Kedar tragedy: a cause and impact assessment. Int J Sci Environ Technol 6(1):493–499

Mishra VN, Rai PK, Singh P (2021) Geo-information technology in earth resources monitoring and management (edited book). Nova Science Publishers, USA, pp 1–395. ISBN: 978-1-53619-669-60

Mishra A (2017) Changing temperature and rainfalls patterns of Uttarakhand. Int J Environ Sci Nat Resour 7(4):90–95

Nair AS, Singh SK (2014) Understanding the causes of Uttarakhand disater of June 2013: a scientific review. In: Paper presented in 2nd disaster risk and vulnerability conference. Department of Geology, University of Kerala, Trivandrum, p 64

NIDM Report (2013) Uttarakhand disaster 2013. Published by National Institute of Disaster Management, Ministry of Home Affairs, Government of India

NIDM (2015) Uttarakhand disaster 2013. Published by National Institute of Disaster Management, Ministry of Home Affairs, Government of India

Rai PK (2023a) River conservation and water resource management. Springer Nature, Singapore, pp 1–352. ISBN No.: 978-981-99-2604-6. eBook ISBN: 978-981-99-2605-3

Rai PK (2023b) Advances in water resource planning & sustainability. Springer Nature, Singapore, pp 1–322. ISBN No. 978-981-99-3659-5

Rai PK, Mohan K, Kumra VK (2014) Landslide hazard and its mapping using remote sensing & GIS techniques. J Sci Res Fac Sci, Banaras Hindu Univ 58:1–13

Rai PK, Mishra VN, Singh P (2021) Recent technologies for disaster management & risk reduction-sustainable community resilience & responses (edited book). Springer Nature, Switzerland, pp 1–471. ISBN: 978-3-030-76116-5. https://doi.org/10.1007/978-3-030-76116-5

Rai PK, Mishra VN, Singh P (2022) Geospatial technology for landscape and environment management: sustainable assessment & planning (edited book). Springer Nature, Singapore, pp 1–431. ISBN: 978-981-16-7373-3. https://doi.org/10.1007/978-981-16-7373-3

Sharma AK et al (2014) Response to Uttarakhand disaster 2013. Int J Sci Eng Res 5(10):1251–1256

Shaw R, Nibanupudi HK (2014) Overview of mountain hazards issues. In: Shaw R, Nibanupudi HK (eds) Mountain hazards and disaster risk reduction. Springer Publication, p 4

Singh RB (2006) Introduction to natural hazards and disasters, conceptual framework, vulnerability, spatial pattern, impacts and mitigation strategy: Indian experience. In: Singh RB (ed) Natural hazards and disasters management (vulnerability and mitigation). Rawat Publication, Jaipur, pp 3–39

Singh P, Sharma A, Sur U, Rai PK (2021) Comparative landslide susceptibility assessment using statistical information value and index of entropy model in Bhanupali Beri region Himachal Pradesh, India. Environ Dev Sustain 23:5233–5250. https://doi.org/10.1007/s10668-020-00811-0

Singh P, Sur U, Rai PK, Singh SK (2023) Landslide susceptibility prediction using frequency ratio model: a case study of Uttarakhand, Himalaya (India). Proc Indian Natl Sci Acad 1–13. https://doi.org/10.1007/s43538-023-00171-z

Sur U, Singh P, Rai PK (2021) Landslide probability mapping by considering fuzzy numerical risk factor (FNRF) and landscape change for road corridor of Uttarakhand, India. Environ Dev Sustain (Springer) 23:13526–13554. https://doi.org/10.1007/s10668-021-01226-1

Subramanian R (2018) Disaster management. Vikas Publication House, India, p 2

Chapter 20
Leaf to Life: An Overview of the Socio-economic Status of Tea Tribes in Assam

Archina Siddika, Praveen Kumar Rai, and Ashutosh Singh

Abstract The review paper titled "Leaf to Life: An Overview of Socio-economic Status of Tea Tribes in Assam" provides an insightful examination of the socio-economic status of the tea tribes in Assam, India. Starting with a historical account of the discovery of tea plantations in Assam, the paper focuses on the tea tribes of Assam. It scrutinized the socio-economic challenges faced by the community, such as low wages, lack of education, poor housing and living conditions, poor healthcare, and women's rights challenges. In this review paper, some measures have been suggested to ameliorate the predicaments faced by tea tribes in Assam.

Keywords Tea plantation · Tea tribes · Socio-economic challenge · Poor healthcare

20.1 Introduction

Tea tribes in India, found primarily in Assam, have a unique socio-economic background shaped by their historical relationship with tea plantations. This community, commonly called Adivasis, migrated from the Chota Nagpur Plateau to Assam during British colonialism, particularly from present-day Jharkhand, Bihar, and Odisha. This migration was to meet the increasing labor requirements of the growing tea industry in Assam in the nineteenth century. The tea tribe community is one of the backward

A. Siddika
Department of Geography and RM, Mizoram University, Aizawl, India

P. K. Rai
Department of Geography, Khwaja Moinuddin Chishti Lucknow, Lucknow, India

A. Singh (✉)
Department of Geography, HNB Garhwal University (A Central University), SRT Campus, Tehri, India
e-mail: singh4as@gmail.com

P. K. Rai et al. (eds.), *Natural Resources Management and Sustainable Livelihoods in the Mountainous Region*, Advances in Geographical and Environmental Sciences,
https://doi.org/10.1007/978-981-97-2100-9_20

communities of Assam. Although there has been progress since the colonial era, they still face problems.

20.1.1 Study Area

Assam sits roughly 24.4° N to 28.2° N and 89.5° E to 96° E (Fig. 20.1). Assam, a state in northeastern India, boasts diverse topography ranging from plains to hills. The Brahmaputra River, an important geographical feature, flows through the region, shaping its landscape. The fertile Brahmaputra Valley, famous for its tea cultivation, is home to tea tribes, including communities like the Adivasis, known for their integral role in the tea industry. Lush tea plantations cover vast expanses, creating a unique visual identity for Assam and contributing significantly to its economy. Moving away from the plains, the state gradually ascends into the picturesque hills of the Karbi Anglong and North Cachar Hills, where diverse flora, fauna, and distinct cultures thrive, offering a rich tapestry of natural beauty and ethnic heritage in this northeastern Indian state.

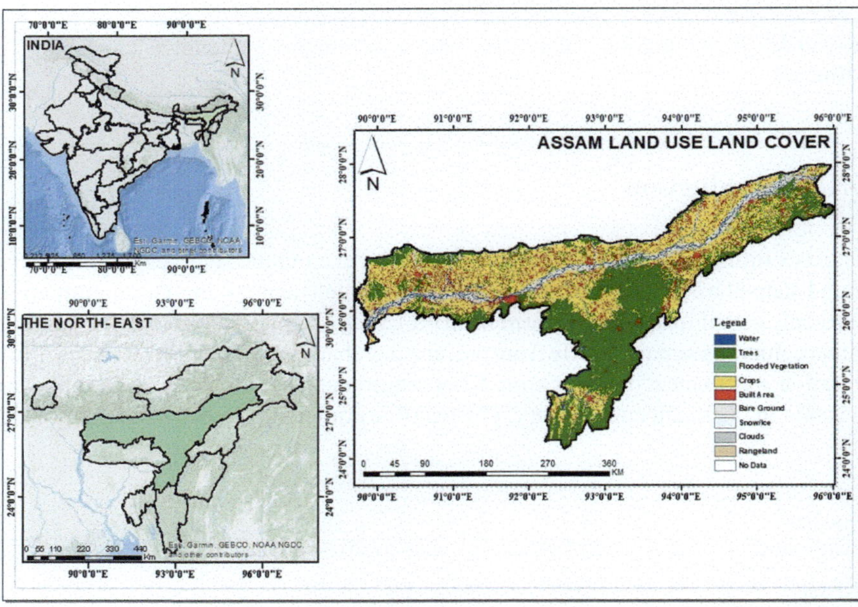

Fig. 20.1 Location map of the study area

20.2 Objective

1. To examine the historical context and origin of tea tribes in Assam.
2. To analyze the major socio-economic challenges faced by the tea tribes in Assam.
3. To propose and discuss potential measures and solutions aimed at addressing the identified challenges and improving the socio-economic conditions of the tea tribes in Assam.

20.2.1 Discovery of Tea Plantation in Assam

Plantations represent one of the most systematic systems of agricultural production. The tea plantation industry is recognized as one of the most well-organized sectors in India. It was a momentous political and economic development for Britain when the indigenous tea plant was discovered in Assam (Kalita 2019; Mishra et al. 2021; Rai et al. 2021, 2022; Singh et al. 2021; Brototi et al. 2022). In 1823, the discovery of tea plants in upper Assam is attributed to Robert Bruce, who was introduced to them by Singphow Chief Bisa Gam. Interestingly, tea had been noted by Mandbelo in 1640, and Sir Joseph Bank attempted to grow it in 1788 using foreign seeds. Additionally, the indigenous habit of drinking 'Fanap,' which was indeed tea, was observed by Col. Letter in 1815 among the people of Assam. After the East India Company annexed Assam in 1826, the possibility of tea cultivation became more evident. In 1834, Lord William Bentinck, the Governor General, allowed experimental tea cultivation under government initiative. The taste of the tea produced in Assam was found to be satisfactory when a few boxes were sent to London in 1839. Consequently, European planters started rushing into Assam for tea cultivation by 1852. There were 20 tea gardens in Assam in 1851, and this number surged to 883 within the next 50 years. This marked a significant transformation in the region's tea industry (Sarmah 2018).

20.2.2 Tea Tribes of Assam

The industrious workforce, consisting of 99% migrant laborers, is the backbone of the tea industry. The migrant laborers who moved to Assam, in northeastern India, became known as the "tea community" or "tea tribe" over time due to their close ties to the tea plantations (Gogoi 2022). The tea tribes in Assam have diverse origins, hailing from various parts of India, including Madhya Pradesh, Bihar, Orissa, Andhra Pradesh, Bengal, Maharashtra, and Tamil Nadu (Akhtar 2013). They are currently predominantly found in districts such as Dibrugarh, Tinsukia, Jorhat, Golaghat, Morigaon, and Sonitpur. According to the 2001 Census, they make up about 20% of the state's population. These communities play a significant role in Assam's organized workforce and contribute significantly to its economy. It's worth noting that they are unique in the state as they are named based on their occupational (Borah and Talukdar

2014). Within their communities, a wide range of castes and sub-castes exist, totaling one hundred thirty-one. Notable sub-castes among them include Arya Male, Asar, Bhumihar, Mali, Basar, Basphor, Bania, Banzara, Chamar, Chara, Dhonwar, Dhopi, Dhanger, Ganda, Garhwali, Hari, Kawar, Kalhandi, Karmakar, Kharia, Lodhi, Majhi, Nag, Orang, Pradhan, Rautia, Sanatal, Tanti, Teli, and more (Akhtar 2013). People from various linguistic backgrounds in several Indian states make up the tea community, which has found comfort in Assam's shared culture and togetherness (Konwar et al. 2023).

The Tea tribes' rich cultural heritage has significantly enriched the greater Assamese culture. They possess a wealth of folk songs, folktales, and folk talks that are unique to their community. These cultural elements serve as a testament to their vibrant traditions (Sarmah 2018). Today, the majority of the Tea tribe community in Assam is predominantly Hindu, with Hindus comprising the larger portion of the population. In contrast, a smaller section, approximately 10% of the total Tea and ex-Tea tribe population, consists of individuals who follow the Christian faith (Magar and Kar 2016). The primary language of the Tea tribes is known as Bagania or Sadani language (Sarmah 2018). Odia, Saora, Kuruk, Gondi, Kui, Kharia, Santhi, and Mundari are different languages spoken by tea tribes in Assam (Sharma 2022).

20.3 Socio-economic Challenges of Tea Tribes in Assam

20.3.1 Low Wages

Income level is the best indicator of the economic well-being of a person, family, and society as a whole (Sharma 2022). The low wages of tea tribes in Assam have been a long-standing issue. The tea industry in Assam is one of the largest in India, but historically, tea tribes have been subjected to low wages, long working hours, and minimal social security benefits. The wages often do not meet the labor-intensive work they perform in the tea gardens, leading to economic hardship and poverty. The lack of land rights and job insecurity further exacerbate their socioeconomic vulnerability.

20.3.2 Lack of Education

Education is a lifetime, ongoing process of empowerment that serves as the foundation for growth and transformation (Akhtar 2013). The tea-tribe community of Assam continues to face educational challenges, which, in turn, hinder their ability to attain higher social status (Sahu and Bhuyan 2022). The rights to education that tea garden laborers lack awareness of. The Indian government added a new basic duty to Indian citizens by revising the 86th constitutional amendment, which instructed guardians to send their children under the age of six to fourteen to school and to view this as a moral obligation (Hazarika and Boruah 2020). Despite numerous efforts by the government to improve education, the tea tribes of Assam remain among the communities with low levels of education (Saikia 2020). The literacy level within the community stands at a mere 46%, which is notably lower when compared to Assam's overall literacy rate of 72% (Sharma 2022). Lack of education negatively impacts tribal society's economic condition, leading to food crises and child marriages, exacerbated by parents' lack of value for education (Bishwasarma and Borah 2019). Boys are frequently driven by their financial situation and desire for partners in their businesses where they may make money (Saikia 2020). In recent times, they have received modern education. But even today, girls of tea tribes are deprived of education compared to males (Gogoi and Handique 2014).

There has always been a problem with the tea tribes in Assam not having access to education. Limited access to quality education and a high dropout rate among children due to economic pressure and lack of proper schools hinder their socio-economic progress. This perpetuates a cycle of illiteracy and limited opportunities for better employment.

20.3.3 Poor Housing and Living Conditions

The ability to live comfortably is thought to be one of humanity's most basic needs. Tea tribes comprise about 20% of Assam's population, and according to housing circumstances, most reside in battered dwellings. Structures without the provision of basic and necessary dwelling facilities, including power and sanitation, the availability of restrooms, drinking water, smokeless cooking fuel, etc. The most underprivileged social group in the state of Assam, the tea tribes, live in appalling conditions without even the most basic necessities for survival (Mishra 2019).

Housing and living conditions pose significant challenges for the tea tribes in Assam, a marginalized and historically disadvantaged community primarily engaged in the state's tea plantation. One major issue is the substandard housing provided to them by tea estate owners.

Many people live in overcrowded, dilapidated, and unhygienic quarters. Those houses lack basic amenities like proper sanitation, clean drinking water, and electricity, making daily life difficult for tea tribe families. The inadequate housing situation directly impacts the community's health and well-being. Overcrowding contributes to the spread of diseases, while the absence of sanitation facilities can lead to poor hygiene and sanitation-related illnesses. Moreover, the lack of clean drinking water exacerbates health concerns, as tea tribe members often rely on contaminated sources. The subpar living conditions also have social implications. They perpetuate a cycle of poverty and low socio-economic status among the tea tribe population. Inadequate housing and living conditions further limit children's access to education, as they often lack a conducive environment for studying.

20.3.4 Poor Healthcare

It is noteworthy that the tea garden population is often situated within the lower socioeconomic strata, residing in unhygienic environments where access to fundamental necessities like clean drinking water and proper sanitation is sorely lacking. This confluence of factors, including impoverished economic circumstances, low levels of literacy, and overcrowded living conditions in their residential colonies, ultimately renders these communities highly susceptible to a myriad of communicable diseases (Bora et al. 2018).

However, over the years, government initiatives and non-governmental organizations have been working towards improving healthcare access for these communities. Efforts have been made to establish more healthcare facilities, provide better healthcare education, and promote awareness about hygiene and preventive healthcare practices. While progress has been made, addressing the healthcare needs of the tea tribes in Assam remains an ongoing challenge that requires continued attention and investment in infrastructure and community health programs.

20.3.5 Women's Rights Challenge

Women account for over fifty percent of the total workforce employed in the tea plantation industry in Assam (Duara and Mallick 2019). From a socioeconomic, cultural, and political standpoint, women are viewed as being weak. Women's rights are increasingly important worldwide and at the regional and national levels (Tamuli 2015). For equivalent work, women are not paid equally (Hazarika and Boruah 2020). Moreover, the tea gardens society in Assam believes in spirits and evil practices, with sickness and disease seen as evil effects of supernatural powers. They also strongly believe in witches, who can bring calamities to individuals, families, or the villagers. The term "*daini*" refers to witches, and women members (Sometimes male) identified as witches are often tortured or killed by co-villagers. Witch hunting is a serious issue in Assam (Kalita 2018).

Women's rights in the Tea tribes of Assam have historically faced challenges, but efforts have been made to address these issues. In recent years, there has been a growing awareness and advocacy for gender equality and women's rights within these communities. Various organizations and government initiatives, such as the Assam Tea Tribes Women's Association and the Assam Tea Corporation's efforts to provide maternity benefits and education to women, have significantly empowered women in the tea tribes of Assam (Gohain 2017).

20.4 Suggestive Measures

To promote the prosperity of tea tribes in Assam socio-economically, several measures and recommendations can be considered:

i. Fair Wages: Implement fair wage policies, ensure fair remuneration, and facilitate skill development programs to increase employability and earning potential within the tea industry.
ii. Education Access: There is a need to build additional schools, provide scholarships, and address the prevalent issue of high dropout rates due to economic barriers to accessing quality education for Tea Tribe's children. Furthermore, it is important to promote adult literacy programs to empower the entire community, ensuring a holistic approach towards educational development among the entire community.
iii. Improving Housing and Living Conditions: Steps should be taken to initiate housing development projects, ensure access to basic amenities like clean water and sanitation, and promote community-driven efforts to raise overall living standards.
iv. Enhancing Healthcare Facilities: Enhancing the healthcare infrastructure in tea-growing regions is necessary to address the health problems that tea tribes face, including better access to medical treatment and implementing health awareness initiatives.
v. Women's Empowerment: Women in tea tribes should be empowered by providing education, vocational treatment, and support for their rights, and by this, they will encourage their active participation in decision-making processes and improve their status within the community by ensuring equal opportunities.
vi. Government and NGO Support: Even though NGOs and the Government help the tea tribes, some things still need fixing. There are some areas where they have not fully solved the problems faced by these communities. They will have to take more steps to solve the problem of tea tribes. The government should create more programs for education, healthcare, and job training. On the other hand, NGOs can support tea tribe communities by providing the resources they need and helping to make their lives better. Simply put, they have the power to impact the welfare of the tea tribes significantly.
vii. Awareness and Advocacy: It is important to raise awareness of the problems faced by tea tribes and advocate for their rights and welfare on local, national, and international levels.

These measures and recommendations, when implemented collectively, can contribute to the prosperity and improved socio-economic status of the tea tribes in Assam (Chatterjee et al. 2017).

20.5 Conclusion

In conclusion, the socio-economic conditions of tea tribes in Assam reflect a complex history rooted in their relationship with tea gardens. These communities' challenges are multidimensional, affecting various aspects of their lives. However, through concerted efforts and suggested measures, there is hope to improve vibrant societies and promote sustainable development.

References

Akhtar PR (2013) Need of education among the tea tribe women in the Udalguri district of Assam, India. Clarion Int Multidiscip J 2(1):167–173

Bishwasarma A, Borah B (2019) Social and educational condition of tea tribes of Assam: a study. J Emerg Technol Innov Res (JETIR) 6(6):343–344

Bora PJ, Das BR, Das N (2018) Availability and utilization of sanitation facilities amongst the tea garden population of Jorhat district, Assam. Int J Commun Med Public Health 5(6):2506–2511

Borah J, Talukdar B (2014) Civil societies and plight of marginalised communities—a case of the tea tribes of Assam. Int J Phys Soc Sci 4(6):207

Brototi B, Singh A, Rai PK, Kumar J, Walker S (2022) GIS based study of reclamation of degraded semi-arid soil: a case study of Rajasthan from India. Indian J Environ Prot 42(3):302–315

Chatterjee P, Choudhury S, Kumar A (2017) A study on education status and social life of tea tribe community of Assam. IOSR J Hum Soc Sci 22(9):17–24

Duara M, Mallick S (2019) Women workers & industrial relations in tea estates of Assam. Indian J Ind Relat 55(1):15–26

Gogoi D, Handique M (2014) Girl child education among tea tribes: a case study of Rajgarh tea estate of Dibrugarh district of Assam. Vidyawarta Interdiscip Multiling Ref J 5(8):0143–0147

Gogoi N (2022) Jhumur folk tradition: a socio-cultural identity of tea community in Assam. Int J Health Sci (6S3):5059–5064

Gohain H (2017) Empowerment of tea garden women workers: a study on two tea gardens of Assam. J Indian Anthropol Soc 52(1)

Hazarika B, Boruah D (2020) Rights of marginalized and role of state: a study on tea garden labourer in Assam. J Crit Rev 7(4):2150–2156

Kalita BJ (2018) Socio-cultural life of the tea plantation workers in Assam: a study from historical perspective. Res Guru Online J Multidiscip Subjects 12(2):160–168

Kalita BJ (2019) Historical background of the tea industry in Assam. Res Rev Int J Multidiscip 04(01):548–550

Konwar P, Das B, Saikia D (2023) Seasonal festivals ceremonies of the tea tribal society of Assam—a study. Int J Sci Res (IJSR) 12(8):2056–2058

Magar CK, Kar BK (2016) Tea plantations and socio-cultural transformation: the case of Assam (India). Space Culture India 4(1):25–39

Mishra D (2019) Status of housing and household amenities of tea tribe population—a case study in the Biswanath district, Assam. Int J Interdiscip Res Arts Hum 4(1):93–99

Mishra VN, Rai PK, Singh P (eds) (2021) Geo-information technology in earth resources monitoring and management. Nova Science Publishers, U.S.A., pp 1–395. ISBN: 978-1-53619-669-6

Rai PK, Mishra VN, Singh P (eds) (2021) Recent technologies for disaster management & risk reduction-sustainable community resilience & responses. Springer Nature, Switzerland, pp 1–471. ISBN: 978-3-030-76116-5. https://doi.org/10.1007/978-3-030-76116-5

Rai PK, Mishra VN, Singh P (eds) (2022) Geospatial technology for landscape and environment management: sustainable assessment & planning. Springer Nature, Singapore, pp 1–431. ISBN: 978-981-16-7373-3. https://doi.org/10.1007/978-981-16-7373-3

Sahu E, Bhuyan S (2022) A study on the educational status of tea-tribe community in Assam. J Positive School Psychol 6(2):5385–5389

Saikia J (2020) Education and tea tribe community women—a case study in Dibrugarh district of Assam. Int J Psychosoc Rehabil 24(06):5613–5621

Sarmah P (2018) Tea tribes of Assam: their migration & assimilation to Assamese society. Int J Res Anal Rev (IJRAR) (E-ISSN 2348-1269, P-ISSN 2349-5138). www.Ijrar.Org

Sharma R (2022) A study on socio-economic and educational status of tea tribe community with special reference to Thowra tea estate, Dibrugarh, Assam. J Positive School Psychol 6(3):6768–6772

Singh A, Rai PK, Deka G, Biswas B, Prasad D, Rai VK (2021) Management of natural resources through integrated watershed management in Nana Kosi micro watershed, district Almora, India. Ecol Environ Conserv 27(February Suppl. Issue):S260–S268

Tamuli D (2015) Socio-economic condition and rights of the tea-garden women in Assam: a study on Negheriting tea estate area of Golaghat district. Labour 53(20):33

The manufacturer's authorised representative in the EU is Springer Nature Customer Service Centre GmbH, Europaplatz 3, 69115 Heidelberg, Germany. If you have any concerns regarding our products, please contact ProductSafety@springernature.com

Printed and bound by CPI Group (UK) Ltd, Croydon, CR0 4YY
30/09/2026
02225253-0012